METHODS IN MOLECULAR BIOLOGY

Series Editor
John M. Walker
School of Life Sciences
University of Hertfordshire
Hatfield, Hertfordshire, AL10 9AB, UK

For further volumes:
http://www.springer.com/series/7651

Plant Isoprenoids

Methods and Protocols

Edited by

Manuel Rodríguez-Concepción

Centre for Research in Agricultural Genomics (CRAG), CSIC-IRTA-UAB-UB, Barcelona, Spain

Humana Press

Editor
Manuel Rodríguez-Concepción
Centre for Research in Agricultural Genomics (CRAG)
CSIC-IRTA-UAB-UB
Barcelona, Spain

ISSN 1064-3745 ISSN 1940-6029 (electronic)
ISBN 978-1-4939-5391-2 ISBN 978-1-4939-0606-2 (eBook)
DOI 10.1007/978-1-4939-0606-2
Springer New York Heidelberg Dordrecht London

Printed on acid-free paper

Humana Press is a brand of Springer
Springer is part of Springer Science+Business Media (www.springer.com)

Preface

Plant isoprenoids form one the most diverse family of metabolites in nature, with tens of thousands of structures known to date. Among them, some are essential for plant photosynthesis (carotenoids and the side chain of chlorophylls, plastoquinone, and phylloquinones), respiration (ubiquinone), and development (brassinosteroids, cytokinins, gibberellins, abscisic acid, strigolactones), whereas others have a great economic interest as drugs (artemisinin, paclitaxel), polymers (rubber), phytonutrients (phytosterols, carotenoids), or even biofuels (limonene, farnesene, or bisabolene).

Because isoprenoids are such a diverse family and they participate in a large variety of processes, the collection of detailed techniques and protocols included in the volume should be a useful tool for a wide range of plant biologists as well as for scientists of other fields with an interest in plant isoprenoids. Rather than being exhaustive, my intention has been that the protocols in this volume would cover strategic areas in plant isoprenoid research. Thus, this volume focuses on four major areas: (1) measurement of core enzyme activities involved in the production of isoprenoid precursors, (2) targeted analysis of major groups of isoprenoid metabolites, (3) isoprenoid profiling in specialized organs such as trichomes and oil glands, and (4) genetic, pharmacological, and bioinformatic tools that are particularly useful for plant molecular biologists.

Thanks to the excellent work of the contributing authors, the protocols provide step-by-step guidance and are easy to follow even for users with little or no experience in the field. At the same time, they can also serve as reference materials that could be adapted to develop customized methods for different needs. I would like to thank all the authors for agreeing to participate and for their generous effort to produce this issue on Plant Isoprenoids, a badly needed resource that will contribute to make the world of plant isoprenoids more accessible for all researchers. I would also like to thank Rosa Rodriguez for her help in editing and adjusting the format of the chapters and acknowledge John M. Walker for his invitation to write this volume for Methods in Molecular Biology and for his useful advices for the preparation of the issue.

Barcelona, Spain *Manuel Rodríguez-Concepción*

Contents

Preface . . . . *v*
Contributors . . . . *ix*

1 Plant Isoprenoids: A General Overview 1
Manuel Rodríguez-Concepción

Part I Measurement of Core Enzyme Activities

2 Measuring the Activity of 1-Deoxy-D-Xylulose 5-Phosphate Synthase, the First Enzyme in the MEP Pathway, in Plant Extracts 9
Louwrance P. Wright and Michael A. Phillips

3 Determination of 3-Hydroxy-3-methylglutaryl CoA Reductase Activity in Plants 21
Narciso Campos, Montserrat Arró, Albert Ferrer, and Albert Boronat

4 Farnesyl Diphosphate Synthase Assay 41
Montserrat Arró, David Manzano, and Albert Ferrer

Part II Targeted Analysis of Isoprenoid Metabolites

5 Metabolite Profiling of Plastidial Deoxyxylulose-5-Phosphate Pathway Intermediates by Liquid Chromatography and Mass Spectrometry 57
Edward E.K. Baidoo, Yanmei Xiao, Katayoon Dehesh, and Jay D. Keasling

6 Analysis of Carotenoids and Tocopherols in Plant Matrices and Assessment of Their In Vitro Antioxidant Capacity 77
Antonio J. Meléndez-Martínez, Carla M. Stinco, Paula Mapelli Brahm, and Isabel M. Vicario

7 Simultaneous Analyses of Oxidized and Reduced Forms of Photosynthetic Quinones by High-Performance Liquid Chromatography 99
Masaru Shibata and Hiroshi Shimada

8 Determination of Sterol Lipids in Plant Tissues by Gas Chromatography and Q-TOF Mass Spectrometry 115
Vera Wewer and Peter Dörmann

9 Analysis of Plant Polyisoprenoids 135
Katarzyna Gawarecka and Ewa Swiezewska

10 Analysis of Diterpenes and Triterpenes from Plant Foliage and Roots 149
Qiang Wang, Reza Sohrabi, and Dorothea Tholl

11 Gas Chromatography–Mass Spectrometry Method for Determination of Biogenic Volatile Organic Compounds Emitted by Plants 161
Astrid Kännaste, Lucian Copolovici, and Ülo Niinemets

12 Analysis of Steroidal Alkaloids and Saponins in *Solanaceae* Plant Extracts Using UPLC-qTOF Mass Spectrometry. 171
Uwe Heinig and Asaph Aharoni

PART III ISOPRENOID PROFILING IN SPECIALIZED ORGANS

13 Isoprenoid and Metabolite Profiling of Plant Trichomes 189
Gerd U. Balcke, Stefan Bennewitz, Sebastian Zabel, and Alain Tissier

14 Sample Preparation for Single Cell Transcriptomics: Essential Oil Glands in *Citrus* Fruit Peel as an Example 203
Siau Sie Voo and Bernd Markus Lange

15 Prenylquinone Profiling in Whole Leaves and Chloroplast Subfractions 213
Felix Kessler and Gaetan Glauser

16 Confocal Laser Scanning Microscopy Detection of Chlorophylls and Carotenoids in Chloroplasts and Chromoplasts of Tomato Fruit 227
Lucio D'Andrea, Montse Amenós, and Manuel Rodríguez-Concepción

PART IV GENETIC, PHARMACOLOGICAL, AND BIOINFORMATIC TOOLS

17 Heterologous Expression of Triterpene Biosynthetic Genes in Yeast and Subsequent Metabolite Identification Through GC-MS 235
Ery Odette Fukushima, Hikaru Seki, and Toshiya Muranaka

18 High-Throughput Testing of Terpenoid Biosynthesis Candidate Genes Using Transient Expression in *Nicotiana benthamiana* 245
Søren Spanner Bach, Jean-Étienne Bassard, Johan Andersen-Ranberg, Morten Emil Møldrup, Henrik Toft Simonsen, and Björn Hamberger

19 Heterologous Stable Expression of Terpenoid Biosynthetic Genes Using the Moss *Physcomitrella patens* 257
Søren Spanner Bach, Brian Christopher King, Xin Zhan, Henrik Toft Simonsen, and Björn Hamberger

20 Quantification of Plant Resistance to Isoprenoid Biosynthesis Inhibitors 273
Catalina Perelló, Manuel Rodríguez-Concepción, and Pablo Pulido

21 A Flexible Protocol for Targeted Gene Co-expression Network Analysis 285
Diana Coman, Philipp Rütimann, and Wilhelm Gruissem

Index *301*

Contributors

ASAPH AHARONI • *Department of Plant Sciences, Weizmann Institute of Science, Rehovot, Israel*
MONTSE AMENÓS • *Centre for Research in Agricultural Genomics (CRAG), CSIC-IRTA-UAB-UB, Barcelona, Spain*
JOHAN ANDERSEN-RANBERG • *Plant Biochemistry Laboratory, Department of Plant and Environmental Sciences, Copenhagen Plant Science Centre, University of Copenhagen, Copenhagen, Denmark; Center for Synthetic Biology, University of Copenhagen, Copenhagen, Denmark*
MONTSERRAT ARRÓ • *Center for Research in Agricultural Genomics (CRAG), CSIC-IRTA-UAB-UB, Barcelona, Spain; Faculty of Pharmacy, Department of Biochemistry and Molecular Biology, University of Barcelona, Barcelona, Spain*
SØREN SPANNER BACH • *Plant Biochemistry Laboratory, Department of Plant and Environmental Sciences, Copenhagen Plant Science Centre, University of Copenhagen, Copenhagen, Denmark*
EDWARD E.K. BAIDOO • *Physical Biosciences Division, Lawrence Berkeley National Laboratory, Berkeley, CA, USA; Joint BioEnergy Institute, Emeryville, CA, USA*
GERD U. BALCKE • *Department of Cell and Metabolic Biology, Institute of Plant Biochemistry, Halle, Germany*
JEAN-ÉTIENNE BASSARD • *Plant Biochemistry Laboratory, Department of Plant and Environmental Sciences, Copenhagen Plant Science Centre, University of Copenhagen, Copenhagen, Denmark; Center for Synthetic Biology, University of Copenhagen, Copenhagen, Denmark*
STEFAN BENNEWITZ • *Department of Cell and Metabolic Biology, Institute of Plant Biochemistry, Halle, Germany*
ALBERT BORONAT • *Center for Research in Agricultural Genomics (CRAG), CSIC-IRTA-UAB-UB, Barcelona, Spain; Department of Biochemistry and Molecular Biology, Faculty of Biology, University of Barcelona, Barcelona, Spain*
PAULA MAPELLI BRAHM • *Food Colour & Quality Laboratory, Department of Nutrition & Food Science, Facultad de Farmacia, Universidad de Sevilla, Sevilla, Spain*
NARCISO CAMPOS • *Center for Research in Agricultural Genomics (CRAG), CSIC-IRTA-UAB-UB, Barcelona, Spain; Department of Biochemistry and Molecular Biology, Faculty of Biology, University of Barcelona, Barcelona, Spain*
DIANA COMAN • *Department of Biology, Plant Biotechnology, ETH Zurich, Zurich, Switzerland*
LUCIAN COPOLOVICI • *Institute of Agricultural and Environmental Sciences, Estonian University of Life Sciences, Tartu, Estonia; Institute of Technical and Natural Sciences Research-Development of "Aurel Vlaicu" University, Arad, Romania*
LUCIO D'ANDREA • *Centre for Research in Agricultural Genomics (CRAG), CSIC-IRTA-UAB-UB, Barcelona, Spain*
KATAYOON DEHESH • *Department of Plant Biology, University of California, Davis, CA, USA*

PETER DÖRMANN • *Institute of Molecular Physiology and Biotechnology of Plants (IMBIO), University of Bonn, Bonn, Germany*
ALBERT FERRER • *Faculty of Pharmacy, Department of Biochemistry and Molecular Biology, University of Barcelona, Barcelona, Spain; Center for Research in Agricultural Genomics (CRAG) CSIC-IRTA-UAB-UB, Campus UAB Bellaterra, Barcelona, Spain*
ERY ODETTE FUKUSHIMA • *Department of Biotechnology, Graduate School of Engineering, Osaka University, Osaka, Japan; Frontier Research Base for Global Young Researchers, Graduate School of Engineering, Osaka University, Osaka, Japan*
KATARZYNA GAWARECKA • *Institute of Biochemistry and Biophysics, Polish Academy of Sciences, Warsaw, Poland*
GAETAN GLAUSER • *Chemical Analytical Service of the Swiss Plant Science Web, University of Neuchâtel, Neuchâtel, Switzerland*
WILHELM GRUISSEM • *Department of Biology, Plant Biotechnology, ETH Zurich, Zurich, Switzerland; Functional Genomics Center Zurich, ETH Zurich, Zurich, Switzerland*
BJÖRN HAMBERGER • *Plant Biochemistry Laboratory, Department of Plant and Environmental Sciences, Copenhagen Plant Science Centre, University of Copenhagen, Copenhagen, Denmark; Center for Synthetic Biology, University of Copenhagen, Copenhagen, Denmark*
UWE HEINIG • *Department of Plant Sciences, Weizmann Institute of Science, Rehovot, Israel*
ASTRID KÄNNASTE • *Institute of Agricultural and Environmental Sciences, Estonian University of Life Sciences, Tartu, Estonia*
JAY D. KEASLING • *Physical Biosciences Division, Lawrence Berkeley National Laboratory, Berkeley, CA, USA; Joint BioEnergy Institute, Emeryville, CA, USA; Department of Chemical Engineering, University of California, Berkeley, CA, USA; Department of Bioengineering, University of California, Berkeley, CA, USA*
FELIX KESSLER • *Laboratory of Plant Physiology, University of Neuchâtel, Neuchâtel, Switzerland*
BRIAN CHRISTOPHER KING • *Plant Biochemistry Laboratory, Department of Plant and Environmental Sciences, Copenhagen Plant Science Centre, University of Copenhagen, Copenhagen, Denmark*
BERND MARKUS LANGE • *Institute of Biological Chemistry and M.J. Murdock Metabolomics Laboratory, Washington State University, Pullman, WA, USA*
DAVID MANZANO • *Center for Research in Agricultural Genomics (CRAG), CSIC-IRTA-UAB-UB, Barcelona, Spain; Department of Biochemistry and Molecular Biology, Faculty of Pharmacy, University of Barcelona, Barcelona, Spain*
ANTONIO J. MELÉNDEZ-MARTÍNEZ • *Food Colour & Quality Laboratory, Department of Nutrition & Food Science, Facultad de Farmacia, Universidad de Sevilla, Sevilla, Spain*
MORTEN EMIL MØLDRUP • *Section for Molecular Plant Biology, DynaMo Center of Excellence, University of Copenhagen, Copenhagen, Denmark*
TOSHIYA MURANAKA • *Department of Biotechnology, Graduate School of Engineering, Osaka University, Osaka, Japan*
ÜLO NIINEMETS • *Institute of Agricultural and Environmental Sciences, Estonian University of Life Sciences, Tartu, Estonia*
CATALINA PERELLÓ • *Centre for Research in Agricultural Genomics (CRAG), CSIC-IRTA-UAB-UB, Barcelona, Spain*
MICHAEL A. PHILLIPS • *Centre for Research in Agricultural Genomics (CRAG), CSIC-IRTA-UAB-UB, Barcelona, Spain*

PABLO PULIDO • *Centre for Research in Agricultural Genomics (CRAG), CSIC-IRTA-UAB-UB, Barcelona, Spain*
MANUEL RODRÍGUEZ-CONCEPCIÓN • *Centre for Research in Agricultural Genomics (CRAG), CSIC-IRTA-UAB-UB, Barcelona, Spain*
PHILIPP RÜTIMANN • *Seminar for Statistics, ETH Zurich, Zurich, Switzerland*
HIKARU SEKI • *Department of Biotechnology, Graduate School of Engineering, Osaka University, Osaka, Japan*
MASARU SHIBATA • *Faculty of Education, Biological institute, Yamaguchi University, Yamaguchi, Japan*
HIROSHI SHIMADA • *Laboratory of Molecular Plant Biology, Department of Mathematical and Life Sciences, Graduate School of Sciences, Hiroshima University, Hiroshima, Japan*
HENRIK TOFT SIMONSEN • *Plant Biochemistry Laboratory, Department of Plant and Environmental Sciences, Copenhagen Plant Science Centre, University of Copenhagen, Copenhagen, Denmark*
REZA SOHRABI • *Department of Biological Sciences, Virginia Tech, Blacksburg, VA, USA; Molecular, Cellular, and Developmental Biology, University of Michigan, Ann Arbor, MI, USA*
CARLA M. STINCO • *Food Colour & Quality Laboratory, Department of Nutrition & Food Science, Facultad de Farmacia, Universidad de Sevilla, Sevilla, Spain*
EWA SWIEZEWSKA • *Institute of Biochemistry and Biophysics, Polish Academy of Sciences, Warsaw, Poland*
DOROTHEA THOLL • *Department of Biological Sciences, Virginia Tech, Blacksburg, VA, USA*
ALAIN TISSIER • *Department of Cell and Metabolic Biology, Institute of Plant Biochemistry, Halle, Germany*
ISABEL M. VICARIO • *Food Colour & Quality Laboratory, Department of Nutrition & Food Science, Facultad de Farmacia, Universidad de Sevilla, Sevilla, Spain*
SIAU SIE VOO • *Institute of Biological Chemistry and M.J. Murdock Metabolomics Laboratory, Washington State University, Pullman, WA, USA*
QIANG WANG • *Department of Biological Sciences, Virginia Tech, Blacksburg, VA, USA; Department of Plant Physiology, Sichuan Agricultural University, Wenjiang, Chengdu, China*
VERA WEWER • *Institute of Molecular Physiology and Biotechnology of Plants (IMBIO), University of Bonn, Bonn, Germany*
LOUWRANCE P. WRIGHT • *Department of Biochemistry, Max Planck Institute for Chemical Ecology, Jena, Germany*
YANMEI XIAO • *Department of Plant Biology, University of California, Davis, CA, USA*
SEBASTIAN ZABEL • *Department of Cell and Metabolic Biology, Institute of Plant Biochemistry, Halle, Germany*
XIN ZHAN • *Plant Biochemistry Laboratory, Department of Plant and Environmental Sciences, Copenhagen Plant Science Centre, University of Copenhagen, Copenhagen, Denmark*

Chapter 1

Plant Isoprenoids: A General Overview

Manuel Rodríguez-Concepción

1 Introduction

Isoprenoids, also known as terpenoids, are a group of metabolites with an astounding functional and structural diversity [1–6]. Although they are produced in all free-living organisms, their abundance and variety is highest in plants. Despite the huge structural and functional diversity found among plant isoprenoids, they all derive from the same five-carbon (C_5) precursors: isopentenyl diphosphate (IPP) and its double-bond isomer dimethylallyl diphosphate (DMAPP), also called isoprene units (Fig. 1). Until the last decade of the past century, it was believed that IPP was synthesized from acetyl-CoA via mevalonic acid (MVA) and then isomerized to DMAPP in all living organisms [7]. Soon after the discovery of the MVA pathway in the 1950s, it was found to also function in plant cells [8]. However, experimental evidence supporting the presence of a MVA-independent pathway in bacteria and plant plastids grew stronger until the early 1990s, when compelling new results demonstrated the existence of a completely novel pathway for the production of IPP and DMAPP [9, 10]. The new pathway is currently known as the methylerythritol 4-phosphate (MEP) pathway [11]. With the key contribution of emerging genomic approaches, all the genes encoding MEP pathway enzymes in the model plant *Arabidopsis thaliana* were identified by 2002 [12]. The availability of the Arabidopsis genome also allowed to construct a comprehensive catalogue of genes potentially encoding enzymes of the MVA pathway as well as downstream pathways for the production of the main groups of isoprenoid end-products [13]. Current efforts are focused on understanding how the metabolic flux through the MVA pathway and the MEP pathway is regulated (*see* Chapters 2–5).

Manuel Rodríguez-Concepción (ed.), *Plant Isoprenoids: Methods and Protocols*, Methods in Molecular Biology, vol. 1153, DOI 10.1007/978-1-4939-0606-2_1, © Springer Science+Business Media New York 2014

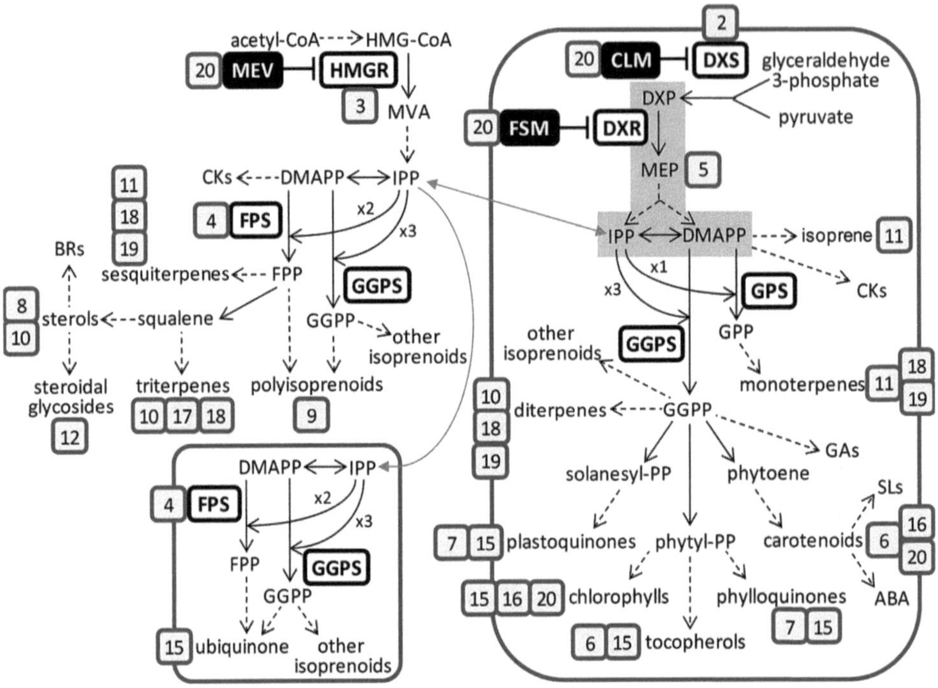

Fig. 1 Isoprenoid biosynthetic pathways in plants. *Gray arrows* represent transport between cell compartments. *Dashed arrows* represent multiple steps. Abbreviations for metabolites are as follows: *ABA* abscisic acid, *BRs* brassinosteroids, *CKs* cytokinins, *DMAPP* dimethylallyl diphosphate, *DXP* deoxyxylulose 5-phosphate, *FPP* farnesyl diphosphate, *GAs* gibberellins, *GGPP* geranylgeranyl diphosphate, *GPP* geranyl diphosphate, *HMG-CoA* hydroxymethylglutaryl CoA, *IPP* isopentenyl diphosphate, *MEP* methylerythritol 4-phosphate, *MVA* mevalonate, *SLs* strigolactones. Enzymes are boxed in *white* (*DXR* DXP reductoisomerase, *DXS* DXP synthase, *FPS* FPP synthase, *GGPP* GGPP synthase, *GPS* GPP synthase, *HMGR* HMG-CoA reductase). Inhibitors are boxed in *black* (*CLM* clomazone, *FSM* fosmidomycin, *MEV* mevinolin). The numbers boxed in *gray* shown next to particular metabolites, enzymes, or inhibitors refer to the chapters that cover them in this issue

Once IPP and DMAPP are produced, prenyltransferase reactions involving the head-to-tail condensation of one or several IPP units to a DMAPP molecule generate geranyl diphosphate (GPP, C_{10}), farnesyl diphosphate (FPP, C_{15}), geranylgeranyl diphosphate (GGPP, C_{20}), and other less abundant prenyldiphosphate molecules of increasing chain length such as octaprenyl diphosphate (C_{40}) and nonaprenyl (solanesyl) diphosphate (C_{45}). These are the starting points for the production of most plant isoprenoids (Fig. 1). Isoprenoids can actually be classified based on the number of isoprene units that form their isoprenoid moiety. Hemiterpenes (C_5) contain a single isoprene unit. Monoterpenes (C_{10}) derive of GPP and consist of two units. Sesquiterpenes (C_{15}) usually derive from FPP and they have three isoprene units. Diterpenes (C_{20}) are GGPP-derived isoprenoids with four C_5 units. Sesterpenes (C_{25}), with five C_5 units, are rare relative to the rest of isoprenoid groups. Triterpenes (C_{30}) are composed of six isoprene units derived from the coupling of two FPP molecules.

Sesquaterpenes (C_{35}) consist of seven C_5 units and they are typically found in microbial organisms. Tetraterpenes (C_{40}) contain eight isoprene units that are normally produced by the condensation of two GGPP molecules. And polyterpenes are formed by more than eight C_5 units.

Plant isoprenoids can also be classified according to their functions in two major groups. The first group is formed by a reduced number of isoprenoid compounds that play essential functions in all plant species and can therefore be considered as "primary" metabolites. This group includes plant hormones such as cytokinins, brassinosteroids, gibberellins, abscisic acid, and strigolactones, sterols (regulators of plant development and membrane architecture), ubiquinone (required for respiration), and photosynthesis related compounds such as carotenoids, chlorophylls, tocopherols, phylloquinones, and plastoquinones (*see* Chapters 6–8). The second group includes tens of thousands of isoprenoid compounds that function as "secondary" metabolites, i.e., nonessential metabolites whose biosynthesis is usually restricted to specific plant families or even to particular plant species, organs, tissues, or developmental stages (*see* Chapters 9–12). They protect plants against herbivores and pathogens, attract pollinators and seed-dispersing animals, and act as allelochemicals that influence competition among plant species. This group includes hemiterpenes (isoprene, prenol), monoterpenes (citral, geraniol, limonene, linalool, menthol, myrcene, pinene, thymol), sesquiterpenes (capsidiol, caryophyllene, farnesene, germacrene, humulene, nerolidol), diterpenes (abietadiene, cafestol, casbene, ferruginol, kaurene, labdane, steviol, rhizathalene, taxadiene), triterpenes (amyrin, arabidiol, lupine, oleanane), and polyterpenes (polyprenols, dolichol, rubber). A large number of secondary isoprenoid metabolites have a commercial value as flavors, pigments, polymers, or drugs. In many cases, plants develop specialized structures for the biosynthesis and storage of very high levels of these secondary isoprenoid metabolites (*see* Chapters 13–16).

An important feature of plant isoprenoid biosynthesis is compartmentalization. Different steps of a particular isoprenoid biosynthetic pathway can take place in several subcellular compartments, cell types, or tissues. Even the production of IPP and DMAPP occurs in different cell compartments (Fig. 1). It is well established that the MEP pathway enzymes are encoded by the nuclear genome and imported into plastids. The MVA pathway enzymes, however, are found in different subcellular compartments, including cytosol, endoplasmic reticulum, and peroxisomes [6]. Prenyltransferase enzymes can also be found in different compartments. In Arabidopsis, most enzymes producing GPP and GGPP are plastidial, but active GGPP synthases are also present in the endoplasmic reticulum and mitochondria [14]. In the case of FPP synthases, cytosolic and mitochondrial forms of the enzyme are present in Arabidopsis (Fig. 1), but peroxisomal and plastidic isoforms have also been reported in other plants (*see* Chapter 4). As represented in Fig. 1, MVA-derived isoprene units are mainly used for the production of triterpenes (including sterols and brassinosteroids), sesquiterpenes, and polyisoprenoids, as well as for the biosynthesis of the polyterpene side chain of ubiquinone upon their transport to mitochondria. And MEP-derived IPP and DMAPP precursors typically support the production of hemiterpenes such as isoprene and some cytokinins, monoterpenes, diterpenes (including gibberellins and the phytol chain of chlorophylls, tocopherols, and phylloquinones), tetraterpenes like carotenoids and derived hormones (abscisic acid and strigolactones), and some polyterpenes like the side chain of plastoquinone. However, labeling experiments have demonstrated that a limited exchange of common isoprenoid precursors between cell compartments takes place, resulting in the production of isoprenoid metabolites with both MVA-derived and MEP-derived isoprene units [15, 16]. Although the exchange rate can be increased by feeding of intermediates of the MVA or the MEP pathways and by metabolic, environmental, or developmental cues, the complete genetic or pharmacological block of either the MVA pathway or the MEP pathway in null mutants or wild type plants treated with inhibitors specifically targeting individual enzymes of each pathway (Fig. 1) is lethal, indicating that the loss of one of the two pathways cannot be compensated by the remaining pathway. A major challenge now is to understand how the production of common isoprenoid precursors in different subcellular locations conveys information between compartments in order to coordinate metabolic fluxes not only between the pathways producing the universal isoprene units but also with downstream pathways leading to the plethora of isoprenoid end-products synthesized in plant cells. Unfortunately, the enzymes involved in most of the pathways leading to secondary isoprenoids still remain to be identified, and the nature of the mechanisms that coordinate the different isoprenoid biosynthesis pathways is little known [4–6, 17]. With the help of new tools and technologies (*see* Chapters 17–21), however, it is expected that many of these pathways will be elucidated and that our knowledge of isoprenoid biosynthesis in plants will be significantly improved in the next few years.

Acknowledgments

Work in my group is supported by grants from Catalan (AGAUR 2009SGR-26), Spanish (DGI BIO2011-23680 and PIM2010IPO-00660), European (FP7 TiMet), and Ibero-American (CYTED 112RT0445-IBERCAROT) agencies.

References

1. Croteau R, Kutchan T, Lewis N (2000) Natural products (secondary metabolites). In: Buchanan B, Gruissem W, Jones R (eds) Biochemistry and molecular biology of plants. American Society of Plant Biologists, Rockville, MD, pp 1250–1268
2. Bouvier F, Rahier A, Camara B (2005) Biogenesis, molecular regulation and function of plant isoprenoids. Prog Lipid Res 44:357–429
3. Tholl D, Lee S (2011) Terpene specialized metabolism in Arabidopsis thaliana. Arabidopsis Book 9:e0143
4. Hemmerlin A, Harwood JL, Bach TJ (2012) A raison d'etre for two distinct pathways in the early steps of plant isoprenoid biosynthesis? Prog Lipid Res 51:95–148
5. Vranova E, Coman D, Gruissem W (2013) Network analysis of the MVA and MEP pathways for isoprenoid synthesis. Annu Rev Plant Biol 64:665–700
6. Pulido P, Perello C, Rodriguez-Concepcion M (2012) New insights into plant isoprenoid metabolism. Mol Plant 5:964–967
7. Chappell J (1995) Biochemistry and molecular biology of the isoprenoid biosynthetic pathway in plants. Annu Rev Plant Physiol Plant Mol Biol 46:521–547
8. Goldstein JL, Brown MS (1990) Regulation of the mevalonate pathway. Nature 343:425–430
9. Lichtenthaler HK (1999) The 1-deoxy-D-XYLULOSE-5-phosphate pathway of isoprenoid biosynthesis in plants. Annu Rev Plant Physiol Plant Mol Biol 50:47–65
10. Rohmer M (1999) The discovery of a mevalonate-independent pathway for isoprenoid biosynthesis in bacteria, algae and higher plants. Nat Prod Rep 16:565–574
11. Phillips MA, Leon P, Boronat A, Rodriguez-Concepcion M (2008) The plastidial MEP pathway: unified nomenclature and resources. Trends Plant Sci 13:619–623
12. Rodríguez-Concepción M, Boronat A (2002) Elucidation of the methylerythritol phosphate pathway for isoprenoid biosynthesis in bacteria and plastids. A metabolic milestone achieved through genomics. Plant Physiol 130:1079–1089
13. Lange BM, Ghassemian M (2003) Genome organization in Arabidopsis thaliana: a survey for genes involved in isoprenoid and chlorophyll metabolism. Plant Mol Biol 51:925–948
14. Beck G, Coman D, Herren E, Ruiz-Sola MA, Rodriguez-Concepcion M, Gruissem W, Vranova E (2013) Characterization of the GGPP synthase gene family in Arabidopsis thaliana. Plant Mol Biol 82:393–416
15. Kasahara H, Hanada A, Kuzuyama T, Takagi M, Kamiya Y, Yamaguchi S (2002) Contribution of the mevalonate and methylerythritol phosphate pathways to the biosynthesis of gibberellins in *Arabidopsis*. J Biol Chem 277:45188–45194
16. Flores-Perez U, Perez-Gil J, Closa M, Wright LP, Botella-Pavia P, Phillips MA, Ferrer A, Gershenzon J, Rodriguez-Concepcion M (2010) PLEIOTROPIC REGULATORY LOCUS 1 (PRL1) integrates the regulation of sugar responses with isoprenoid metabolism in Arabidopsis. Mol Plant 3:101–112
17. Rodríguez-Concepción M, Campos N, Ferrer A, Boronat A (2013) Biosynthesis of isoprenoid precursors in Arabidopsis. In: Bach TJ, Rohmer M (eds) Isoprenoid synthesis in plants and microorganisms: new concepts and experimental approaches. Springer, New York, pp 436–456

Part I

Measurement of Core Enzyme Activities

Chapter 2

Measuring the Activity of 1-Deoxy-D-Xylulose 5-Phosphate Synthase, the First Enzyme in the MEP Pathway, in Plant Extracts

Louwrance P. Wright and Michael A. Phillips

Abstract

The first enzyme in the methylerythritol phosphate (MEP) pathway is 1-deoxy-D-xylulose 5-phosphate (DXP) synthase (DXS). As such this enzyme is considered to be important in the control of plastidial isoprenoid production. Measuring the activity of DXS in plant extracts is therefore crucial to understanding the regulation of the MEP pathway. Due to the relatively low amounts of DXS, the activity of this enzyme can only be measured using highly sensitive analytical equipment. Here, a method is described to determine the DXS enzyme activity in a crude plant extract, by measuring DXP production directly using high performance liquid chromatography linked to a tandem triple quadrupole mass spectrometry detector (LC-MS/MS).

Key words DXS, Enzyme assay, Isoprenoid biosynthesis, LC-MS/MS

1 Introduction

All isoprenoids are produced from the same C5 isoprene units, isopentenyl diphosphate and dimethylallyl diphosphate. In the cytosol these isoprenoid building blocks are biosynthesized through the well-known mevalonate pathway, whereas these same precursors are biosynthesized in the plastids by the recently discovered 2-C-methyl-erythritol 4-phosphate (MEP) pathway [1, 2]. In this pathway, DXS condenses pyruvate and glyceraldehyde 3-phosphate to form 1-deoxy-D-xylulose 5-phosphate (DXP) in the first step towards the formation of plastidic isoprenoids. The MEP pathway provides the precursors for synthesizing products with diverse roles in plant energy metabolism, photosynthesis and plant–insect interactions. Additionally, the MEP pathway provides the precursors for the biosynthesis of many end products with

Manuel Rodríguez-Concepción (ed.), *Plant Isoprenoids: Methods and Protocols*, Methods in Molecular Biology, vol. 1153, DOI 10.1007/978-1-4939-0606-2_2, © Springer Science+Business Media New York 2014

considerable economic value, including the anticancer drugs paclitaxel, vincristine, and vinblastine, common flavor and fragrance compounds such as geraniol, linalool, and menthone, cosmetics such as shikonin, and other industrial raw materials such as the monoterpenoid olefinic hydrocarbons used to make turpentine. A detailed knowledge of the regulation of the early steps which provide common precursors for diverse downstream isoprenoid pathways may lead to increased production of commercially valuable isoprenoids. However, the processes by which this pathway is regulated are still poorly understood.

The amount of end products produced by a metabolic pathway depends on its flux, which in turn is dependent on the rate at which the individual enzymatic steps convert their respective intermediates. Identifying the enzymatic steps most important for metabolic flux will aid the elucidation of the regulatory mechanisms responsible for isoprenoid biosynthesis. To achieve this, it is crucial to measure the activities of the individual enzymes in plant extracts. As the first committed enzyme, DXS is considered to be an important enzyme in the regulation of the MEP pathway. Measurement of DXS activity was first accomplished using recombinant proteins following the initial isolation of their cDNAs from *E. coli* [3] and mint [4]. In the former case, ^{14}C-labelled pyruvate was spiked into the reaction mixture, and the radioactive products were separated by TLC, a technique adopted elsewhere for analyzing DXS activity [5]. In the latter case, the DXP product of mint DXS was derivatized for gas chromatographic analysis by removal of the 5-phosphate and trimethylsilylation of the free hydroxyls. The activity of DXS has also been confirmed on a qualitative level by bacterial complementation through the transformation of *E. coli* strains deficient in DXS activity [6–8].

Measuring the activity of DXS in plant tissue, however, is another matter. Pyruvate and glyceraldehyde 3-phosphate are intermediates of a multitude of other metabolic pathways, and their respective enzymes will also be present in plant protein extracts, competing with DXS for exogenous substrates. Furthermore, the concentration and activity of DXS are very low on a cellular scale, necessitating highly sensitive methods to detect its activity. To overcome these problems methods were developed to derivatize DXP with 3,5-diaminobenzoic acid and measure the fluorescence emitted by the quinaldine product [9]. Another approach uses isotope ratio mass spectrometry to detect the $^{13}CO_2$ emitted by the DXS reaction when using labelled pyruvate as substrate [10]. A further solution to low sensitivity was proposed involving the reductive amination of the DXP product with anthranilic acid. The fluorescent product is then separated by HPLC and detected by a fluorescence detector [11]. However, its sensitivity is limited to measuring recombinant proteins. Other methods overcome the problem of low enzyme activities in plant extracts by increasing the DXS enzyme concentrations by isolating plastids [12, 13], methods that are themselves prone to high variability. Thus far, no published method of measuring DXS activity is capable of reproducibly quantifying the low levels of DXS in plant extracts.

Here we describe a fast and simple method to measure DXS activity. A crude plant protein extract is prepared which is added to an enzyme reaction mixture consisting of the glyceraldehyde 3-phosphate and pyruvate substrates as well as the thiamine diphosphate and Mg^{2+} cofactors. After the reaction is stopped, the DXP product is detected using the sensitivity provided by LC-MS/MS.

2 Materials

Use analytical grade reagents and ultrapure deionized water. Extra care should be taken that the water used for liquid chromatography triple quadrupole mass spectrometry (LC-MS/MS) analyses should be purified to a resistivity of 18 MΩ at 25 °C and the acetonitrile should be at least HPLC grade. All LC-MS/MS solvent additives used must be LC-MS grade. All reagents and plant material should be kept on ice during the enzyme extraction and enzyme assay preparation procedure (unless indicated otherwise). Waste disposal regulations must be meticulously followed.

2.1 Crude Enzyme Extraction

1. Stock extraction buffer: 50 mM Tris–HCl pH 8.0 (*see* **Note 1**), 10 % glycerol (*see* **Note 2**), 0.5 % Tween 20, 1 % polyvinylpyrrolidone (PVP) (average molecular weight 360,000), 2 mM imidazole, 1 mM NaF, and 1.15 mM molybdate (*see* **Note 3**). The stock extraction buffer is made to a volume of 100 mL and can be stored for 1 week at 4 °C.
2. Prepare fresh a small quantity of 1 M dithiothreitol (DTT), a small quantity of 1 M ascorbic acid and a small quantity of 100 mM thiamine pyrophosphate (TPP). These reagents can be prepared in a final volume of 1 mL or scaled up for larger numbers of extractions (*see* **Note 4**). Keep the solutions on ice.
3. Protease inhibitor cocktail for plant cell extracts (Sigma-Aldrich). Store at −20 °C (*see* **Note 5**).
4. Microcentrifuge tubes of 2 mL capacity.
5. Vertical rotator (Stuart rotator SB3, VWR, or equivalent).
6. Refrigerated microcentrifuge (Eppendorf centrifuge 5415R, or equivalent).

2.2 DXS Enzyme Assay

1. Assay buffer: 100 mM Tris–HCl, pH 8.0 (*see* **Note 1**), 20 % glycerol (*see* **Note 2**), 20 mM $MgCl_2$ (*see* **Note 6**). Store for up to a week at 4 °C.
2. Prepare fresh a small quantity of 1 M sodium pyruvate (*see* **Note 4**). Dissolve 110.04 mg in 1 mL water and keep on ice.

3. Freshly prepared 1 M DTT and 100 mM TPP (*see* Subheading 2.1, **item 2**).
4. DL-Glyceraldehyde 3-phosphate (GAP) supplied as a 45–55 mg/mL solution (Sigma-Aldrich). Store as aliquots at -20 °C.
5. Water bath set at 25 °C.
6. Chloroform. Store at room temperature in a solvent cabinet.

2.3 DXP Detection

1. 5 M ammonium acetate stock solution: Dissolve 19.3 g LC-MS grade ammonium acetate in 50 mL water and store at 4 °C.
2. Liquid chromatography (LC) solvents: 20 mM ammonium acetate, pH 10.0 (solvent A), 80 % acetonitrile containing 20 mM ammonium acetate, pH 10.0 (solvent B) (*see* **Note 7**).
3. 1 mg/mL DXP (Sigma-Aldrich) dissolved in 5 mM ammonium acetate, 50 % acetonitrile (*see* **Note 8**).
4. 1 mg/mL ^{13}C isotopically labelled DXP dissolved in water (*see* **Note 9**).
5. XBridge BEH Amide column (150×2.1 mm, 3.5 μm, Waters) with an XBridge BEH Amide Sentry guard cartridge (10×2.1 mm, 3.5 μm, Waters) and a high pressure pre-column filter (SSI High Pressure Pre-column Filter, Sigma) or equivalent.
6. Agilent 1200 HPLC system (Agilent Technologies) or equivalent.
7. API 3200 triple quadrupole mass spectrometer (Applied Biosystems) or equivalent.

3 Methods

All steps used for extraction of enzymes should be carried out on ice, unless otherwise specified. This method has been optimized for measuring DXS activity in leaf material of *Arabidopsis thaliana* and has also been tested on leaf material from *Populus tremuloides*. Measuring DXS activity in other plant species or tissues might require additional optimization.

3.1 Crude Enzyme Extraction

1. Prepare the extraction buffer by adding 100 μL of 1 M DTT, 100 μL of protease inhibitor cocktail, 10 μL of 1 M ascorbic acid and 10 μL of 100 mM TPP to a graduated cylinder and make up to 10 mL with the stock extraction buffer (*see* **Note 10**). Keep the extraction buffer on ice.
2. Homogenize plant material to a fine powder in liquid nitrogen in a pre-cooled mortar and pestle. Transfer the homogenized plant material to a plastic tube kept frozen in liquid nitrogen or on dry ice.

3. Weigh between 20 and 25 mg fresh weight plant material in a 2 mL microcentrifuge tube pre-cooled in liquid nitrogen. It is crucial that the plant material does not thaw; work quickly and return the weighed material to liquid nitrogen or dry ice (*see* **Note 11**). Note the exact weight.
4. Add 1 mL of extraction buffer to each microcentrifuge tube and mix gently on a vertical rotator at 20 rpm for 15 min at 4 °C (*see* **Note 12**). Transfer tubes to a pre-cooled microcentrifuge and centrifuge at 16,000 × *g* for 20 min at 4 °C. The supernatant can now be used in the enzyme assays.

3.2 DXS Enzyme Assay

1. Prepare the enzyme reaction mixture by adding 0.25 μL of 1 M DTT, 1 μL of 100 mM TPP, 1 μL of 1 M pyruvate, 1 μL of the phosphatase inhibitor cocktail (*see* **Note 3**), 1 μL of protease inhibitor cocktail, and 1 μmol of the GAP solution (*see* **Note 13**) to 50 μL assay buffer. Add enough water to reach a final volume of 70 μL (Table 1).
2. Add 30 μL of the enzyme extract to the reaction mixture and incubate in a 25 °C water bath for 2 h (*see* **Note 14**).
3. Stop the enzyme reaction by adding one volume of chloroform and vortexing vigorously (*see* **Note 15**). Centrifuge at maximum velocity in a microcentrifuge to achieve phase separation and

Table 1
Overview of the amounts of the different reagents necessary to prepare a single DXS enzyme reaction with a final volume of 100 μL

Reagent	Stock concentration	Assay concentration	Vol. added
Assay buffer	100 mM Tris–HCl 20 mM $MgCl_2$ 20 % Glycerol	50 mM Tris–HCl 10 mM $MgCl_2$ 10 % Glycerol	50 μL
DTT	1 M	2.5 mM	0.25 μL
TPP	100 mM	1 mM	1 μL
PhI[a]	200 mM imidazole 100 mM NaF 115 mM sodium molybdate	2 mM imidazole 1 mM NaF 1.15 mM sodium molybdate	1 μL
PrI[b]	Solution	1 %	1 μL
Sodium pyruvate	1 M	10 mM	1 μL
GAP	Solution	10 mM	3.4 μL [c]
Water			12.35 μL
Enzyme			30 μL

[a]Phosphatase inhibitors
[b]Protease inhibitor cocktail
[c]Volume of GAP used depends on the concentration of the solution purchased (*see* **Note 13**)

transfer 45 μL of the aqueous upper phase to a HPLC vial fitted with a 200 μL insert. Add 5 μL of a 10 ng/μL [3,4,5-$^{13}C_3$] DXP internal standard (*see* **Note 16**). Dilute this solution with one volume (50 μL) of methanol to approximate the initial mobile phase and improve peak shape.

3.3 DXP Detection with LC-MS/MS

1. Optimize the instrument parameters for detecting DXP by infusing a 100 μg/mL solution of DXP dissolved in 5 mM ammonium acetate in 50 % acetonitrile into the mass spectrometer (*see* **Note 17**) using the automatic optimization procedure under negative ionization mode. The optimized parameters will be different for different LC-MS/MS instruments, *see* Table 2 for the optimized parameters of an API 3200 LC-MS/MS.
2. Liquid chromatographic separation of DXP is achieved with a HILIC column (XBridge BEH Amide) using the following LC parameters: column temperature = 25 °C, injection volume = 20 μL, flow rate = 0.5 mL/min, total run time = 20 min. Use a solvent gradient program starting with a linear gradient from 0 to 20 % solvent A over 0.5 min, isocratic separation at 20 % A until 10 min, a linear increase to 30 % A by 11 min, hold at 30 % A until 15 min, return to initial conditions of 0 % A at 15.1 min and equilibrate at 0 % A until 20 min.
3. Ionization was achieved using eletcrospray ionization with a Turbospray ion source operating under negative ionization mode. The ion spray voltage was maintained at −4,500 eV and the turbo gas temperature at 700 °C. The nebulizing gas was set at 70 psi, the heating gas at 30 psi, the curtain gas at 30 psi and the collision gas at 10 psi. The ionization parameters also need to be optimized for different LC-MS/MS systems.

Table 2
Selected reaction monitoring transitions and conditions used in the API 3200 LC-MS/MS

Analyte	Precursor *m/z*	Product *m/z*	EP[a] (V)	CEP[b] (V)	CE[c] (V)	CXP[d] (V)
DXP	312.9	138.9	−8	−12	−18	−4
		78.9	−7	−28	−42	0
DXP-$^{13}C_3$	215.9	140.9	−8	−12	−18	−4
		78.9	−7	−28	−42	0

The dwell times used were 0.15 s with a declustering potential (DP) of −20 V. Both Q1 and Q3 quadrupoles were operated at unit resolution

[a]Entrance potential

[b]Cell entrance potential

[c]Collision energy

[d]Cell exit potential

4. The optimized parameters of the triple quadrupole mass spectrometer are shown in Table 2. DXP is detected as a mass-to-charge ratio (m/z) of 138.9, the product ion of the DXP [M – 1] precursor ion (m/z 212.9). The labelled internal standard is detected as precursor ion → quantifier ion: m/z 215.9 → 140.9 (*see* Fig. 1). The identities of DXP and its internal standard can be verified using the precursor ion → qualifier ion combinations m/z 212.9 → 78.9 and m/z 215.9 → 78.9, respectively (*see* **Note 18**). The product ion and qualifier ion obtained after fragmentation are shown in Fig. 1.
5. Construct a calibration curve with DXP in the range of 0.1–10 ng/μL.

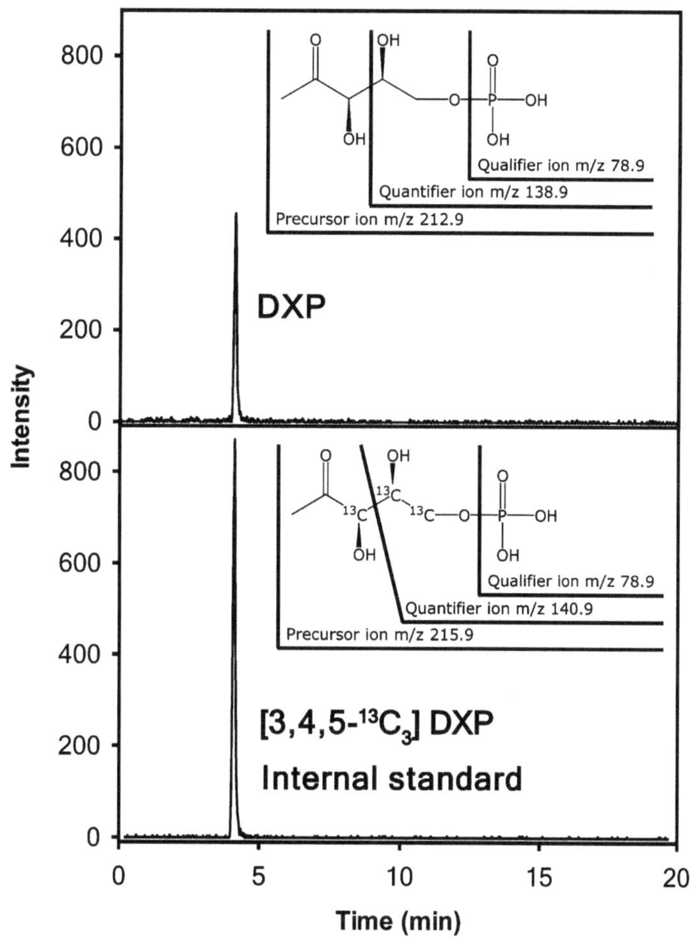

Fig. 1 Representative chromatogram for the DXS assay product, DXP, and the internal standard, [3,4,5-$^{13}C_3$] DXP. The fragmentation of DXP and the internal standard to produce the quantifier ions m/z 138.9 and m/z 140.9, as well as the qualifier ion m/z 78.9, is also shown

6. Analyst 1.5 software (Applied Biosystems) was used for data acquisition and processing. Quantify the DXP produced by the DXS enzyme reaction using the calibration curve and normalize it to the detected internal standard to correct for any ion suppression effects (*see* **Note 19**). Lastly adjust the DXP amounts for the dilution due to the added acetonitrile and internal standard (*see* **Note 20**).

4 Notes

1. Prepare a 1 M Tris–HCl pH 8.0 buffer that can be stored at room temperature and later diluted to the desired concentration when preparing buffers. We usually make a 1 L solution by dissolving 121.1 g Tris in 600 mL water in a graduated cylinder or glass beaker. Adjust the pH by adding HCl, starting with concentrated acid at first and switching to more dilute acid (e.g., 1 N) when nearing the desired pH. When diluting concentrated acids it is important to remember to add the acid to the water. After the buffer was adjusted to pH 8.0, bring the final volume to 1 L.
2. Glycerol is stored as a 50 % solution. This makes it much easier to handle than the viscous undiluted reagent.
3. The phosphatase inhibitor cocktail consisting of imidazole, NaF and molybdate are prepared as a 100× stock solution and stored at 4 °C. Dissolve 1.36 g imidazole, 0.42 g NaF, and 2.78 g molybdate in 100 mL water. Use protective clothing, gloves and a face mask when weighing NaF. Due to its toxicity, protective clothing and gloves should be used when handling any of the reagents containing NaF.
4. All the reagents that are more labile are prepared fresh and added to the extraction buffer shortly before use. To save on consumables and expenses, small quantities are prepared by weighing an approximate amount in 2 mL eppendorf tubes and then adding the appropriate amount of water to obtain the desired concentration. This approach does not take into consideration the volume of the reagent in the final solution, but the slight decrease in reagent concentration so achieved will have no effect on the final enzyme reaction. For example, to make a 1 M solution of DTT of about 1 mL, weigh between 100 and 200 mg in a 2 mL eppendorf tube and write down the exact weight. Now divide the weighed amount of DTT by 154.2 and multiply by 1,000 to get the volume of water (in μL) to add to the 2 mL eppendorf tube to get a 1 M solution. In other words, dissolve 154.2 mg DTT in 1 mL to prepare a 1 M solution, dissolve 176.12 mg L-ascorbic acid in 1 mL to prepare a 1 M solution, and dissolve 46.08 mg TPP in 1 mL to prepare a 100 mM solution.

5. The protease inhibitor cocktail is supplied as a solution in dimethyl sulfoxide (DMSO) which solidifies when kept on ice. Thaw prepared aliquots at room temperature and return to −20 °C as soon as possible to minimize degradation.
6. Prepare a 1 M $MgCl_2$ solution to dilute into the assay buffer to achieve the desired concentration. Weigh out 9.5 g anhydrous $MgCl_2$, dissolve in 100 mL water and store at room temperature.
7. Ammonium acetate is highly hygroscopic and should be kept under argon. To ease preparation of the solvents used for LC a stock solution of 5 M is prepared and stored at 4 °C. This is then diluted to obtain a 10× concentration and adjusted to the correct pH using LC-MS grade ammonium hydroxide. Add 4 mL of the 5 M ammonium acetate stock solution to approximately 80 mL water and add concentrated ammonium hydroxide to reach pH 10. Finally adjust the volume to 100 mL. The ammonium acetate buffer is then diluted with either water or acetonitrile to prepare the respective LC solvents. Add acetonitrile to the aqueous ammonium acetate solution for easy mixing of the salt with the organic solvent.
8. Prepare a 10 mM ammonium acetate solution by diluting the 5 M ammonium acetate stock by adding 200 μL of the stock to a graduated cylinder and make up to 100 mL with water. Then add 1 volume 10 mM ammonium acetate to 1 volume acetonitrile.
9. ^{13}C isotopically labelled DXP is not commercially available, but could be prepared enzymatically as described [14]. The simplest method is to use recombinant *Escherichia coli* DXS to produce [1,2-$^{13}C_2$] DXP by using [2,3-$^{13}C_2$] pyruvate as substrate.
10. Adjust the volumes according to how many extractions are planned, allowing 1 mL of extraction buffer per extract.
11. When the enzyme activity is normalized to protein content, it is not necessary to determine the exact weight, and only an approximate amount of plant material need to be transferred to a pre-cooled microcentrifuge tube. Another alternative, which works well for *Arabidopsis thaliana*, is to freeze dry the plant material and then weigh 5 mg of the dried material in a microcentrifuge tube at room temperature. The dried material should, however, still be kept on ice and be stored at -20 °C before weighing.
12. When adding the extraction buffer to the microcentrifuge tubes pre-cooled in liquid nitrogen, the extraction buffer will freeze on the tube surface, hindering proper mixing of the plant material with the extraction buffer. To circumvent this problem, put the tubes on ice for exactly 2 min before adding the extraction buffer. This allows the microcentrifuge tubes to

warm sufficiently to minimize frozen extraction buffer but still keep the plant material in a frozen state.

13. Use the following equation to calculate the amount of GAP to use in each 100 µl enzyme assay to obtain a 10 mM concentration: µL GAP to be used = 170.06/(concentration of GAP reagent in units of mg/mL). For example, use 170.06/50 = 3.4 µL GAP for a 50 mg/mL solution.
14. Although the maximum temperature for the *A. thaliana* DXS assay is at approximately 40 °C, the technical variation between enzyme assays is lower at room temperature.
15. The chloroform extraction is actually done to remove hydrophobic compounds from the reaction mixture, which otherwise interfere with the separation of the DXP on a HILIC column. The chloroform extraction was consequently also found to be a sufficient procedure for stopping the DXS enzyme reaction.
16. Although LC-MS/MS is a very sensitive analytical technique, it suffers from the disadvantage of ion-suppression. For a compound to be detected with a mass spectrometer it needs to contain a charge. In the case of LC-MS/MS, the compound is ionized through electro spray ionization in the ion source by applying electrical charge to the LC eluent. The presence of other molecules in the eluent, competing with the compound of interest for the available charge, will influence the ionization efficiency of the compound, and hence its detection by the mass spectrometer. It is thus usually not possible to measure the absolute quantity of a compound in a complex mixture, such as an enzyme reaction of a crude plant extract, when using an external standard curve to calculate the amount of the analyte. To compensate for the ion-suppression occurring, a known amount of ^{13}C isotopically labelled DXP is added to the enzyme reaction product. The labelled DXP will have the same ionization efficiency of the unlabelled DXP, but can be distinguished by its mass difference. The labelled DXP can thus be used as internal standard to quantify the absolute amount of DXP. However, when no internal standard is used, it will still be possible to measure the relative DXS activities of different plant extracts, as long as the plant tissues used have a similar matrix composition.
17. Dissolving DXP in 50 % acetonitrile containing 5 mM ammonium acetate significantly increases the ionization efficiency and also more closely represents the ionization conditions during a LC-MS/MS run.
18. Although the qualifier ion (m/z 78.9) gives a more sensitive signal, the less sensitive quantifier ion (m/z 138.9, and m/z 140.9 for the internal standard) is used for quantification to minimize potential background signals.

19. Use the following equation to normalize the detected DXP relative to the internal standard: DXP = ((DXP measured) × (Internal standard added))/(Internal standard measured). For example, if 1 ng internal standard was injected and 0.55 ng DXP and 0.75 ng internal standard was detected, the absolute amount of DXP injected will be (0.55 × 1)/0.75 = 0.73 ng.
20. To account for the dilution due to the addition of acetonitrile and the internal standard, multiply the amount of DXP detected by 2 and divides this by 0.9. Using the example of **Note 19**, this means that the undiluted amount of DXP will be (0.73 × 2)/0.9 = 1.62 ng/μL if 1 μL was injected.

Acknowledgments

We thank Bettina Raguschke for technical assistance and Felix Rohdich, Adelbert Bacher and Wolfgang Eisenreich for the kind gift of ^{13}C labelled DXP. The project was funded by the Max Planck Gesellschaft.

References

1. Rodriguez-Concepcion M, Boronat A (2002) Elucidation of the methylerythritol phosphate pathway for isoprenoid biosynthesis in bacteria and plastids. A metabolic milestone achieved through genomics. Plant Physiol 130: 1079–1089
2. Lichtenthaler HK (1999) The 1-deoxy-D-xylulose-5-phosphate pathway of isoprenoid biosynthesis in plants. Annu Rev Plant Physiol Plant Mol Biol 50:47–65
3. Lois LM, Campos N, Putra SR et al (1998) Cloning and characterization of a gene from *Escerichia coli* encoding a transketolase-like enzyme that catalyzes the synthesis of D-1-deoxyxylulose 5-phosphate, a common precursor for isoprenoid, thiamine, and pyridoxol biosynthesis. Proc Natl Acad Sci U S A 95: 2105–2110
4. Lange BM, Wildung MR, McCaskill D et al (1998) A family of transketolases that directs isoprenoid biosynthesis via a mevalonate-independent pathway. Proc Natl Acad Sci U S A 95:2100–2104
5. Walter MH, Hans J, Strack D (2002) Two distantly related genes encoding 1-deoxy-D-xylulose 5-phosphate synthase: differential regulation in shoots and apocarotenoid-accumulating mycorrhizal roots. Plant J 31: 243–254
6. Phillips MA, Walter MH, Ralph SG et al (2007) Functional identification and differential expression of 1-deoxy-D-xylulose 5-phosphate synthase in induced terpenoid resin formation of Norway spruce (*Picea abies*). Plant Mol Biol 65:243–257
7. Sauret-Gueto S, Uros EM, Ibanez E et al (2006) A mutant pyruvate dehydrogenase E1 subunit allows survival of *Escherichia coli* strains defective in 1-deoxy-D-xylulose 5-phosphate synthase. FEBS Lett 580:736–740
8. Cordoba E, Porta H, Arroy A et al (2011) Functional characterization of the three genes encoding 1-deoxy-D-xylulose 5-phosphate synthase in maize. J Exp Bot 62:2023–2038
9. Querol J, Besumbes O, Lois LM et al (2001) A fluorometric assay for the determination of 1-deoxy-D-xylulose 5-phosphate synthase activity. Anal Biochem 296:101–105
10. Ghirardo A, Zimmer I, Brueggemann N et al (2010) Analysis of 1-deoxy-D-xylulose 5-phosphate synthase activity in grey poplar leaves using isotope ratio mass spectrometry. Phytochemistry 71:918–922
11. Han Y-S, Sabbioni C, van der Heijden R (2003) High-performance liquid chromatography assay for 1-deoxy-D-xylulose 5-phosphate synthase activity using fluorescence detection. J Chromatogr A 986:291–296

12. Fraser PD, Enfissi EMA, Halket JM et al (2007) Manipulation of phytoene levels in tomato fruit: effects on isoprenoids, plastids, and intermediary metabolism. Plant Cell 19:3194–3211

13. Flores-Perez U, Perez-Gil J, Closa M et al (2010) PLEIOTROPIC REGULATORY LOCUS 1 (PRL1) integrates the regulation of sugar responses with isoprenoid metabolism in *Arabidopsis*. Mol Plant 3:101–112

14. Hecht S, Kis K, Eisenreich W et al (2001) Enzyme-assisted preparation of isotope-labeled 1-deoxy-D-xylulose 5-phosphate. J Org Chem 66:3948–3952

Chapter 3

Determination of 3-Hydroxy-3-methylglutaryl CoA Reductase Activity in Plants

Narciso Campos, Montserrat Arró, Albert Ferrer, and Albert Boronat

Abstract

The enzyme 3-hydroxy-3-methylglutaryl CoA (HMG-CoA) reductase catalyzes the NADPH-mediated reductive deacylation of HMG-CoA to mevalonic acid, which is the first committed step of the mevalonate pathway for isoprenoid biosynthesis. In agreement with its key regulatory role in the pathway, plant HMG-CoA reductase is modulated by many diverse external stimuli and endogenous factors and can be detected to variable levels in every plant tissue. A fine determination of HMG-CoA reductase activity levels is required to understand its contribution to plant development and adaptation to changing environmental conditions. Here, we report a procedure to reliably determine HMG-CoA reductase activity in plants. The method includes the sample collection and homogenization strategies as well as the specific activity determination based on a classical radiochemical assay.

Key words 3-Hydroxy-3-methylglutaryl CoA reductase, HMG-CoA reductase, HMGR, MVA, Mevalonate pathway, Isoprenoid, Terpenoid

1 Introduction

1.1 Molecular Properties of Plant HMG-CoA Reductase

HMG-CoA reductase (HMGR) (EC 1.1.1.34) catalyzes the first committed step of the mevalonate pathway for isoprenoid biosynthesis, consisting in the NADPH-mediated reductive deacylation of HMG-CoA to mevalonic acid [1] (Fig. 1). The enzyme exerts a key regulatory role on the flux of the mevalonate pathway in all eukaryotes [2–4] and in plants is critical not only for normal growth and development but also for the adaptation to diverse challenging conditions [4]. Plant HMGR is controlled at transcriptional and posttranslational levels in response to many developmental and environmental signals such as phytohormones, calcium, calmodulin, light, wounding, elicitor treatment, and pathogen attack [5, 6]. In all plants studied so far, HMGR is encoded by a multigene family [7]. Some HMGR genes participate in general house-keeping roles such as sterol biosynthesis, whereas others are required for more specific developmental or adaptive

Manuel Rodríguez-Concepción (ed.), *Plant Isoprenoids: Methods and Protocols*, Methods in Molecular Biology, vol. 1153, DOI 10.1007/978-1-4939-0606-2_3, © Springer Science+Business Media New York 2014

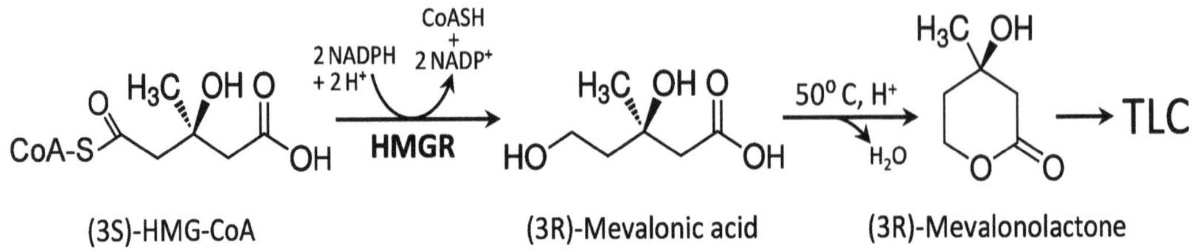

Fig. 1 The HMGR activity assay. Eukaryotic HMGR catalyzes the stereospecific NADPH-dependent reductive deacylation of (3S)-HMG-CoA to (3R)-mevalonic acid [1]. The HMGR assay reaction product is subsequently converted to mevalonolactone by heating in acid medium. The heat treatment also hydrolyses HMG-CoA to free HMG acid and CoASH. The mevalonolactone is more hydrophobic than the HMG acid or any remaining HMG-CoA and separates well from these compounds in the TLC system

processes [4]. This scenario anticipates a complex transcriptional control by multiple regulatory circuits [8]. In addition, posttranslational control of plant HMGR has been proposed to occur in response to light [9, 10], salt stress [11], or alteration of the metabolic flux through sterol or sphingolipid biosynthetic pathways [12, 13]. Protein degradation [10, 11], inhibition [14] or activation by calcium [15], and phosphorylation at a conserved site of the catalytic domain [16–18] are mechanisms by which plant HMGR is posttranslationally modulated. Protein phosphatase 2A (PP2A) has been identified both as a transcriptional and a posttranslational regulator of HMGR in *Arabidopsis* [11]. The interaction with HMGR occurs through a B" PP2A regulatory subunit, which is also a calcium-binding protein [11]. These observations uncovered the potential of PP2A to integrate developmental and Ca^{2+}-mediated environmental signals in the control of plant HMGR [19]. Because plant HMGR may be affected by many diverse stimuli, a careful control of the plant growth and sample collection conditions is critical to attain reproducibility in its activity determination.

Plant HMGR is an endoplasmic reticulum (ER) protein of about 63–70 kDa. *Arabidopsis* and tomato HMGR have been shown to span the ER membrane twice [20–22]. Both the N-terminal region and the highly conserved catalytic domain are in the cytosol, whereas only a short stretch of the protein is in the ER lumen. Insertion in the ER membrane is mediated by the Signal Recognition Particle (SRP) that recognizes the two hydrophobic sequences which will become membrane spanning segments [20]. Since these two sequences are highly conserved, it was proposed that all plant HMGR variants are primarily targeted to the ER [20]. However, in *Arabidopsis* and tobacco cells, HMGR also localizes in still uncharacterized bodies that range between 0.2 and 2.0 μm in diameter [23, 24]. HMGR was purified as a dimer or tetramer with subunits of 55–45 kDa from potato [25], *Hevea* [26], and radish [27] which, according to the corresponding nucleotide sequences [28–30], may correspond to the protease-released catalytic portion. This is in agreement with the previous crystallization of human HMGR catalytic domain as a tetramer, with each active site (four in total) formed in the interphase of two neighboring HMGR subunits [31]. Thus, the available data indicate that not only the sequence of the catalytic domain of HMGR but also its quaternary structure is conserved in high eukaryotes.

1.2 Setting Up the HMG-CoA Reductase Activity Assay

Plant HMGR activity was first detected in 1967, in *Hevea brasiliensis* latex [32], and demonstrated at high levels in the same system 2 years later [33]. In the 1970s, different methods to determine HMGR activity were set up in pea seedlings [34], sweet potatoes [35], anise cell line [36], tobacco seedlings [37] and etiolated radish seedlings [38]. This pioneer work uncovered some common biochemical properties of plant HMGR, that have allowed the detection of its enzymatic activity in many different organs, tissues, cell lines or extracts of 40 plant species (Table 1). When plants are submitted to homogenization and centrifugation, the HMGR activity is detected in the final microsomal pellet (about 100,000 × *g*), as it would be expected from its ER localization, but also in the sediment obtained at low centrifugation forces (1,500–16,000 × *g*) [5, 39] (Table 1). Therefore, a crude extract, instead of a more elaborated subfraction, may be required to measure total HMGR activity. Alternatively, plant HMGR has been released from the insoluble fraction by extraction with a nonionic detergent [27, 40] or by intended proteolytic digestion [25]. Higher levels of potato HMGR activity were obtained in the presence of cysteine, serine and threonine peptidase inhibitors [6, 10], indicating that these should be included in the homogenization buffer to recover the total HMGR activity.

The catalytic activity of plant HMGR depends on free thiol groups and a reducing agent has been used to protect their reduced state [41, 42]. It was reported that DTT is better than 2-mercaptoethanol or glutathione for this purpose [34, 43]. Maximal HMGR activity occurs at pH 7.3–7.5 in radish [27] and sweet potato [42], and about pH 6.8 in Hevea latex [40, 41]. In pea [14] and guayule [44], two pH optima were found, corresponding to HMGR from the heavy or the light fractions: 7.9 and 6.9, respectively, in the case of pea, and 7.5 and 7.0, respectively, in the case of guayule. HMGR activity was assayed in phosphate buffer in all the above systems, but in *Arabidopsis* we found that the Hepes-KOH buffer system is also suitable. Ethylenediaminetetraacetic acid (EDTA) is used in most of the above-mentioned methods. It was proposed to inhibit the subsequent reactions of the mevalonate pathway in Hevea latex [33] and was found to increase the apparent HMGR activity in sweet potato extracts [42].

In most cases, HMGR activity has been detected in plant extracts by a [^{14}C]HMG-CoA-based radiochemical method, which

Table 1
List of plants where HMGR activity has been measured. Only the earliest report for each particular case (plant species, organ, tissue, cell culture and subcellular fraction) is indicated. Pellet (P) and supernatant (S) fractions obtained by differential centrifugation appear with the corresponding gravitational force (g), as a suffix

Species	Organ/tissue	Fraction	References
Arabidopsis thaliana	Rosette leaf	S_{200}	[58]
	Fully expanded leaf	S_{200} and $P_{16,000}$	[57]
	Green seedling	S_{200}	[12]
	Dry seed	S_{200}	[56]
	Seedling aerial part and root	S_{200}	[11]
Arachis hypogaea	Green seedling	$P_{105,000}$	[60]
Artemisia annua	Leaf	$P_{105,000}$	[61]
Bixa orellana	Callus, leaf	$P_{100,000}$	[62]
Brassica napus	Developing seed	P_{1200} and S_{1200}	[63]
Cannabis sativa	Leaf	$S_{24,000}$	[50]
Cucumis melo	Fruit pericarp	$P_{50,000}$	[64]
Daucus carota	Cell culture	$P_{10,000}$ and $P_{105,000}$	[65]
Dunaliella salina	Cell culture (entire organism)	Low speed pellet (?)	[66]
Euphorbia lathyris	Latex, leaf, stem	S_{500} and $P_{18,000}$	[67]
	Latex	$S_{5,000}$ and $P_{5,000}$	[68]
Glycine max	Seedling apical part, cotyledon, hypocotyl and root, cell culture	$P_{5,000}$ and $P_{100,000}$	[69]
Gossypium barbadense	Stele tissue	$P_{105,000}$	[70]
Gossypium hirsutum	Stele tissue	$P_{105,000}$	[70]
Helianthus tuberosus	Tuber explants	$P_{15,000}$ and $P_{105,000}$	[71]
Hevea brasiliensis	Latex	$S_{20,000}$	[32]
	Latex	P_{600}	[33]
Hordeum vulgare	Seedling	$P_{16,000}$ and $P_{105,000}$	[72]
Ipomoea batatas	Root	$P_{105,000}$	[35]
Lithospermum erythrorhizon	Cell culture	$P_{10,000}$ and $P_{100,000}$	[73]
	Hairy root	$S_{100,000}$ and $P_{100,000}$	[74]
Malus x domestica	Fruit skin	$S_{105,000}$ and $P_{105,000}$	[75]
Medicago sativa	Hairy root	$P_{100,000}$	[53]
Nepeta cataria	Leaf, callus tissue	$P_{2,000}$ and $P_{100,000}$	[76]
Nicotiana benthamiana	Leaf	$S_{13,000}$	[77]

(continued)

Table 1 (continued)

Species	Organ/tissue	Fraction	References
Nicotiana tabacum	Aerial part of seedling	$P_{200,000}$	[37]
	Cell culture KY-14	$P_{10,000}$ and $P_{100,000}$	[78]
	Seedling, fully expanded leaf, Callus	$P_{100,000}$	[79]
	Cell culture BY-2	$P_{100,000}$	[13]
	Developing seed	$P_{1,200}$ and $S_{1,200}$	[63]
Ochromonas malhamensis	Cell culture (entire organism)	$P_{145,000}$	[80]
Parthenium argentatum	Fully expanded leaf	$S_{45,000}$ and $P_{2,500}$	[44]
	Bark of lower stem	$P_{105,000}$	[81]
Persea americana	Fruit mesocarp	$P_{27,000}$	[82]
	Seed	$P_{27,000}$	[83]
Phaseolus radiatus	Leaf	$P_{2,500}$	[84]
Picea abies	Seedling, leaf, callus, cell culture	$P_{18,000}$ and $P_{105,000}$	[85]
Pimpinella anisum	Cell culture	Particulate fraction (?)	[36]
Pisum sativum	Etiolated seedling, green seedling	Crude extract, $P_{2,000}$, $P_{10,000}$ and $P_{105,000}$	[34]
Raphanus sativus	Seedling	$P_{16,000}$ and $P_{105,000}$	[38]
Salvia miltiorrhiza	Hairy root	(?)	[48]
Sinapis alba	Green seedling	$S_{200,000}$ and $P_{120,000}$	[86]
Solanum lycopersicum	Fruit, leaf	$S_{105,000}$ and $P_{3,500}$	[87]
Solanum tuberosum	Tuber tissue	$P_{105,000}$	[88]
Solanum xantocarpum	Cell suspension culture	$S_{4,000}$	[89]
Spinacia oleracea	Leaf	$P_{120,000}$	[90]
Stevia rebaudiana	Leaf	$P_{1,500}$ and $P_{105,000}$	[91]
Taraxacum brevicorniculatum	Latex	Crude extract	[92]
Zea mays	Etiolated seedling	$P_{140,000}$	[93]

has thus become classical in this context (Table 1). The assay involves the separation of the reaction product from the labeled substrate by thin layer chromatography (TLC) [43, 45, 46] or organic solvent extraction [47]. To optimize separation, mevalonic acid is first converted to its less hydrophilic derivative mevalonolactone (Fig. 1). Alternative non-radiochemical methods for HMGR activity determination have also been reported. They include the far less sensitive, but simpler, spectrophotometric assay based on monitoring the conversion of NADPH into $NADP^+$ [48–51] and others, still less sensitive than the radiochemical method, using HPLC [52, 53] or liquid chromatography-tandem mass spectrometry [54]. Only the later has not been applied to plants. Analysis done in *Arabidopsis* showed a direct correlation between the HMGR activity levels determined by the radiochemical method and the percentage of seedlings that develop true leaves (achieve seedling establishment) in the presence of the HMGR-specific inhibitor mevinolin [11, 55]. Thus, the HMGR activity measured in plant extracts faithfully represents the HMGR catalytic potential existing in vivo [55].

The above studies allow a well-reasoned design of protocols for plant HMGR extraction and activity determination. The method we report was set up in *Arabidopsis* [11, 56, 58] (Table 1), but should be applicable with minor variations to other plants.

2 Materials

2.1 Reagents and Solutions

Prepare all solutions with ultrapure water (17 MΩ resistivity at 25 °C). Unless otherwise stated, use analytical grade reagents.

1. HCl: 25 % in H_2O. Dilute the 37 % commercial stock 25:12 with H_2O. Store at room temperature.
2. Radioactive ink. Dilute about 1 μCi [^{14}C]-HMG-CoA (*see* **item 18**) in 100 μL of India ink. Store at room temperature.
3. Cytoscint scintillation cocktail (MP Biomedicals). Store at room temperature.
4. Triton X-100: 20 % (v/v) in H_2O. Store at 4 °C.
5. Ethylenediaminetetraacetic acid (EDTA): 500 mM in H_2O, pH 7.5. Add EDTA powder to H_2O and dissolve by adding NaOH pellets to the stirring suspension. Finish pH adjustment with NaOH 1 M and bring to the final volume. Store at 4 °C.
6. Tris–HCl, pH 7.5: 20 mM in H_2O. Store at 4 °C.
7. Bio-Rad Protein Assay dye reagent concentrate (Bio-Rad). Store at 4 °C.
8. HKS buffer: 40 mM Hepes, pH 7.2 (*see* **Note 1**), 50 mM KCl and 100 mM sucrose. Dissolve reagents in H_2O, equilibrate pH with KOH and adjust the volume. Aliquot and store at –20 °C. Keep the thawed aliquot at 4 °C up to 2 weeks.
9. Aprotinin (Sigma): 3 mg/mL in HKS buffer. Store at –20 °C.
10. Leupeptin (Sigma): 4 mg/mL in HKS buffer. Store at –20 °C.
11. Trans-epoxysuccinyl-L-leucylamindo-(4-guanidino)-butane (E-64) (Sigma): 3 mg/mL in HKS buffer. Store at –20 °C.

12. Pepstatin (Sigma): 3 mg/mL in methanol. Store at −20 °C.
13. Bovine Serum Albumin (BSA) protein concentration standards (0.1–0.6 mg/mL). Store at −20 °C.
14. DTT: 1 M in H_2O. Store at −80 °C.
15. DL-3-Hydroxy-3-methylglutaryl coenzyme A (HMG-CoA) (Sigma) (*see* **Note 2**): 4 mM in 50 mM KH_2PO_4 pH 4.5. Store at −80 °C.
16. Yeast glucose 6-phosphate dehydrogenase (G6P-DH) (Sigma): 1 U/μL in HKS buffer. Aliquot and store at −80 °C; thaw only once.
17. TGNB solution: 158 mM glucose 6-phosphate (G6P) (Roche), 7.9 mM NADP (Roche) and 1.58 mg/mL BSA. Weight and dissolve reagents in 658 mM Tris–HCl, pH 7.2. Aliquot and store at −80 °C. The TGNB solution, together with G6P-DH (*see* **item 16**), will constitute the NADPH regeneration system.
18. Hydroxy-3-methylglutaryl Coenzyme A, DL-3-[Glutaryl-3-^{14}C]-, 50 μCi (1.85 MBq) (0.02 mCi/mL, 40–60 mCi/mmol) ([^{14}C]-HMG-CoA, CAS Number: 103476-21-7) (PerkinElmer) (*see* **Note 2**). Aliquot and store at −80 °C.
19. Phenylmethylsulphonyl fluoride (PMSF) (Sigma). Aliquot in microcentrifuge tubes (about 5–15 mg per tube) and store at room temperature. Just before use, dissolve the aliquot in isopropanol at 20 mg/mL.
20. HM buffer (homogenization buffer): 40 mM Hepes-KOH, pH 7.2, 50 mM KCl, 100 mM sucrose, 16 mM EDTA, 0.2 % (v/v) Triton X-100, 10 mM DTT, 15 μg/mL aprotinin, 20 μg/mL leupeptin, 4.2 μg/mL E-64, 1.8 μg/mL pepstatin A, and 100 μg/mL PMSF. Prepare just before use. For 5 mL of HM buffer, add the following stocks in the indicated order: 4.655 mL HKS buffer, 160 μL EDTA, 50 μL Triton X-100, 50 μL DTT, 25 μL aprotinin, 25 μL leupeptin, 7 μL E-64, 3 μL pepstatin A, and 25 μL PMSF. Mix immediately after PMSF addition and keep on ice.
21. Reaction cocktail (RC): 547 mM Tris–HCl pH 7.2, 131 mM G6P, 6.6 mM NADP, 1.3 mg/mL BSA, 15.6 mM DTT, 87.5 μM HMG-CoA, 38.36 μM (1,295 Bq) [3-^{14}C]-HMG-CoA, and 0.35 U yeast G6P-DH. Just before use, prepare sufficient volume of RC for the total number of samples to be assayed (16 μL per assay), keeping the following proportions: 13.3 μL TGNB, 0.25 μL DTT, 0.35 μL HMG-CoA, 1.75 μL [^{14}C]-HMG-CoA, and 0.35 μL G6P-DH. Mix and keep on ice.

2.2 Equipment

1. Polypropylene tubes (50 mL).
2. Racks and appropriate containers for liquid nitrogen.
3. Eppendorf Safe-Lock micro test tubes™. Alternatively, any other good quality microcentrifuge tube that withstand freezing

in liquid nitrogen and shaking in TissueLyser (Qiagen) (*see* **item 6**).

4. Cane for 5 NUNC cryo vials (MTG—Medical Technology Vertriebs).
5. Stainless steel beads 5 mm (Qiagen). Alternatively, stainless steel beads of similar diameter, from any local supplier. Clean the beads twice with H_2O, several times with acetone, until no dirt is released, and twice with 96 % ethanol. Dry the beads at 200 °C.
6. TissueLyser (Qiagen). Alternatively, use mortar and pestle.
7. Refrigerated microcentrifuge (*see* **Note 3**).
8. TLC plates (20×20 cm) made of silica gel 60 on plastic support (Merck).
9. TLC chamber with vertical grooves on the transverse walls.
10. Scotch® Magic™ Removable Tape 811.
11. Fujifilm™ BAS-MS Imaging Plate 20×40 cm (Fisher Scientific). Alternatively, any phosphor imaging screen of similar size to detect ^{14}C isotope emission.
12. Cassette for the 20×40 cm imaging plate. It should contain a separate plastic board, to stick TLC plates.
13. A storage phosphor imaging system (PhosphorImager™, Storm™, Typhoon™ or similar) to scan the exposed imaging plate.
14. Long, fine scissors (*see* **Note 4**).
15. Plastic 20 mL scintillation counting vials.
16. Liquid Scintillation β-counter.
17. ELISA plates and ELISA plate reader.

3 Methods

3.1 Sample Collection and Storage

1. Establish sample collection conditions and adhere to these conditions to keep reproducibility. In particular, collect plants or plant organs at a fixed time of the day and freeze them immediately after removal from their growing place. Be particularly expeditive when collecting in the dark growth period, since this may imply plant exposure to light.
2. Fill 50 mL polypropylene tubes with liquid nitrogen and place them in a stainless steel rack inside a proper container filled with liquid nitrogen.
3. Deep freeze the samples by placing them into the 50 mL polypropylene tubes. Let the liquid nitrogen to evaporate, till only few milliliters are left. Then, break plant samples into

small pieces with a spatula. Store the tubes unfastened at –80 °C until their liquid nitrogen evaporates completely.

4. Transfer 100–200 mg of the crumbled tissue to a pre-weighed microcentrifuge tube with a deep frozen spatula (*see* **Note 5**). Be fast to avoid sample thawing. Add a nitrogen-frozen stainless steel bead and close the microcentrifuge tube. Repeat this sequence for all samples (*see* **Note 6**).
5. Crush samples in a TissueLyser to obtain a fine powder (*see* **Notes** 7 and **8**). Knock the tubes softly with a spatula or forceps to bring the powder to the bottom. Process samples immediately or store them at –80 °C (*see* **Note 9**).

3.2 Extract Preparation

1. Add 2 μL of homogenization buffer per mg of frozen tissue. Knock down the tube on the bench to facilitate a fast penetration of the buffer. Invert the tube several times to thaw the sample completely and put it on ice immediately afterwards. Do not process more than two tubes at once to avoid sample thawing in the absence of buffer. Keep the samples on ice until all of them are ready.
2. Invert all tubes once again (*see* **Note 10**). Centrifuge samples at 200×*g* and 2 °C for 10 min. Brake down slowly to avoid sediment resuspension. Recover the supernatant (S_{200}) carefully with a micropipette and transfer it to a new tube.
3. Centrifuge again, in the same conditions, to remove the sediment completely. Keep the tubes with the clean S_{200} extract on ice.

3.3 HMGR Activity Assay

1. Dispense 26 μL of the extract at the bottom of a fresh microcentrifuge tube and 16 μL of RC (*see* **Note 11**) in the inside side of the lid. Pipette with care such that all the whole volume is released. Close the tube carefully. No RC liquid should fall down at this step. Let the tube on ice. Blanks should contain 26 μL of HM buffer instead of plant extract.
2. When all samples are ready, start the reaction by a short centrifuge pulse, just enough to bring the RC to the bottom, avoiding pellet formation (about 6,000–8,000 rpm in a microcentrifuge for few seconds). Immediately after the pulse, mix each sample by gentle vortexing. Incubate at 37 °C for 30–60 min (*see* **Notes 12** and **13**).
3. Stop the reaction by precipitation with acid (*see* **Note 14**). Pipette 7 μL of 25 % HCl solution in the inside side of the lid and close carefully. When all tubes are ready, centrifuge for few seconds to bring the HCl to the bottom. Vortex immediately after the pulse (*see* **Note 15** for **steps 1–3**).
4. Incubate at 50 °C for 10 min, to lactonize mevalonate.
5. Complete precipitation by incubating on ice for at least 10 min (*see* **Note 16**).

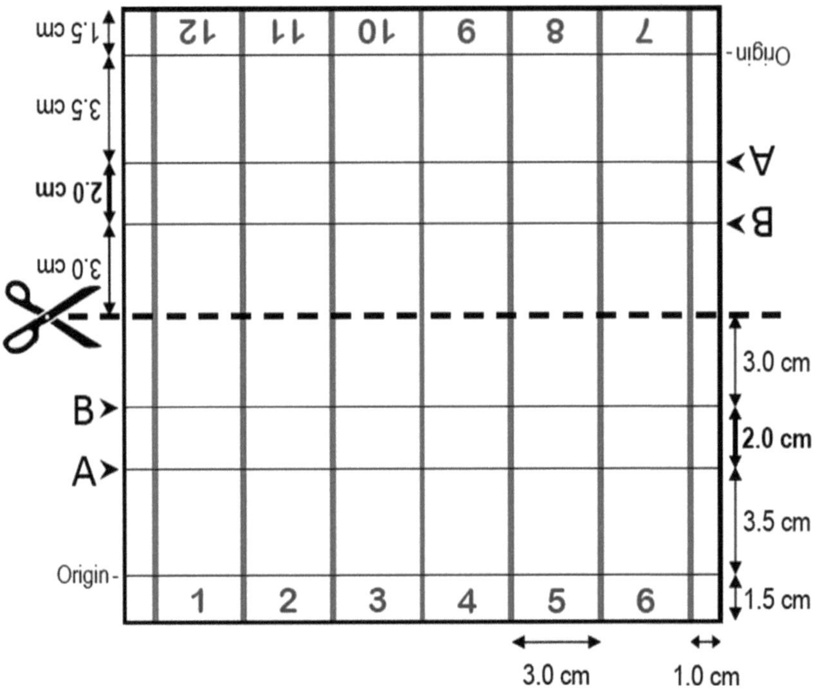

Fig. 2 Preparation of TLC plates for HMGR activity assays. A design to analyze up to twelve samples per 20 × 20 cm silica gel plate is shown. First, *horizontal* and *vertical lines* are drawn with a soft-tip pencil, as indicated in the figure. Next, a separation between lanes (*vertical gray lines*) are made by scoring the plate from top to bottom, displacing a pipette tip along a ruler. Removal of the silica gel avoids sample cross-contamination during the run. Finally, the plate is cut in two halves 10 × 20 cm in size. After the run, the mevalonolactone band will be positioned between *lines A* and *B* of the half-plate

6. Centrifuge at maximum speed for 5 min, at room temperature, to obtain a protein-free supernatant containing the mevalonolactone.

3.4 TLC and Radioactive Counting

1. Prepare TLC plates as indicated in Fig. 2.
2. Place the TLC plate on a double-block heater at 85 °C under the hood, with the silica gel facing up, and put a heavy flat object (i.e., a plastic rack) on the plate bottom to avoid curling and ensure full contact with the metal blocks (*see* **Notes 17** and **18**).
3. Streak the sample supernatant along the origin line. Do several applications for a total of 40 μL per sample and let the liquid dry before re-pipetting in the same surface. Keep the application band as thin as possible. It should not exceed 4–5 mm wide.
4. When all samples have been applied, let the plate cool down to room temperature. Make sure that all samples dry completely.
5. Prepare 100 mL of the eluent by adding 50 mL each of acetone and benzene (1:1) into the TLC chamber. Close the chamber and mix softly (*see* **Note 19**).

6. Place the plates into the chamber, with their lateral borders between the vertical grooves (*see* **Note 20**). Be fast and close the chamber immediately. Run chromatography for 18–20 min, just until the eluent reaches the upper border.
7. Remove plates from the chamber and let them dry for 5 min.
8. Mark plate lines A and B near the plate edges (outside the chromatography tracks, Fig. 2), with a spot of radioactive ink, using a needle or toothpick (*see* **Note 21**).
9. Fix the plates to the cassette board with removable tape (stuck behind the plates) and cover the plates with transparent plastic foil wrapped behind the board (*see* **Note 22**).
10. Fit the sandwich in an autoradiography cassette, together with a sensitive screen. Expose at room temperature for 12–36 h.
11. Read the imaging plate in a phosphor imaging system and print a copy in its true dimensions. Evaluate whether the mevalonolactone band is between lines A and B and whether the closely migrating by-product is outside this region (Fig. 3) (*see* **Note 23**). Otherwise, correct the line positions upwards or downwards, using the image of the radioactive ink marks as a reference.
12. Cut rectangular pieces of the TLC plate (3 cm wide, 2 cm high) containing the mevalonolactone band of each lane with long scissors.
13. Place each silica gel piece into a 20 mL counting vial by bending its plastic support. Fit the rectangular piece to the bottom

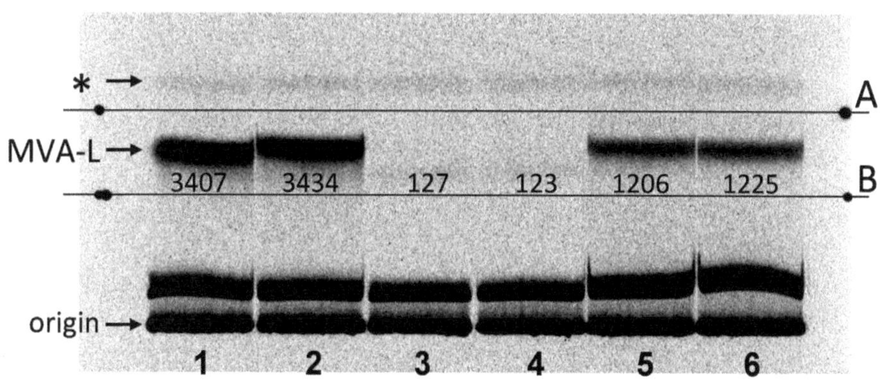

Fig. 3 Chromatogram of the HMGR assay reaction products. Two replicas from two samples (*lanes 1, 2* and *5, 6*, respectively) and a blank (*lanes 3, 4*) were applied at the bottom of the plate (origin). After migration, the TLC plate was exposed to a phosphor imaging screen, for 18 h. The radioactive spots drawn in the plate borders (**step 8** of Subheading 3.4) were used to confirm that the mevalonolactone band (MVA-L) ($R_f = 0.57$) was between the reference *lines A* and *B* (**step 11** of Subheading 3.4). The plate fragments delimited by these lines were processed for radioactivity counting. Notice that a weak side reaction product ($R_f = 0.77$, *asterisk*) migrating ahead of the mevalonolactone band was not included in the cut fragment. The number below each mevalonate band indicates the corresponding radioactive counts (cpm). Under our conditions, the assay replicas did not usually differ more than a 5 %

and add 10 mL of scintillation cocktail. The liquid should cover the TLC plate fragment completely.

14. Prepare a counting vial with 16 μL of RC, in duplicate, to determine the total number of counts per assay.
15. Keep vials in the dark at room temperature for 12–24 h to completely extract the mevalonolactone from the silica gel. Count samples with an appropriate ^{14}C program in a liquid scintillation β-counter (*see* **Note 24**).

3.5 Specific Activity Determination

1. Set an ELISA plate with 90 μL of H_2O per well (*see* **Note 25**).
2. Predilute plant extract with 20 mM Tris–HCl pH 7.5 (*see* **Note 26**).
3. Pipette 10 μL of the diluted plant extracts or BSA standards in the corresponding wells (*see* **Note 27**).
4. Predilute the concentrated Bradford reagent 2.5-fold with H_2O.
5. Dispense 100 μL of the diluted Bradford reagent to the ELISA plate wells with a multichannel pipette. Mix by pipetting up and down a few times avoiding bubble formation, until solution is homogenous, (*see* **Note 28**).
6. Measure absorbance at 595 nm in an ELISA plate reader, 30 min after protein and Bradford reagent mixing (*see* **Note 29**).
7. Calculate the HMGR specific activity in units per mg of protein extract. One HMGR unit is defined as the activity that converts 1 pmol of HMG-CoA into MVA per min at 37 °C. The HMGR specific activity can be calculated with the following formula:

$$\frac{(\text{Cpm}_s - \text{Cpm}_b)\cdot T_H \cdot V}{\text{Cpm}_{RC}\cdot V' \cdot t \cdot V_{ext} \cdot P_c},$$

where:

Cpm_s: Counts per min incorporated into MVA in the sample.

Cpm_b: Counts per min determined in the blank.

T_H: Total HMG-CoA substrate (labeled plus unlabeled) per assay, in pmol.

V: Total assay volume after HCl addition (49 μL in the above protocol).

Cpm_{RC}: Total counts per min present in the assay (sample prepared in **step 14** of Subheading 3.4).

V': Assay volume applied to the TLC plate (40 μL in the above protocol).

t: Incubation time, in min.

V_{ext}: Plant extract volume added to the assay, in μL (26 μL in the above protocol).

P_c: Protein concentration of the plant extract, in mg per μL.

4 Notes

1. Buffer 40 mM H_2KPO_4 pH 7.2 can be used, instead of 40 mM Hepes-KOH pH 7.2, to determine HMGR activity in *Arabidopsis*.
2. Note that the HMG-CoA and [^{14}C]-HMG-CoA available stocks are racemic mixtures of the two possible stereoisomers. Since only the (3S)-HMG-CoA isomer will be processed by HMGR (Fig. 1), the effective concentration of the labeled and the unlabelled HMG-CoA will be half of the one indicated in the corresponding vials. This should be kept in mind in any calculation of substrate-dependent kinetic constants. The total HMGR activity of the plant sample will be also likely affected by the presence of the (3R)-HMG-CoA isomer, which was shown to be a competitive inhibitor of rat liver HMGR [59]. The rat liver HMGR activity was 1.8- to 2.0-fold higher with pure (3S)-HMG-CoA than with (3RS)-HMG-CoA racemic mixture [59].
3. A slow-down braking option in the centrifuge is important to have a good separation between supernatant and pellet in **steps 2** and **3** of Subheading 3.2.
4. The longer the scissors, the easier to cut the silica gel plate without border scrapping.
5. A spatula with capacity to transfer about 100 mg of tissue in a single step is advisable. We built such a device by bending the bottom end of an aluminum cryopreservation cane (*see* Subheading 2.2, **item 4**). The modified end just fitted into the round opening of the microcentrifuge tube.
6. The microcentrifuge tubes can be kept frozen by standing inside a metal block (of a common dry heater) immersed in liquid nitrogen. This system can be placed beside a precision balance, such that each deep frozen tube can be tared to zero just before sample transfer.
7. Two runs for 1 min, at 30 beats per second in a TissueLyser, are sufficient to crush *Arabidopsis* seedlings. Tubes should be refrozen in liquid nitrogen between runs. Invert tubes in the second run.
8. Some plant organs can not be crushed completely with TissueLyser. Grind them in a mortar under liquid nitrogen. Mortar grinding requires longer time and at least 400–600 mg per sample.
9. The crushed samples can be stored at −80 °C at least 1 month, without reduction of HMGR activity.
10. Optionally, incubate the samples for 10–20 min at 4 °C in a rotating wheel, at about 15 rpm. This may be required to

completely extract HMGR protein from certain tissues, but long incubation periods should be avoided to prevent HMGR activity loses. We routinely skipped this incubation step with *Arabidopsis* samples.

11. In advance, prepare sufficient volume of RC for the total number of samples to be assayed. Each plant extract and a blank should be assayed in duplicate and two more aliquots of RC (twice 16 μL) should be left to determine the total counts per assay in **step 14** of Subheading 3.4.
12. The final reaction mix contains 24.8 mM Hepes-KOH pH 7.2, 31 mM KCl, 62 mM sucrose, 9.9 mM EDTA, 0.12 % (v/v) Triton X-100, 12.1 mM DTT, 9.3 μg/mL aprotinin, 12.4 μg/mL leupeptin, 2.6 μg/mL E-64, 1.11 μg/mL pepstatin A, 62 μg/mL PMSF, 208 mM Tris–HCl pH 7.2, 50 mM glucose 6-phosphate, 2.5 mM NADP, 0.5 mg/mL BSA, 33.3 μM HMG-CoA, 14.6 μM (77,700 dpm) [3-^{14}C]-HMG-CoA, 0.35 U yeast glucose 6-phosphate dehydrogenase, in a final volume of 42 μL.
13. Incubation time should be adapted to the expected activity. A too long incubation may cause HMGR degradation or substrate shortage. Make sure that mevalonate synthesis is linear with pilot experiments. Mevalonolactone radioactivity higher than 10 % of total counts (about 8,000 cpm in the assayed conditions) likely indicates that the reaction was not linear.
14. After the reaction, open the tubes under the hood. Production of volatile compounds has not been demonstrated in the HMGR activity assays, but is always advisable when running radioactive reactions in complex enzymatic extracts.
15. To start and stop reaction, alternatively to **steps 1–3** (Subheading 3.3), pipette 16 μL of RC to the bottom of the tube (containing 26 μL of extract) at time intervals, close the tube, mix immediately and transfer to a 37 °C water bath. To stop the reaction, add 7 μL of HCl at the same time intervals and mix immediately.
16. Alternatively, freeze the samples down at −20 °C. The process can be interrupted at this point.
17. The flat object should hold the TLC plate below the origin line, such that it does not interfere with the plate loading nor damages the running area.
18. The whole TLC process, including sample application, eluent preparation and plate and chamber handling (**steps 3–7** of Subheading 3.4) should be done under the hood.
19. Prepare the eluent 10–20 min in advance to the run, to allow chamber saturation.

20. Up to three plates can be run at once in a 20 × 20 cm chamber containing seven lateral grooves.
21. Lines A and B delimit the plate area that will contain the mevalonolactone band after migration.
22. Direct contact between the plates and the imaging screen should be avoided. Otherwise, the screen could become contaminated by radioactive material or could be damaged by still remaining HCl.
23. In the recommended TLC system (benzene/acetone 1:1 on silica gel 60), mevalonolactone migrates to an R_f of 0.56–0.58 [45].
24. Counting can be done immediately after adding the scintillation liquid, but this will give lower incorporation and higher variability between replicas.
25. ELISA plates are preferred over individual spectrophotometric cuvettes because less sample volume is required and measurement is done simultaneously for all samples.
26. Dilution should be about 1:10 for *Arabidopsis* seedling extracts obtained as indicated above (2 μL of extraction buffer per mg of tissue). For other plant tissues or other buffer to tissue ratios, appropriate dilutions should be established empirically.
27. We determined plant extract concentration in quadruplicate, with two replicas of the dilution (**step 2** of Subheading 3.5) and two replicas of the measurement of each dilution (**step 3** of Subheading 3.5). Standards were measured in duplicate.
28. The final concentration of the Bradford reagent in the ELISA plate will be one fifth of that in the original stock.
29. According to the Protein Assay manufacturer, absorbance reading can be done 5–60 min after mixing, but we observed that the A_{595} values usually decrease along time, and this can happen at a different rate in the plant extracts and standards. The incubation time should be fixed to better compare the HMGR specific activity between experiments.

Acknowledgment

This work was supported by grants of the Spanish Ministerio de Economía y Competitividad and the Spanish Ministerio de Ciencia e Innovación (BFU2011-24208 to N.C., BIO2009-06984 to A.F. and M.A., and BIO2009-09523 to A.B., including FEDER funds), the Spanish Consolider-Ingenio Program (CSD2007-00036 Centre for Research in Agrigenomics), and the Generalitat de Catalunya (2009SGR0026).

References

1. Rogers DH, Panini SR, Rudney H (1983) Properties of HMGCoA reductase and its mechanism of action. In: Sabine JR (ed) 3-hydroxy-3-methylglutaryl coenzyme A reductase. CRC, Boca Raton, FL, pp 57–75
2. Burg JS, Espenshade PJ (2011) Regulation of HMG-CoA reductase in mammals and yeast. Prog Lipid Res 50:403–410
3. Goldstein JL, Brown MS (1990) Regulation of the mevalonate pathway. Nature 343: 425–430
4. Rodríguez-Concepción M, Campos N, Ferrer A et al (2013) Biosynthesis of isoprenoid precursors in Arabidopsis. In: Bach TJ, Rohmer M (eds) Isoprenoid synthesis in plants and microorganisms: new concepts and experimental approaches. Springer Science + Business Media, New York, pp 459–456
5. Bach TJ (1987) Synthesis and metabolism of mevalonic acid in plants. Plant Physiol Biochem 25:163–178
6. Stermer BA, Bianchini GM, Korth KL (1994) Regulation of HMG-CoA reductase activity in plants. J Lipid Res 35:1133–1140
7. Chappell J (1995) Biochemistry and molecular biology of the isoprenoid biosynthetic pathway in plants. Annu Rev Plant Physiol Plant Mol Biol 46:521–547
8. Vranová E, Coman D, Gruissem W (2013) Network analysis of the MVA and MEP pathways for isoprenoid synthesis. Annu Rev Plant Biol 64:665–700. doi:10.1146/annurev-arplant-050312-120116
9. Brooker JD, Russell DW (1979) Regulation of microsomal 3-hydroxy-3-methylglutaryl coenzyme A reductase from pea seedlings: Rapid posttranslational phytochrome-mediated decrease in activity and in vivo regulation by isoprenoid products. Arch Biochem Biophys 198:323–334
10. Korth KL, Jaggard DAW, Dixon RA (2000) Developmental and light-regulated post-translational control of 3-hydroxy-3-methylglutaryl-CoA reductase levels in potato. Plant J 23:507–516
11. Leivar P, Antolín-Llovera M, Ferrero S et al (2011) Multilevel control of *Arabidopsis* 3-hydroxy-3-methylglutaryl coenzyme A reductase by protein phosphatase 2A. Plant Cell 23: 1494–1511
12. Nieto B, Forés O, Arró M et al (2009) Arabidopsis 3-hydroxy-3-methylglutaryl-CoA reductase is regulated at the post-translational level in response to alterations of the sphingolipid and the sterol biosynthetic pathways. Phytochemistry 70:53–59
13. Wentzinger LF, Bach TJ, Hartmann MA (2002) Inhibition of squalene synthase and squalene epoxidase in tobacco cells triggers an up-regulation of 3-hydroxy-3-methylglutaryl coenzyme A reductase. Plant Physiol 130: 334–346
14. Russell DW, Knight JS, Wilson TM (1985) Pea seedling HMG-CoA reductases: regulation of activity in vitro by phosphorylation and Ca^{2+}, and posttranslational control in vivo by phytochrome and isoprenoid hormones. In: Randall DD, Blevins DG, Larson RL, Kagawa T (eds) Current topics in plant biochemistry and physiology. University of Missouri-Columbia, Missouri, CO, pp 191–206
15. Wititsuwannakul R, Wititsuwannakul D, Dumkong S (1990) Hevea calmodulin: regulation of the activity of latex 3-hydroxy-3-methylglutaryl coenzyme A reductase. Phytochemistry 29:1755–1758
16. Dale S, Arró M, Becerra B et al (1995) Bacterial expression of the catalytic domain of 3-hydroxy-3-methylglutaryl-CoA reductase (isoform HMGR1) from *Arabidopsis thaliana*, and its inactivation by phosphorylation at Ser577 by *Brassica oleracea* 3-hydroxy-3-methylglutaryl-CoA reductase kinase. Eur J Biochem 233:506–513
17. Hemmerlin A (2013) Post-translational events and modifications regulating plant enzymes involved in isoprenoid precursor biosynthesis. Plant Sci 203:41–54
18. Sugden C, Donaghy PG, Halford NG et al (1999) Two SNF1-Related protein kinases from spinach leaf phosphorylate and inactivate 3-hydroxy-3-methylglutaryl-coenzyme A reductase, nitrate reductase, and sucrose phosphate synthase in vitro. Plant Physiol 120:257–274
19. Antolín-Llovera M, Leivar P, Arró M et al (2011) Modulation of plant HMG-CoA reductase by protein phosphatase 2A: positive and negative control at a key node of metabolism. Plant Signal Behav 6:1127–1131
20. Campos N, Boronat A (1995) Targeting and topology in the membrane of plant 3-hydroxy-3-methylglutaryl coenzyme A reductase. Plant Cell 7:2163–2174
21. Denbow CJ, Lang S, Cramer CL (1996) The N terminal domain of tomato 3-hydroxy-3-methylglutaryl-CoA reductases – sequence, microsomal targeting, and glycosylation. J Biol Chem 271:9710–9715
22. Re EB, Brugger S, Learned M (1997) Genetic and biochemical analysis of the transmembrane domain of Arabidopsis 3-hydroxy-3-methylglutaryl coenzyme A reductase. J Cell Biochem 65:443–459

23. Leivar P, González VM, Castel S et al (2005) Subcellular localization of Arabidopsis 3-hydroxy-3-methylglutaryl-coenzyme A reductase. Plant Physiol 137:57–69
24. Merret R, Cirioni JR, Bach TJ et al (2007) A serine involved in actin-dependent subcellular localization of a stress-induced tobacco BY-2 hydroxymethylglutaryl-CoA reductase isoform. FEBS Lett 581:5295–5299
25. Kondo K, Oba K (1986) Purification and characterization of 3-hydroxy-3-methylglutaryl CoA reductase from potato tubers. J Biochem 100:967–974
26. Wititsuwannakul R, Wititsuwannakul D, Suwanmanee P (1990) 3-Hydroxy-3-methylglutaryl coenzyme A reductase from the latex of *Hevea brasiliensis.* Phytochemistry 29:1401–1403
27. Bach TJ, Rogers DH, Rudney H (1986) Detergent-solubilization, purification, and characterization of membrane-bound 3-hydroxy-3-methylglutaryl-coenzyme-A reductase from radish seedlings. Eur J Biochem 154:103–111
28. Choi D, Ward BL, Bostock RM (1992) Differential induction and suppression of potato 3-hydroxy-3-methylglutaryl coenzyme A reductase genes in response to *Phytophthora infestans* and to its elicitor arachidonic acid. Plant Cell 4:1333–1344
29. Chye ML, Kush A, Tan CT et al (1991) Characterization of cDNA and genomic clones encoding 3-hydroxy-3-methylglutaryl-coenzyme A reductase from *Hevea brasiliensis.* Plant Mol Biol 16:567–577
30. Vollack KU, Dittrich B, Ferrer A et al (1994) Two radish genes for 3-hydroxy-3-methylglutaryl-CoA reductase isozymes complement mevalonate auxotrophy in a yeast mutant and yield membrane-bound active enzyme. J Plant Physiol 143:479–487
31. Istvan ES, Palnitkar M, Buchanan SK et al (2000) Crystal structure of the catalytic portion of human HMG-CoA reductase: insights into regulation of activity and catalysis. EMBO J 19:819–830
32. Lynen F (1967) Biosynthetic pathways from acetate to natural products. Pure Appl Chem 14:137–167
33. Hepper CM, Audley BG (1969) Biosynthesis of rubber from β-hydroxy-β-methylglutaryl-coenzyme A in Hevea brasiliensis latex. Biochem J 114:379–386
34. Brooker JD, Russell DW (1974) Some properties of 3-hydroxy-3-methylglutaryl coenzyme A reductase from *Pisum sativum*. Bull Roy Soc New Zeal 12:365–370
35. Suzuki H, Oba K, Uritani I (1974) Occurrence of 3-hydroxy-3-methyl-glutaryl coenzyme A reductase in sweet-potato. Agric Biol Chem 38:2053–2055
36. Huber J, Rudiger W (1978) Subcellular localization and properties of 3-hydroxy-3-methylglutaryl coenzyme A reductase in plant cell suspension cultures of anise. Hoppe-Seylers Z Physiol Chem 359:277–277
37. Douglas TJ, Paleg LG (1978) AMO 1618 effects on incorporation of ^{14}C-MVA and ^{14}C-acetate into sterols in *Nicotiana* and *Digitalis* seedlings and cell-free preparations from *Nicotiana*. Phytochemistry 17:713–718
38. Grumbach KH, Bach TJ (1979) Effect of PSII herbicides, amitrol and SAN 6706 on the activity of 3-hydroxy-3-methylglutaryl-coenzyme A reductase and the incorporation of [2-^{14}C]acetate and [2-^{3}H]mevalonate into chloroplast pigments of radish seedlings. Z Naturforsch (C) 34:941–943
39. Bach TJ, Wettstein A, Boronat A et al (1991) Properties and molecular cloning of plant HMG-CoA reductase. In: Patterson GW, Nes WD (eds) Physiology and biochemistry of plant sterols. American Oils Chemical Society, Champaign, IL, pp 29–49
40. Sipat A (1985) 3-Hydroxy-3-methylglutaryl-CoA reductase in the latex of *Hevea brasiliensis.* Methods Enzymol 110:40–51
41. Sipat AB (1982) Hydroxymethylglutaryl CoA reductase (NADPH) in the latex of *Hevea brasiliensis.* Phytochemistry 21:2613–2618
42. Suzuki H, Oba K, Uritani I (1975) The occurrence and some properties of 3-hydroxy-3-methylglutaryl coenzyme A reductase in sweet potato roots infected by *Ceratocystis fimbriata.* Physiol Plant Pathol 7:265–276
43. Russell DW (1985) 3-Hydroxy-3-methylglutaryl-CoA reductases from pea seedlings. Methods Enzymol 110:26–40
44. Reddy AR, Das VSR (1986) Partial purification and characterization of 3-hydroxy-3-methylglutaryl coenzyme A reductase from the leaves of guayule (*Parthenium argentatum*). Phytochemistry 25:2471–2474
45. Shapiro DJ, Imblum RL, Rodwell VW (1969) Thin-layer chromatographic assay for HMG-CoA reductase and mevalonic acid. Anal Biochem 31:383
46. Young NL, Berger B (1981) Assay of S-3-hydroxy-3-methylglutaryl-CoA reductase. Methods Enzymol 71(Pt C):498–509
47. Chappell J, Wolf F, Proulx J et al (1995) Is the reaction catalyzed by 3-hydroxy-3-methylglutaryl coenzyme A reductase a rate-limiting step for

isoprenoid biosynthesis in plants ? Plant Physiol 109:1337–1343
48. Dai Z, Cui G, Zhou S-F et al (2011) Cloning and characterization of a novel 3-hydroxy-3-methylglutaryl coenzyme A reductase gene from Salvia miltiorrhiza involved in diterpenoid tanshinone accumulation. J Plant Physiol 168:148–157
49. Douglas P, Pigaglio E, Ferrer A et al (1997) Three spinach leaf nitrate reductase-3-hydroxy-3-methylglutaryl-CoA reductase kinases that are required by reversible phosphorylation and/or Ca^{2+} ions. Biochem J 325:101–109
50. Mansouri H, Asrar Z, Mehrabani M (2009) Effects of gibberellic acid on primary terpenoids and D^9-tetrahydrocannabinol in Cannabis sativa at flowering stage. J Integr Plant Biol 51:553–561
51. Toroser D, Huber SC (1998) 3-Hydroxy-3-methylglutaryl-coenzyme A reductase kinase and sucrose-phosphate synthase kinase activities in cauliflower florets: Ca^{2+} dependence and substrate specificities. Arch Biochem Biophys 355:291–300
52. Mozzicafreddo M, Cuccioloni M, Eleuteri AM et al (2010) Rapid reverse phase-HPLC assay of HMG-CoA reductase activity. J Lipid Res 51:2460–2463
53. Soto G, Stritzler M, Lisi C et al (2011) Acetoacetyl-CoA thiolase regulates the mevalonate pathway during abiotic stress adaptation. J Exp Bot 62:5699–5711
54. Waldron J, Webster C (2011) Liquid chromatography-tandem mass spectrometry method for the measurement of serum mevalonic acid: a novel marker of hydroxymethylglutaryl coenzyme A reductase inhibition by statins. Ann Clin Biochem 48:223–232
55. Rodríguez-Concepción M, Forés O, Martínez-García JF et al (2004) Distinct light-mediated pathways regulate the biosynthesis and exchange of isoprenoid precursors during *Arabidopsis* seedling development. Plant Cell 16:144–156
56. Closa M, Vranová E, Bortolotti C et al (2010) The *Arabidopsis thaliana* FPP synthase isozymes have overlapping and specific functions in isoprenoid biosynthesis, and complete loss of FPP synthase activity causes early developmental arrest. Plant J 63:512–525
57. Manzano D, Fernàndez-Busquets X, Schaller H et al (2004) The metabolic imbalance underlying lesion formation in *Arabidopsis thaliana* overexpressing farnesyl diphosphate synthase (isoform 1S) leads to oxidative stress and is triggered by the developmental decline of endogenous HMGR activity. Planta 219: 982–992
58. Masferrer A, Arró M, Manzano D et al (2002) Overexpression of *Arabidopsis thaliana* farnesyl diphosphate synthase (FPS1S) in transgenic Arabidopsis induces a cell death/senescence-like response and reduced cytokinin levels. Plant J 30:123–132
59. Pastuszyn A, Havel CM, Scallen TJ et al (1983) (R)3-Hydroxy-3-methylglutaryl coenzyme A: the not-so-innocent bystander in the enzymatic reduction of (S)HMG-CoA to (R) mevalonate. J Lipid Res 24:1411–1411
60. Lalitha R, George R, Ramasarma T (1989) Mevalonate-metabolizing enzymes in *Arachis hypogaea*. Mol Cell Biochem 87:161–170
61. Ram M, Khan MA, Jha P et al (2010) HMG-CoA reductase limits artemisinin biosynthesis and accumulation in *Artemisia annua* L. plants. Acta Physiol Plant 32:859–866
62. Narváez JA, Flores-Pérez P, Herrera-Valencia V et al (2001) Development of molecular techniques for studying the metabolism of carotenoids in Bixa orellana L. Hortscience 36:982–986
63. Harker M, Hellyer A, Clayton JC et al (2003) Co-ordinate regulation of sterol biosynthesis enzyme activity during accumulation of sterols in developing rape and tobacco seed. Planta 216:707–715
64. Kato-Emori S, Higashi K, Hosoya K et al (2001) Cloning and characterization of the gene encoding 3-hydroxy-3-methylglutaryl coenzyme A reductase in melon (Cucumis melo L. reticulatus). Mol Genet Genomics 265:135–142
65. Nishi A, Tsuritani I (1983) Effect of auxin on the metabolism of mevalonic acid in suspension-cultured carrot cells. Phytochemistry 22: 399–401
66. Golbeck JH, Maurey KM, Newberry GA (1985) Detection of 3 hydroxy-3-methylglutaryl coenzyme A reductase activity in *Dunaliella salina*. Plant Physiol 77:48
67. Skrukrud CL, Taylor SE, Hawkins DR et al (1987) Triterpenoid biosynthesis in *Euphorbia lathyris*. In: Stumpf PK, Mudd JB, Nes WD (eds) The metabolism, structure, and function of plant lipids. Plenum, New York, pp 115–118
68. Skrukrud CL, Taylor SE, Hawkins DR et al (1988) Subcellular fractionation of triterpenoid biosynthesis in *Euphorbia lathyris* latex. Physiol Plant 74:306–316
69. Leube J, Grisebach H (1983) Further studies on induction of enzymes of phytoalexin synthesis in soybean and cultured soybean cells. Z Naturforsch (C) 38:730–735

70. Joost O, Bianchini G, Bell AA et al (1995) Differential induction of 3-hydroxy-3-methylglutaryl CoA reductase in two cotton species following inoculation with Verticillium. Mol Plant Microbe Interact 8:880–885
71. Ceccarelli N, Lorenzi R (1984) Growth inhibition by competitive inhibitors of 3-hydroxymethylglutarylcoenzyme A reductase in *Helianthus tuberosus* tissue explants. Plant Sci Lett 34:269–276
72. Bach TJ (1981) Untersuchungen zur Charakterisierung und Regulation der 3-Hydroxy-3-methylglutaryl-Coenzym-A-Reduktase [Hydroxy-methylglutaryl-Coenzym-A Reduktase] (Mevalonat : NADP+Oxidoreduktase, CoA acylierend, E.C. 1.1.1.34) in Keimlingen von Raphanus sativus. Ph.D. Universität Karlsruhe
73. Srinivasan V, Ryu DDY (1992) Enzyme activity and shikonin production in *Lithospermum erythrorhizon* cell cultures. Biotechnol Bioeng 40:69–74
74. Köhle A, Sommer S, Yazaki K et al (2002) High level expression of chorismate pyruvate-lyase (UbiC) and HMG-CoA reductase in hairy root cultures of *Lithospermum erythrorhizon*. Plant Cell Physiol 43:894–902
75. Rupasinghe HPV, Almquist KC, Paliyath G et al (2001) Cloning of *hmg1* and *hmg2* cDNAs encoding 3-hydroxy-3-methylglutaryl coenzyme A reductase and their expression and activity in relation to alpha-farnesene synthesis in apple. Plant Physiol Biochem 39:933–947
76. Arebalo RE, Mitchell ED (1984) Cellular distribution of 3-hydroxy-3-methylglutaryl coenzyme A reductase and mevalonate kinase in leaves of *Nepeta cataria*. Phytochemistry 23:13–18
77. van Deenen N, Bachmann AL, Schmidt T et al (2012) Molecular cloning of mevalonate pathway genes from *Taraxacum brevicorniculatum* and functional characterisation of the key enzyme 3-hydroxy-3-methylglutaryl-coenzyme A reductase. Mol Biol Rep 39:4337–4349
78. Chappell J, Nable R (1987) Induction of sesquiterpenoid biosynthesis in tobacco cell suspension cultures by fungal elicitor. Plant Physiol 85:469–473
79. Gondet L, Weber T, Maillotvernier P et al (1992) Regulatory role of microsomal 3-hydroxy-3-methylglutaryl-coenzyme A reductase in a tobacco mutant that overproduces sterols. Biochem Biophys Res Commun 186:888–893
80. Maurey K, Wolf F, Golbeck J (1986) 3-Hydroxy-3-Methylglutaryl Coenzyme A reductase activity in *Ochromonas malhamensis*: a system to study the relationship between enzyme activity and rate of steroid biosynthesis. Plant Physiol 82:523–527
81. Ji W, Benedict CR, Foster MA (1993) Seasonal variations in rubber biosynthesis, 3-hydroxy-3-methylglutaryl-coenzyme A reductase, and rubber transferase activities in *Parthenium argentatum* in the chihuahuan desert. Plant Physiol 103:535–542
82. Cowan AK, MooreGordon CS, Bertling I et al (1997) Metabolic control of avocado fruit growth – Isoprenoid growth regulators and the reaction catalyzed by 3-hydroxy-3-methylglutaryl coenzyme A reductase. Plant Physiol 114:511–518
83. Cowan AK (2004) Metabolic control of avocado fruit growth: 3-hydroxy-3-methylglutaryl coenzyme a reductase, active oxygen species and the role of C7 sugars. S Afr J Bot 70:75–82
84. Reddy AR, Das VSR (1987) Chloroplast autonomy for the biosynthesis of isopentenyl diphosphate in guayule (*Parthenium argentatum gray*). New Phytol 106:457–464
85. Boll M, Kardinal A (1990) 3-Hydroxy-3-methylglutaryl coenzyme A reductase from spruce (*Picea abies*). Properties and regulation. Z Naturforsch (C) 45:973–979
86. Lütke-Brinkhaus F, Kleinig H (1987) Formation of isopentenyl diphosphate via mevalonate does not occur within etioplasts and etiochloroplasts of mustard *Sinapis alba* L. seedlings. Planta 171:406–411
87. Narita JO, Gruissem W (1989) Tomato hydroxymethylglutaryl-CoA reductase is required early in fruit development but not during ripening. Plant Cell 1:181–190
88. Oba K, Kondo K, Doke N et al (1985) Induction of 3-hydroxy-3-methylglutaryl CoA reductase in potato tubers after slicing, fungal infection or chemical treatment, and some properties of the enzyme. Plant Cell Physiol 26:873–880
89. Josekutty PC (1998) Selection and characterization of *Solanum xanthocarpum* cell line with augmented steroid metabolism. S Afr J Bot 64:238–243
90. Kreuz K, Kleinig H (1984) Synthesis of prenyl lipids in cells of spinach leaf. Compartmentation of enzymes for formation of isopentenyl diphosphate. Eur J Biochem 141:531–535
91. Kim KK, Yamashita H, Sawa Y et al (1996) A high activity of 3-hydroxy-3-methylglutaryl coenzyme a reductase in chloroplasts of *Stevia rebaudiana* Bertoni. Biosci Biotechnol Biochem 60:685–686
92. Post J, van Deenen N, Fricke J et al (2012) Laticifer-specific cis-prenyltransferase silencing affects the rubber, triterpene, and inulin content

of *Taraxacum brevicorniculatum*. Plant Physiol 158:1406–1417

93. Bach TJ, Weber T, Motel A (1990) Some properties of enzymes involved in the biosynthesis and metabolism of 3-hydroxy-3-methylglutaryl-CoA in plants. In: Towers GHN, Stafford HA (eds) Biochemistry of the mevalonic acid pathway to terpenoids. Plenum Press Div Plenum Publishing Corp, New York, pp 1–82

Chapter 4

Farnesyl Diphosphate Synthase Assay

Montserrat Arró, David Manzano, and Albert Ferrer

Abstract

Farnesyl diphosphate synthase (FPS) catalyzes the sequential head-to-tail condensation of isopentenyl diphosphate (IPP, C_5) with dimethylallyl diphosphate (DMAPP, C_5) and geranyl diphosphate (GPP, C_{10}) to produce farnesyl diphosphate (FPP, C_{15}). This short-chain prenyl diphosphate constitutes a key branch-point of the isoprenoid biosynthetic pathway from which a variety of bioactive isoprenoids that are vital for normal plant growth and survival are produced. Here we describe a protocol to obtain highly purified preparations of recombinant FPS and a radiochemical assay method for measuring FPS activity in purified enzyme preparations and plant tissue extracts.

Key words Farnesyl diphosphate synthase, Dimethylallyl diphosphate, Geranyl diphosphate, Isopentenyl diphosphate, Isoprenoid, Mevalonate, Prenyltransferase, Terpenoid

1 Introduction

Farnesyl diphosphate (FPP) synthase (FPS) (EC 2.5.1.10) is the best-studied member of the family of enzymes called prenyltransferases [1]. This short-chain prenyltransferase catalyzes a two-step reaction consisting of the head-to-tail condensation of one unit of isopentenyl diphosphate (IPP, C_5) with its isomer dimethylallyl diphosphate (DMAPP, C_5) to form the intermediate geranyl diphosphate (GPP, C_{10}) and a second condensation of IPP with GPP to produce FPP (C_{15}) (Fig. 1). The mechanism for chain elongation is a stereo-selective electrophilic alkylation of the carbon–carbon double bond in IPP by a resonance-stabilized cation generated from the allylic substrates DMAPP and GPP leading to production of the all-*trans*-(*E*,*E*) isomer of FPP. The reaction requires divalent metal cations for activity, usually Mg^{2+} or Mn^{2+}, which coordinate to the diphosphate moiety of the allylic substrates and interact with highly conserved aspartate residues in the substrate-binding pocket [2]. FPS has been purified from a variety

Manuel Rodríguez-Concepción (ed.), *Plant Isoprenoids: Methods and Protocols*, Methods in Molecular Biology, vol. 1153, DOI 10.1007/978-1-4939-0606-2_4,

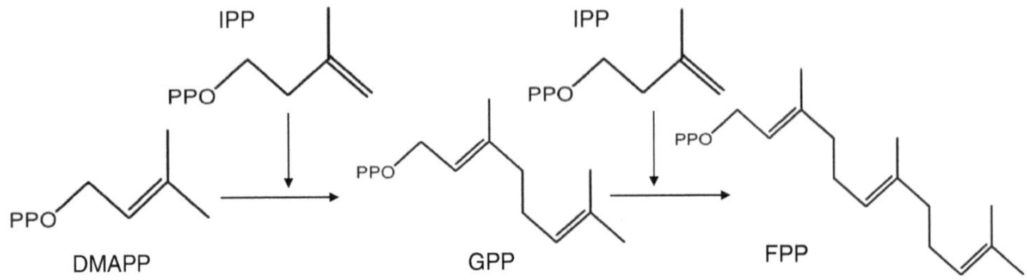

Fig. 1 FPS catalyzes the sequential condensation of two units of isopentenyl diphosphate (IPP; C_5) with the substrate dimethylallyl diphosphate (DMAPP; C_5) and the reaction intermediate geranyl diphosphate (GPP; C_{10}) to form the short-chain prenyl diphosphate FPP (C_{15})

of organisms and functions as a homodimer of 80–84 kDa [3] where each subunit bears one active site, even though it has been proposed that interaction between subunits is required to form a shared active site in the homodimer structure rather than an independent active site in each subunit [4]. In addition to the *trans*-FPSs, some mechanistically similar but genetically unrelated FPSs leading to the synthesis of *cis–trans-*(*Z,E*)-FPP and all-*cis*-(*Z,Z*)-FPP have been reported [5, 6].

FPS is located at a key position in the plant isoprenoid biosynthesis pathway as both the substrates and the reaction product are multiple branch-point intermediates leading to the synthesis of isoprenoids such as cytokinins, sterols, brassinosteroids, ubiquinones, dolichols, and sesquiterpenes. FPP is also used as substrate for protein farnesylation. All these products play pivotal roles in a wide range of essential cellular processes that are vital for plant growth and survival [7]. Plants generally have multiple FPS isozymes that localize in different subcellular compartments including the cytosol [8], mitochondria [9], peroxisomes [10], and chloroplasts [11]. Specifically, the model plant *Arabidopsis thaliana* contains three FPS isozymes (FPS1L, FPS1S, and FPS2) that are encoded by two genes, *FPS1* (At5g47770) and *FPS2* (At4g17190). Isozymes FPS1S and FPS2 both localize in the cytosol [8], whereas isozyme FPS1L is targeted to mitochondria [9]. The multiplicity of FPS isozymes certainly hampers their individual characterization by using conventional purification techniques from plant extracts. This limitation can be overcome by producing the desired isozyme through expression in heterologous systems like *E. coli*. Here we describe a protocol to obtain large amounts of catalytically active FPS isozymes devoid of any affinity-purification tag and a radiochemical assay method to determine FPS activity in both purified enzyme preparations and extracts from tissues of *A. thaliana*. The reported assay method may need to be slightly adapted to suit the particular characteristics of extracts from plant species other than Arabidopsis.

2 Materials

2.1 Cloning

1. Standard equipment and reagents for recombinant DNA work.
2. 10 mM stock solution of forward primers FPS1S-orf-fw (5′-**ATG**GAGACCGATCTCAAGTCAACC-3′), FPS2-orf-fw (5′-**ATG**GCGGATCTGAAATCAACC-3′), and reverse primer FPS-orf-rv (5′-CGCGGATCC**CTA**CTTCTGCCTCTT GTAG-3′) (*see* **Note 1**). Translation start and stop codons are shown in bold. A *Bam*HI restriction site in the sequence of primer FPS-orf-rv is shown underlined.
3. Thermostable DNA polymerase with proof-reading activity.
4. Nuclease S1 from *Aspergillus oryzae* cells (Thermo Scientific).
5. Alkaline phosphatase 1 U/μL (Roche).
6. Ethylene diamine tetraacetic acid (EDTA): 0.5 M solution, pH 8.0, in water.
7. pGEX-3X-NotI vector.
8. pGEX 5′ sequencing primer: 5′-GGGCTGGCAAGCCAC GTTTGGTG-3′.
9. *E. coli* strains: DH5α (F^- endA1 glnV44 thi-1 recA1 relA1 gyrA96 deoR nupG Φ80d*lacZ*ΔM15 Δ(*lacZYA-argF*)U169, hsdR17(r_K^- m_K^+), λ^-), and BL21(DE3) (F^- ompT gal dcm lon hsdS_B(r_B^- m_B^-) λ(DE3 [lacI lacUV5-T7 gene 1 ind1 sam7 nin5] (pUBS520)).

2.2 Protein Expression and Purification

1. Isopropyl thio-β-D-galactopyranoside (IPTG): 1 M solution in water (store at −20 °C).
2. PBS: 80 mM Na_2HPO_4, 20 mM NaH_2PO_4, pH 7.5, 100 mM NaCl.
3. Poly-Prep chromatography columns (0.8 × 4.0 cm) (Bio-Rad).
4. Glutathione Sepharose™ 4B (GE Healthcare Life Sciences).
5. Factor Xa (GE Healthcare Life Sciences): 1 U/mL solution (*see* **Note 2**).
6. Factor Xa cleavage buffer: 50 mM Tris-HCl, pH 7.5, 50 mM NaCl, 1 mM $CaCl_2$.
7. Factor Xa removal resin (Qiagen).
8. Glycerol.
9. Protein assay kit.

2.3 FPS Activity Assay

1. Mortar and pestle.
2. Aprotinin: 3 mg/mL in water (store at −20 °C).
3. Pepstatin: 3 mg/mL in methanol (store at −20 °C).

4. *trans*-Epoxysuccinyl-L-leucylamindo-(4-guanidino)-butane (E64): 3 mg/mL in water (store at −20 °C).
5. Phenylmethylsulfonyl fluoride (PMSF): 100 mM in isopropanol (store at room temperature protected from light).
6. DTT: 1 M in water (store at −20 °C).
7. Homogenization buffer: 50 mM PIPES, pH 7.0, 250 mM sucrose, 10 mM NaCl, 5 mM $MgCl_2$, 5 mM DTT, 15 μg/mL aprotinin, 3 μg/mL E64, 1.5 μg/mL pepstatin, 0.5 mM PMSF. Add DTT and protease inhibitors immediately prior to use.
8. Eppendorf Safe-Lock Tubes™, 1.5 mL capacity and 2 mL capacity.
9. Rotating wheel Rotator SB2 (Stuart Benchtop Science Equipment).
10. Benchtop refrigerated centrifuge (or access to cold room).
11. Bovine serum albumin (BSA): 10 mg/mL in water (store at −20 °C).
12. $MgCl_2$: 25 mM in water.
13. IPP and GPP (Echelon Biosciences, Inc.). Both prenyl diphosphates are provided as the lyophilized tris-ammonium salt. Prepare 4 mM stock solutions in water, dispense in aliquots of 100 μL and store at −80 °C.
14. IPP triammonium salt [4-^{14}C]IPP (Perkin Elmer, Inc.) (40–60 mCi (1.48–2.22 GBq)/mmol): 50 μCi (1.85 MBq) in 2.5 mL of EtOH:0.15 N NH_4OH (1:1). Dispense in aliquots of 150 μL and store at −80 °C (*see* **Note 3**).
15. HCl: 2 N solution.
16. Solid sodium chloride.
17. Hexane analytical grade reagent saturated with water.
18. Scintillation vials (20 mL capacity) with screw caps (National Diagnostics).
19. Scintillation cocktails Ecoscint O and Ecoscint H (National Diagnostics).
20. Beta liquid scintillation counter.

3 Methods

3.1 Expression and Purification of Recombinant Arabidopsis FPS

1. To amplify the cDNA sequences coding for isozymes FPS1S and FPS2, set up a 100 μL PCR reaction containing 50 ng of plasmid DNA, either pcNC3 for FPS1S [9] or pcNC2 for FPS2 [12], 2.5 mM of each dNTP, 1× PCR buffer, 0.2 mM of each primer (*see* **Note 1**), and 2 U of a thermostable DNA

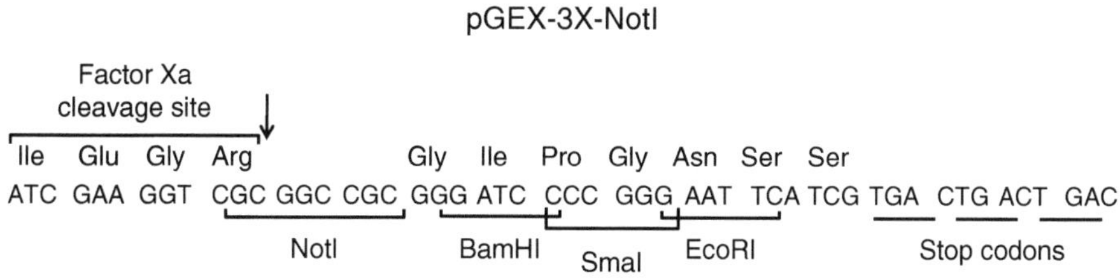

Fig. 2 Multiple cloning site of pGEX-3X-NotI vector. A *NotI* restriction site (*underlined*) was introduced into the multiple cloning site of pGEX-3X plasmid (GE Healthcare Life Sciences) between the sequence coding for the Factor Xa protease cleavage site and the *Bam*HI restriction site. Factor Xa cuts on the carboxy-terminal side of an arginine residue (*arrow*)

polymerase with proof-reading activity. Perform the DNA amplification for 30 cycles under the following conditions: 95 °C for 40 s, 60 °C for 1 min, and 72 °C for 3 min.

2. Check for the success of the PCR by agarose-gel electrophoresis and purify the amplified cDNA fragments with your preferred PCR clean-up method. Digest the PCR products with *Bam*HI and repurify the resulting cDNA fragments.
3. Digest pGEX-3X-NotI vector (*see* **Note 4**) with *Not*I (Fig. 2). Check complete digestion of plasmid by agarose-gel electrophoresis and collect the linearized DNA by ethanol precipitation.
4. Make blunt ends by nuclease S1 treatment (2 U/mg of plasmid DNA) in 1× reaction buffer at 37 °C for 30 min. Stop the reaction by adding EDTA to a final concentration of 15 mM and inactivate the nuclease by heating the reaction mixture at 70 °C for 10 min. Purify the blunt-ended pGEX-3X-NotI vector by either running the DNA on an agarose gel and excising the DNA band or using a spin column. Digest the vector with *Bam*HI and repurify as above.
5. Dephosphorylate the plasmid with alkaline phosphatase to prevent re-ligation. Inactivate the phosphatase by phenol:chloroform:isoamyl alcohol extraction and collect the DNA by ethanol precipitation.
6. Ligate approximately 100 ng of the plasmid DNA with 100 ng of each cDNA fragment (1:5 vector to insert molar ratio). Set up in parallel a control ligation without insert.
7. Transform competent DH5α *E. coli* cells with the ligation mixtures and select the transformed colonies on LB plates supplemented with ampicillin (100 μg/mL). Confirm the presence of recombinant plasmids by digesting plasmid DNA from different colonies with the appropriate restriction enzymes.
8. Select the appropriate recombinant plasmids pGEX-3X-NotI-FPS1 and pGEX-3X-NotI-FPS2, and verify the correct fusion

between the GST and FPS open reading frames by sequencing with the pGEX 5′ primer (*see* **Note 5**).

9. Transform competent BL21(DE3) *E. coli* cells harboring plasmid pUBS520 with the selected recombinant plasmids. Plasmid pUBS520 carries the gene encoding the tRNA of rare arginine codons AGG and AGA [13]. Select the transformed colonies on LB plates supplemented with ampicillin (100 μg/mL) and kanamycin (25 μg/mL).
10. Inoculate 3 mL of LB medium supplemented with ampicillin (100 μg/mL) and kanamycin (25 μg/mL) with a freshly isolated colony of BL21(DE3) cells harboring either pGEX-3X-NotI-FPS1 or pGEX-3X-NotI-FPS2. Grow overnight at 37 °C with constant shaking at 200 rpm.
11. On the next day, inoculate 30 mL of LB medium supplemented with the same antibiotics with 0.2 mL of the overnight culture and incubate at 37 °C with shaking until an optical density at 600 nm of 0.5–0.6 is reached. At this point, induce expression of GST-FPS fusion proteins by adding IPTG to a final concentration of 0.4 mM. Shift cultures to 20 °C and after incubation for additional 16 h chill flasks on ice.
12. Harvest *E. coli* cells by centrifugation at 7,000 ×*g* at 4 °C for 5 min. Pour off the supernatant and thoroughly resuspend cell pellets in 3 mL of ice-cold PBS. Disrupt cells by sonication at 12 μm (0.5 min/mL suspension) at 30–45 s intervals while being chilled in a −10 °C bath.
13. Remove cell debris by centrifugation at 15,000 ×*g* for 30 min at 4 °C and carefully collect the supernatant. Retain a sample for SDS-PAGE analysis to estimate the efficiency of soluble GST-FPS protein expression.
14. Transfer sufficient Glutathione-Sepharose 4B slurry to a disposable plastic column to achieve a 2.5–3 mL bed volume. Equilibrate the resin with minimum ten column volumes of cold PBS and keep the column at 4 °C.
15. Apply the soluble fraction of bacterial lysates to the column of Glutathione-Sepharose 4B. Allow the sample to flow through the column and wash it with minimum ten column volumes of cold PBS to remove unbound proteins. Re-equilibrate the column with Factor Xa cleavage buffer.
16. Apply to the column 50 U of Factor Xa in 1 mL of cleavage buffer to excise the FPS protein from the GST moiety (*see* **Note 6**) while bound to the Glutathione-Sepharose 4B. Allow the column to drain until the Factor Xa solution level is about 1–2 mm above the top of the column matrix and place a bottom cap or seal tightly with Parafilm® to stop the protease solution flow. Keep the column containing the Factor Xa reaction mixture overnight at 20–22 °C.

17. On the next day, elute the excised FPS protein with cleavage buffer at 4 °C and collect fractions of 0.4–0.5 mL (*see* **Note 7**). Keep the fractions at 4 °C while monitoring their protein concentration with a NanoDrop spectrophotometer and the purity of the FPS protein by SDS-PAGE analysis followed by Coomassie blue staining (*see* **Note 8**). Pool fractions enriched in FPS protein (*see* **Note 9**).
18. To remove Factor Xa protease from the pooled FPS fractions after GST-FPS digestion, add sufficient amount of Factor Xa removal resin equilibrated in cleavage buffer (*see* **Note 10**) to bind all Factor Xa protease. One hundred microliter of slurry contains sufficient resin to bind 4 U of Factor Xa protease. Mix gently to resuspend the resin and shake the tube with the slurry for 10 min at 4 °C (*see* **Note 11**) on a rotating wheel at 20 rpm.
19. Centrifuge the tube at 1,000 × *g* for 5 min at 4 °C to pellet the resin. Transfer the supernatant that contains the cleaved FPS protein without any vector-derived amino acids at the N-terminus to a clean tube and repeat centrifugation step twice more. Transfer the supernatant to a clean tube carefully avoiding any of the protease removal resin pellet.
20. Determine the protein content using a commercial protein assay kit and check the purity of the FPS protein by SDS-PAGE followed by Coomassie blue staining (*see* **Note 12**). Add glycerol to a final concentration of 15 % (v/v) and mix gently. Dispense the FPS preparation into 100 μL aliquots and store at −80 °C (*see* **Note 13**).

3.2 Preparation of Arabidopsis Extracts

1. Weigh a minimum of 25 mg of seeds in an Eppendorf tube or 200 mg of any other Arabidopsis freshly sampled tissue using homemade tin foil pouches (*see* **Note 14**). Take care to sample the tissue you are interested in a representative manner. Place the weighed sample in a small mortar (e.g., approximately 5 cm inner diameter) pre-chilled on ice.
2. Add ice-cold tissue homogenization buffer at a ratio of either 20 μL of buffer per mg of seeds or 2 μL of buffer per mg of any other tissue.
3. Keep the mortar on ice and crush tissue with the pestle until the sample becomes completely homogeneous (*see* **Note 15**). Carefully transfer the homogenate into a 2 mL Eppendorf tube with a pipette equipped with a 1 mL plastic tip. Incubate samples for 30 min at 4 °C on a rotating wheel.
4. Centrifuge samples at 200 × *g* for 10 min at 4 °C and carefully transfer the supernatant to a clean Eppendorf tube. Centrifuge at 16,000 × *g* for 20 min at 4 °C (*see* **Note 16**). Remove the supernatant and centrifuge again at 16,000 × *g*. Keep the resulting

supernatant at 4 °C until ready to use in the assay for FPS activity (*see* **Note 17**).

5. Determine extract protein content using a commercial protein assay kit.

3.3 Determination of FPS Activity

1. Dispense in 2 mL Eppendorf tubes the appropriate amount of either freshly prepared 16,000 × *g* supernatant from tissue extracts (*see* **Note 18**) or purified recombinant FPS (*see* **Note 19**). Prepare samples in duplicate. Complete to a total volume of 68 μL with homogenization buffer. Prepare in parallel a blank sample with 68 μL of homogenization buffer. Keep tubes on ice until ready to use in the assay.
2. Add 12 μL of 25 mM $MgCl_2$ to each tube.
3. Prepare sufficient volume of reaction mix for the total number of samples to be assayed (*see* **Note 20**). The assay of FPS activity is based on the quantification by liquid scintillation counting of the ^{14}C radioactivity incorporated into the reaction product FPP using [^{14}C]IPP as radiolabeled substrate and unlabeled GPP as the allylic substrate. For every 20 μL of reaction mix, combine in an Eppendorf tube 2.5 μL of 4 mM GPP and the required volumes of 4 mM IPP and [^{14}C]IPP to achieve a 0.5 mM concentration of [^{14}C]IPP with a specific radioactivity of 7.5 nCi/μmol. Complete to the required final volume with water and keep the tube on ice until ready to use in the assay.
4. Let stand tubes containing samples to be assayed and the reaction mix at room temperature for 5–10 min. At timed intervals add 20 μL of reaction mix to each tube and after each addition vortex to initiate the reaction. Incubate tubes at 37 °C for 30 min.
5. At the same timed intervals as above add 586 μL of ice-cold 2 N HCl to each tube. Vortex to stop the reaction and keep the tubes on ice for at least 5 min.
6. Incubate tubes at 37 °C for 30 min to hydrolyze the phosphate groups (*see* **Note 21**).
7. Add to each tube 200–250 mg of solid NaCl with a stainless-steel small laboratory spoon (approximately three spoons) to saturate the reaction mix (*see* **Note 22**) and 1 mL of hexane saturated with water (*see* **Note 23**). Vortex each tube for 1 min.
8. Centrifuge tubes at 6,000 × *g* for 10 min at 4 °C to enhance separation of the two phases.
9. Collect 500 μL of the upper hexanic phase (*see* **Note 24**) and place into vials containing 10 mL of scintillation cocktail Ecoscint O (*see* **Note 25**). In parallel place an aliquot of 20 μL of the reaction mix into a vial containing 10 mL of scintillation

cocktail Ecoscint H (*see* **Note 25**) to determine the specific radioactivity of the substrate [^{14}C]IPP.

10. Vortex vials for 10 s and allow them to stand for 1 h prior to counting. Quantify the ^{14}C-radioactivity by counting vials in a beta liquid scintillation counter (*see* **Note 26**).
11. Calculate the specific FPS activity (*see* **Note 27**) as units/mg of protein. One unit of FPS activity is defined as the amount of enzyme that catalyzes the incorporation of 1 nmol of [^{14}C]IPP into GPP to produce [^{14}C]FPP per minute at 37 °C.

4 Notes

1. Primers are usually supplied in a non-phosphorylated form so that PCR-amplified products do not contain a phosphate group in their 5′ ends. On the contrary digestion of DNA with restriction endonucleases always produce 5′-phosphorylated ends. For the resulting PCR products to be cloned in a dephosphorylated vector, the amplification products need to be phosphorylated, which can be performed by using T4 polynucleotide kinase. However, since phosphorylation of PCR products is a rather inefficient process, it is advisable to phosphorylate primers prior to PCR or even better to order 5′-phosphorylated primers.
2. Dissolve the Factor Xa lyophilized powder in water to a final concentration of 1 U/mL, dispense into 50 μL aliquots and store at −80 °C.
3. [^{14}C]IPP must only be used and stored in authorized areas by responsible persons who have received sufficient training necessary to safely handle radioactive products. Users of radioactive products must observe the local regulations governing the handling, use, storage, disposal, and transportation of radioactive materials.
4. Plasmid pGEX-3X (GE Healthcare Life Sciences) is a prokaryotic expression vector of approximately 4.9 kb designed to produce recombinant proteins fused to the 26 kDa glutathione *S*-transferase (GST) from *Schistosoma japonicum*. Protein expression from a pGEX plasmid is driven by the *tac* promoter, which is induced using IPTG. Fusion proteins are purified from bacterial lysates by affinity chromatography using Glutathione Sepharose 4B. Liberation of the protein of interest from the GST moiety is achieved using Factor Xa protease whose recognition sequence (Ile-Glu-Gly-Arg↓) is located immediately upstream from the *Bam*HI site of the multiple cloning region on pGEX-3X. After Factor Xa digestion, the protease is removed in a batch procedure with Factor Xa

removal resin followed by centrifugation. It is of primary importance to check the sequence of the protein you wish to express to ensure that it does not contain the Factor Xa protease recognition sequence. In the pGEX-3X-NotI vector (Fig. 2), a *Not*I restriction site has been added to the multiple cloning site of pGEX-3X between the sequence coding for the Factor Xa recognition sequence and the *Bam*HI restriction site. When the gene of interest is cloned blunt ended at the 5′-end using the *Not*I restriction site, cleavage of the recombinant fusion protein with Factor Xa Protease results in a protein product without any vector-derived amino acids at the N-terminus.

5. Even though a proof-reading polymerase is used, it is advisable to confirm that no artifactual sequence variations have been introduced in the amplified fragments by re-sequencing the entire PCR products.
6. So far there is no clear consensus about the interference that extra amino acid residues may have on FPS activity. A GST-tagged FPS of the rubber-producing mushroom *Lactarius chrysorrheus* has been reported to show less activity than the enzyme without the GST moiety and the same FPS N-terminally fused to a 6xHis tag does not show any prenyltransferase activity [14]. On the contrary, the properties of *D. melanogaster* FPS are not affected by the presence of a 6xHis tag fused at its N-terminus [15]. In view of this controversy, we prefer to stay on the safe side and work with recombinant FPS enzymes without any extra amino acid.
7. We collect manually a maximum of 20 fractions.
8. We usually electrophorese 5 μL aliquots of each fraction in a 10 % SDS-PAGE using a Mini PROTEAN® Tetra Cell (Bio-Rad) system.
9. We usually pool those fractions containing a concentration of protein ≥0.25 mg/mL, although this threshold may vary somewhat depending on the yield of recombinant protein expression.
10. Factor Xa removal resin is equilibrated with Factor Xa cleavage buffer (50 mM Tris–HCl, pH 7.5, 50 mM NaCl, 1 mM $CaCl_2$) instead of Factor Xa removal resin buffer (20 mM Tris–HCl, pH 6.5; 50 mM NaCl; 1 mM $CaCl_2$) because FPS proteins eluted from the Glutathione-Sepharose 4B column are dissolved in cleavage buffer. Binding of Factor Xa protease to the removal resin is unaffected by increasing the pH to 7.5 and the presence of 20–100 mM Tris–HCl.
11. Factor Xa binding to the removal resin can be performed at 4 °C without any loss of binding efficiency compared to that at room temperature.

12. A single protein band migrating with an apparent molecular mass of ~40 kDa is observed in the purified Arabidopsis FPS preparations. The Arabidopsis GST-FPS fusion proteins migrate with an apparent molecular mass of ~66 kDa in the bacterial lysates prior to Factor Xa digestion.
13. Both Arabidopsis FPS isozymes retain 90–95 % of their initial activity after 2 months of storage at −80 °C.
14. Tissue samples can also be collected in Eppendorf tubes partially dipped in liquid nitrogen. The frozen tissue is rapidly transferred to a mortar under liquid nitrogen, ground to a fine powder with pestle and transferred back to Eppendorf tubes that are stored at −80 °C until used. Frozen tissue can also be disrupted with a tissue grinder (e.g., TissueLyserII from Qiagen) if available. In any case, do not ever let samples to thaw. To obtain the extracts, take tubes containing ground tissue from the freezer, place them on ice, and add the appropriate volume of ice-cold tissue homogenization buffer. Let samples thaw slowly and incubate tubes at 4 °C as described (*see* Subheading 3.2, **step 3**).
15. Seed homogenates acquire an oily appearance.
16. First centrifugation at 16,000 × *g* can also be performed immediately after the 200 × *g* centrifugation step without removing the supernatant. To collect supernatant after centrifugation of seed homogenates, carefully insert the pipette tip below the floating lipid layer and above the pellet.
17. Determine FPS activity in freshly prepared extracts, as there is a significant loss of enzyme activity in tissue extracts that have been stored frozen.
18. The amount of 16,000 × *g* supernatant to be assayed varies according to the starting material. In the case of Arabidopsis, we usually measure FPS activity in 30 μL of seed extracts, 5 μL of seedling extracts, and 15 μL of leaf extracts.
19. Prepare a set of five serial 1:2 dilutions from a first 1:20 dilution of the purified FPS preparation and assay activity in 10 μL of each dilution to establish the appropriate working dilution. To ensure enzyme stability, purified FPS preparations are diluted with homogenization buffer supplemented with 0.5 mg/mL BSA.
20. It is advisable to prepare sufficient reaction mix for *N* + 3 samples, where *N* is the number of enzyme samples plus the blank sample. A 20 μL aliquot of the remaining reaction mix is used for determining the specific radioactivity of the substrate [^{14}C] IPP (*see* Subheading 3.3, **step 9**).
21. IPP is resistant to treatment with acid at 37 °C, while allylic prenyl diphosphates (DMAPP, GPP, and FPP) are acid-labile

yielding a mixture of alcohols that can be separated from IPP by extraction with a nonpolar solvent like hexane [16].

22. Solid NaCl is added to saturation to facilitate further separation of the phases.
23. When pipetting volatile liquids like hexane it is advisable to use a positive displacement Microman® pipet (Gilson) equipped with the appropriate capillary piston to assure accurate dispensing. To avoid exposure to hexane, wear proper personal protective equipment and use only under a chemical fume hood.
24. Radioactivity measured in the hexanic phase corresponds to that of the alcohols resultant from acid hydrolysis at 37 °C of the [^{14}C]FPP produced in the reaction mixture.
25. Ecoscint O is a scintillation solution for use with non-aqueous samples, whereas Ecoscint H is a scintillation solution for counting of aqueous samples.
26. We re-assay samples when cpm duplicates differ by more than 10 % from each other.
27. As a rule of thumb, enzyme activity values should only be calculated from time points where no more than 15 % of the substrate has been consumed. Specific FPS activity is determined by the formula:

$$\left[(\mathrm{Cs}-\mathrm{Cb})/(atm)\right]\times\left(V/V'\right)\times d$$

where

Cs: Number of counts per min (cpm) in 500 μL of the hexanic phase from FPS sample.

Cb: Number of cpm in 500 μL of the hexanic phase from blank sample.

a: specific activity of the substrate in cpm per nmol of [^{14}C]IPP (*see* Subheading 3.3, **step 9**).

t: time of incubation in min.

m: amount of protein in the assay mixture in mg.

V: total volume of hexanic phase (1 mL).

V′: volume of hexanic phase placed into the vial (0.5 mL).

d: fold of dilution of the FPS sample, if any.

Acknowledgements

This work was supported by grants from the Spanish Ministerio de Ciencia e Innovación (BIO2009-06984, cofinanced by the European Regional Development Funds), the Spanish Consolider-Ingenio 2010 Program (CSD2007-00036, Center for Research in Agricultural Genomics), and the Generalitat de Catalunya (2009SGR0026).

References

1. Kellogg BA, Poulter CD (1997) Chain elongation in the isoprenoid biosynthetic pathway. Curr Opin Chem Biol 1:570–578
2. Poulter CD (2006) Farnesyl diphosphate synthase. A paradigm for understanding structure and function relationships in *E*-polyprenyl diphosphate synthases. Phytochem Rev 5:17–26
3. Dhar MK, Koul A, Kaul S (2013) Farnesyl pyrophosphate synthase: a key enzyme in isoprenoid biosynthetic pathway and potential molecular target for drug development. New Biotechnol 30:114–123
4. Koyama T, Gotoh Y, Nishino T (2000) Intersubunit location of the active site of farnesyl diphosphate synthase: reconstruction of active enzymes by hybrid-type heteromeric dimers of site-directed mutants. Biochemistry 39:463–469
5. Ambo T, Noike M, Kurokawa H, Koyama T (2008) Cloning and functional analysis of novel short-chain *cis*-prenyltransferases. Biochem Biophys Res Commun 375:536–540
6. Sallaud C, Rontein D, Onillon S, Jabès F, Duffé P, Giacalone C, Thoraval S, Escoffier C, Herbette G, Leonhardt N, Causse M, Tissier A (2009) A novel pathway for sesquiterpene biosynthesis from Z,Z-farnesyl pyrophosphate in the wild tomato *Solanum habrochaites*. Plant Cell 21:301–317
7. Tholl D, Sungbeon L (2011) Terpene specialized metabolism in Arabidopsis thaliana. Arabidopsis Book 9:e0143. doi:10.1043/tab.0143
8. Keim V, Manzano D, Fernández FJ, Closa M, Andrade P, Caudepón D, Bortolotti C, Vega MC, Arró M, Ferrer A (2012) Characterization of Arabidopsis FPS isozymes and FPS gene expression analysis provide insight into the biosynthesis of isoprenoid precursors in seeds. PLoS ONE 7:e49109. doi:10.1371/journal.pone.0049109.t001
9. Cunillera N, Boronat A, Ferrer A (1997) The *Arabidopsis thaliana* FPS1 gene generates a novel mRNA that encodes a mitochondrial farnesyl-diphosphate synthase isoform. J Biol Chem 272:15381–15388
10. Thabet I, Guirimand G, Courdavault V, Papon N, Godet S, Dutilleul C, Bouzid S, Giglioli-Guivarc'h N, Clastre M, Simkin AJ (2011) The subcellular localization of periwinkle farnesyl diphosphate synthase provides insight into the role of peroxisome in isoprenoid biosynthesis. J Plant Physiol 168:2110–2116
11. Sanmiya K, Ueno O, Matsuoka M, Yamamoto N (1999) Localization of farnesyl diphosphate synthase in chloroplasts. Plant Cell Physiol 40:348–354
12. Cunillera N, Arró M, Delourme D, Karst F, Boronat A, Ferrer A (1996) *Arabidopsis thaliana* contains two differentially expressed farnesyl-diphosphate synthase genes. J Biol Chem 271:7774–7780
13. Brinkmann U, Mattes RE, Buckel P (1989) High-level expression of recombinant genes in *Escherichia coli* is dependent on the availability of the dnaY gene product. Gene 85:109–114
14. Mekkriengkrai D, Sando T, Hirooka K, Sakdapipanich J, Tanaka Y, Fukusaki E, Kobayashi A (2004) Cloning and characterization of farnesyl diphosphate synthase from the rubber-producing mushroom Lactarius chrysorrheus. Biosci Biotechnol Biochem 68:2360–2368
15. Sen SE, Trobaugh C, Béliveau C, Richard T, Cusson M (2007) Cloning, expression and characterization of a dipteran farnesyl diphosphate synthase. Insect Biochem Mol Biol 37:1198–1206
16. Ogura K, Nishino T, Shinka T, Seto S (1985) Prenyltransferases of pumpkin fruit. Methods Enzymol 110:167–171

Part II

Targeted Analysis of Isoprenoid Metabolites

Chapter 5

Metabolite Profiling of Plastidial Deoxyxylulose-5-Phosphate Pathway Intermediates by Liquid Chromatography and Mass Spectrometry

Edward E.K. Baidoo, Yanmei Xiao, Katayoon Dehesh, and Jay D. Keasling

Abstract

Metabolite profiling is a powerful tool that enhances our understanding of complex regulatory processes and extends to the comparative analysis of plant gene function. However, at present, there are relatively few examples of metabolite profiling being used to characterize the regulatory aspects of the plastidial deoxyxylulose-5-phosphate (DXP) pathway in plants. Since the DXP pathway is one of two pathways in plants that are essential for isoprenoid biosynthesis, it is imperative that robust analytical methods be employed for the characterization of this metabolic pathway. Recently, liquid chromatography-mass spectrometry (LC-MS), in conjunction with traditional molecular biology approaches, established that the DXP pathway metabolite, methylerythritol cyclodiphosphate (MEcPP), previously known solely as an intermediate in the isoprenoid biosynthetic pathway, is a stress sensor that communicates environmental perturbations sensed by plastids to the nucleus, a process referred to as retrograde signaling. In this chapter, we describe two LC-MS methods from this study that can be broadly used to characterize DXP pathway intermediates.

Key words DXP pathway, MEP pathway, Intermediates, Metabolites, Quantification, LC-MS

1 Introduction

The DXP pathway, also known as the methylerythritol 4-phosphate (MEP) pathway, consists of seven enzymes that transform glyceraldehdye-3-phosphate (G3P) and pyruvic acid to intermediates and the final products namely, isopentenol diphosphate (IPP) and dimethylallyl diphosphate (DMAPP) (Fig. 1), from which all isoprenoids are synthesized [1–4]. DXP pathway intermediates can be anionic in nature and, therefore, carry net negative charges. This means that they are detected as deprotonated ions $[M-H]^-$ in the negative mode of mass spectrometry. However, when profiling metabolites, it is better to separate them in real-time prior to their detection. The addition of this separation step

Manuel Rodríguez-Concepción (ed.), *Plant Isoprenoids: Methods and Protocols*, Methods in Molecular Biology, vol. 1153, DOI 10.1007/978-1-4939-0606-2_5,

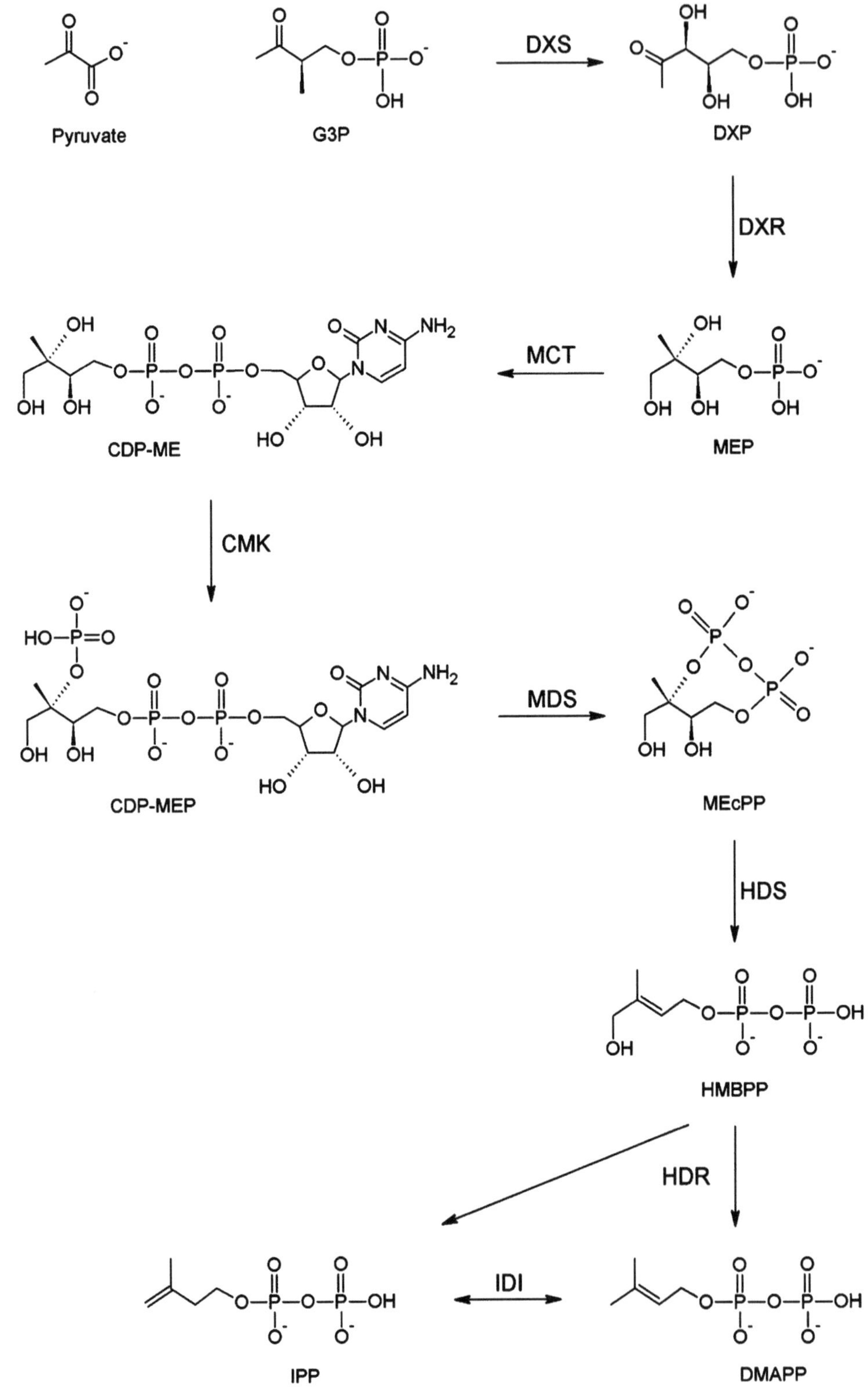

Fig. 1 Overview of the DXP pathway. Intermediates of the DXP pathway are as follows: Pyruvate (pyr) and glyceraldehyde 3-phosphate (G3P) are condensed to DXP with a loss of carbon dioxide, a reaction performed by DXP synthase (DXS); DXP is reduced to methylerythritol phosphate (MEP) by DXP reductoisomerase (DXR); MEP

reduces ion suppression effects (in electrospray ionization, ESI) caused from interfering compounds and, thus, improves upon quantitation.

Separation of hydrophilic compounds such as the intermediates of the DXP pathway is generally best achieved by ion exchange chromatography (IXC), ion exclusion chromatography (IEC), and hydrophilic interaction liquid chromatography (HILIC). In IXC, retention is based on the interaction between the charged analyte and the charged functional groups bound to the stationary phase [5, 6]. This retention property enables anionic metabolites, such as the DXP pathway intermediates, to be retained by their interaction with the positively charged groups on the stationary phase. In IEC, ions with the same charge as the stationary phase (such as those from the mobile phase) are not permitted to penetrate it as they are excluded through repulsion [5, 6]. Here, only the analyte is allowed to interact with the stationary phase. A stationary phase that employs both mechanisms is normally composed of a resin. The charged groups on the resin allow for IXC and IEC to take place, and the polystyrene backbone of the resin allows for hydrophobic interactions. The resin, therefore, permits the retention mechanisms of reversed phase, normal phase partitioning, and size exclusion to occur. These multiple modes of interaction provide resin-based HPLC columns with added selectivity under isocratic conditions [5, 6]. However, a major drawback to using this approach with mass spectrometry (MS) is that analyte separation is normally achieved via aqueous mobile phases.

HILIC retention is achieved through hydrogen bonding between the analyte and water bound to the silica stationary phase [7]. Consequently, analytes are best retained by the stationary phase when the organic content (e.g., acetonitrile) of the mobile phase is high and they are eluted from the column when the aqueous content of the mobile phase is high. Secondary interactions can be achieved through the addition of charged groups (e.g., zwitter ions) to the silica stationary phase to further improve upon the selectivity of the HILIC method [8]. The volatile nature of HILIC mobile phases makes this separation technique ideally suited to electrospray mass spectrometry (ESI-MS).

Once the separated metabolites have been eluted from the column they are delivered to the ion source. Here, they are converted to gaseous ions by the combination of a nebulizer gas and a heated

reacts with cytidine triphosphate to produce diphosphocytidylyl methylerythritol (CDP-ME), a reaction catalyzed by MEP cytidylyltransferase (MCT); CDP-ME is phosphorylated to CDP-ME phosphate (CDP-MEP) by CDP-ME kinase (CMK); CDP-MEP is converted to MEcPP via a loss of cytidine monophosphate, a reaction catalyzed by MEcPP synthase (MDS); MEcPP is converted to hydroxymethylbutenyl diphosphate (HMBPP) by HMBPP synthase (HDS); HMBPP is converted to IPP and DMAPP by HMBPP reductase (HDR); IPP can be reversibly isomerized to DMAPP via IPP isomerase (IDI). MEcPP is thought to undergo reduction and elimination to HMBPP [4]

drying gas during ESI. This ionization technique is established via the application of a strong electric field, under atmospheric pressure, to the eluent passing through a narrow stainless steel capillary, where the accumulation of charge at the liquid surface at the tip of the capillary leads to the formation of highly charged droplets. As the droplets become increasingly smaller the accumulation of destabilizing like-charges, through natural repulsion, causes the release of ions. Consequently a cascade of coulombic repulsions eventually give rise to either an ion contained within a single droplet and/or the emission of solvated ions from charged droplets, which lead to the formation of gas phase ions upon evaporation [9–14].

Ions from the ESI source are electrostatically drawn towards the entrance of the MS, where they pass through a heated sampling capillary until they reach a metal skimmer. From there, they pass through an ion guide (e.g., an octapole) under a specified radio frequency (rf) voltage. The application of a high vacuum at this stage reduces the number of collisions between ions and gas molecules. Before entering the mass analyzer, ions are further focused by a series of lenses [15]. For the purpose of this chapter, only the quadrupole (quad; Fig. 2) and the time-of-flight (TOF; Fig. 3) mass analyzer will be discussed.

A single quad is a low resolution, low cost, highly robust mass spectrometer that consists of four parallel cylindrical or hyperbolic

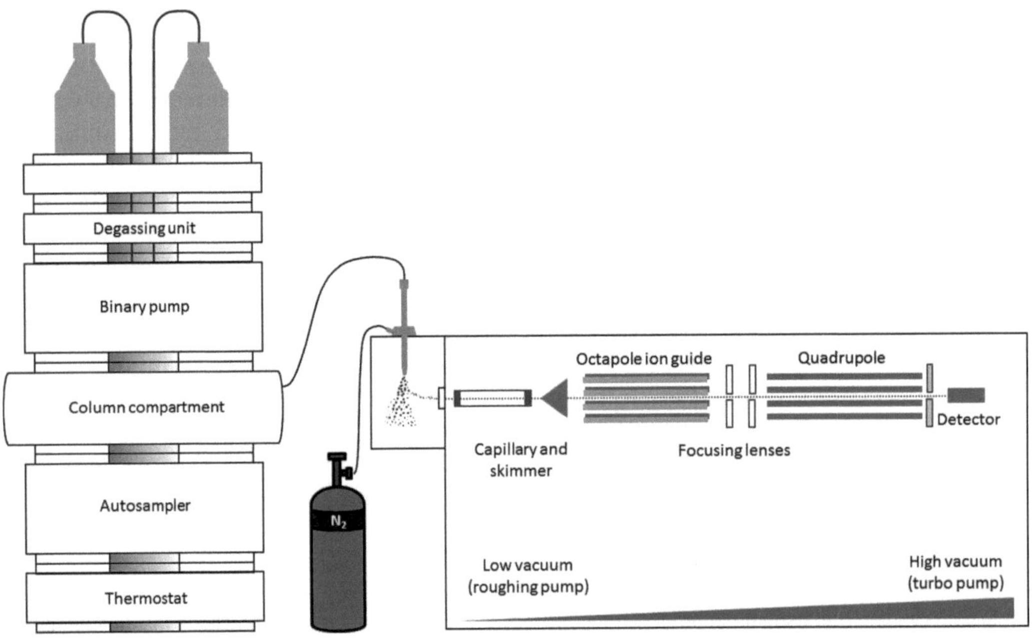

Fig. 2 LC-ESI-quad MS. After analyte ions are delivered to the ion source, ESI facilitates their transfer (from the liquid to the gas phase) to the entrance of the MS. From there they are transmitted to the quadrupole (via a glass capillary, a skimmer, and focusing lenses) and transit this mass analyzer in a stable spiral like trajectory at its center (while dc and ac/rf voltages are alternated), en route to the detector. This illustration is based on an Agilent Technologies LC-ESI-quad MS system

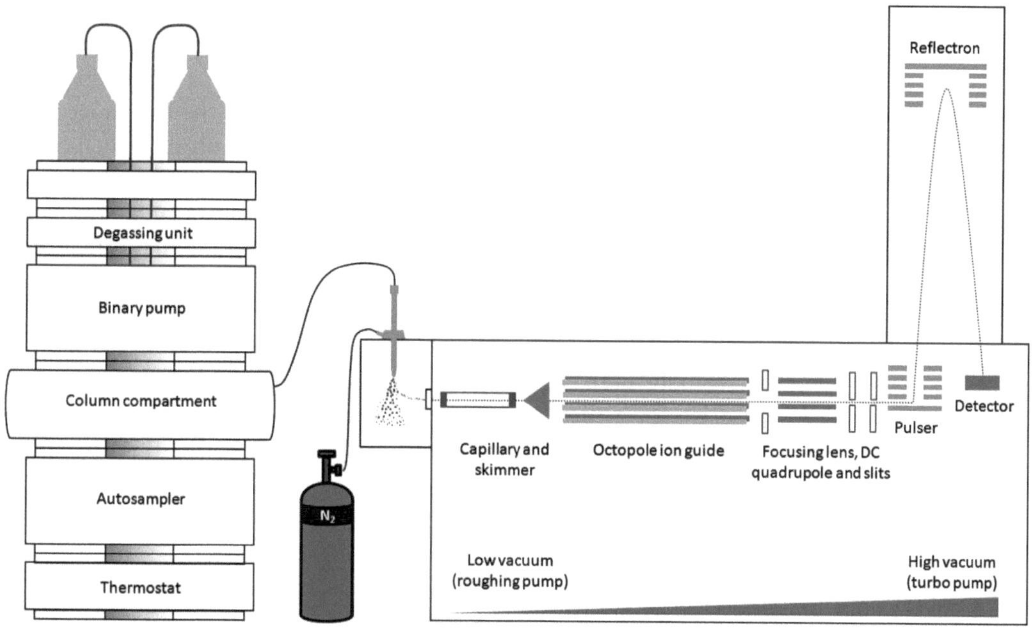

Fig. 3 LC-ESI-TOF MS. After analyte ions are delivered to the ion source, ESI facilitates their transfer (from the liquid to the gas phase) to the entrance of the MS. From there they are transmitted to the TOF (via a glass capillary, a skimmer, an ion guide and focusing lenses) and are pulsed up the flight tube and reflected down (by specific voltages) en route to the detector. Lighter ions arrive at the detector first. This illustration is based on an Agilent Technologies LC-ESI-quad MS system

rods that are equally spaced around a central axis [16], with each rod serving as an electrode. Opposing sets of rods have both direct current (dc) and radio frequency (rf) voltages applied to them. As ions transit the rods both dc and rf voltages are increased at a constant ratio [17]. The application of these voltages creates an area of mutual stability at the center of the rods, where only ions with certain m/z ratios are transmitted through the quadrupole, whereas ions exhibiting different m/z ratios collide with the rods as result of their unstable trajectories [16]. Thus, the quadrupole forms a low and high mass filter. The relationship of m/z to these potentials is described in following equations:

$$U = a_u = \frac{m}{z}\frac{\omega^2 r_0^2}{8e} \qquad V = q_u = \frac{m}{z}\frac{\omega^2 r_0^2}{4e}$$

where a_u and q_u are the stability parameters and e is the charge of an electron [9]. The other parameters ω^2 and r_0^2 are related to the angular frequency and radius, respectively [9]. The output of the dc generator is described by the V voltage and the rf generator is described by the U voltage [9]. To scan a mass spectrum, both voltages are increased at the same time from zero to some maximum value while their ratio is maintained constant [17–19]. A scan is, therefore, achieved in sequential steps, from the lowest to the highest m/z ratio.

In the most sensitive operating mode of the single quadrupole, selected ion monitoring (SIM), the mass analyzer is programmed to allow the passage of ions with specific *m/z* ratios (typically ±0.5 *m/z*) through to the detector. In this way, the mass spectrometer spends more time collecting information on preselected ions rather than scanning across a wide mass range. This leads to a reduction in background noise and, by extension, a significant increase in signal-to-noise (*see* **Note 1**). The quad mass analyzer is ideally suited to targeted metabolite analyses, especially when the SIM mode is employed (Fig. 2).

A TOF is a medium to high resolution mass analyzer with an extremely high scan rate (*see* **Note 2**). The TOF mass analyzer consists of a flight tube under a high vacuum. The principle of TOF is based on the assumption that ions with the same initial kinetic energy (E_{kin}) travel at different velocities ν that are proportional to their *m/z* ratios and inversely proportional to their masses (*m*):

$$\nu = \frac{\sqrt{E_{kin}}}{m}$$

For a given energy (*E*) and distance (*L*), the mass is proportional to the square of the flight time (*t*) [15]:

$$m = t^2\left(\frac{2E}{L^2}\right)$$

The *m/z* ratio of a given ion can, therefore, be calculated by measuring the time it takes to travel a certain distance [15]:

$$\frac{m}{z} = t^2\left(\frac{2eV}{L^2}\right)\frac{1}{2}$$

where ions are selected by a potential *V* and *e* is the charge of an electron [9, 15, 20, 21].

As ions enter the TOF they are subjected to a pulsed electric field, which is applied at the entrance of the TOF. Lighter, multiple-charged ions reach the detector before heavier single-charged ions [15]. Once ions are pulsed up the flight tube, they enter the reflectron region which, as the name suggests, is responsible for repelling ions towards the detector based on their forward kinetic energies (*see* **Notes 3** and **4**). A consequence of this is that the same ions arrive at the flight tube at the same time (*see* **Note 5**). In this way, the reflectron region improves mass resolution and accuracy by minimizing variations in the flight times of ions (resulting from different spatial and kinetic energy distributions) [15]. The reflectron region also helps to double the flight path, giving ions more time to separate and, hence, increases the overall resolution (the mass difference, Δ*m*, between two adjacent masses) of the mass analyzer (*see* **Note 6**).

The monoisotopic mass is used to determine the mass accuracy of analyte ions, which is the difference between the theoretical m/z ratio and the measured (*see* **Note 7**).

$$\text{Mass accuracy}(\text{ppm}) = 1 \times 10^6 \left(\frac{mTheo - mMeas}{mTheo} \right)$$

In the above equation mTheo is the theoretical m/z ratio and mMeas is the measured m/z ratio [22]. The high acquisition rate, high sensitivity, high mass resolution, and high mass accuracy of TOF make this mass analyzer ideally suited to targeted metabolomics analyses (Fig. 3).

Information obtained from MS (i.e., a measure of the characteristics of charged molecules based on the mass-to-charge ratio, m/z, of an ion) is commonly used to deduce elemental composition and elucidate the chemical structure of a compound. A mass spectrometer can, therefore, be used to produce a mass spectrum (a plot of m/z vs. intensity) or used as a detector (a plot of peak intensity vs. time) [6, 16, 17]. An analyte ion is generally quantified by the area under the corresponding chromatographic peak. The TOF utilizes extracted ion chromatograms (EIC) for accurate peak integration, whereas the quad utilizes either EIC (in the full scan mode) or SIM chromatograms, with each data point across the chromatographic peak corresponding to the m/z ratio of the analyte ion (*see* **Note 8**). The added selectivity that the mass spectrometer brings allows LC-MS to be used to quantify metabolites from complex biological matrices such as plant biomass. For a more comprehensive study of IXC, IEC, HILIC, ESI, quad and TOF mass spectrometry, the reader is kindly referred to other sources for this information [5–21]. In this chapter we describe IXC-IEC-ESI-TOF MS and HILIC-ESI-quad MS methods for the measurement of plastidial DXP pathway intermediates.

2 Materials

It is recommended that solvents used for LC-MS experiments are of HPLC grade (99.9 % chemical purity) or greater, and that the chemicals used are of analytical grade (with a chemical purity >90 %). Chemical standard and sample solutions are prepared in acetonitrile–water (1:1, v/v) [1]. Alternatively, they can also be prepared in the appropriate reconstitution medium (i.e., the initial mobile phase solvent composition) (*see* **Note 9**). Filter and degas all HPLC eluents (i.e., with a 0.2- or 0.45-μm membrane pore size filter) prior to use. It is also recommended that a guard column is used as it can prolong the lifetime of the analytical column. Use nitrogen as both the nebulizer and heated drying gas for ESI (*see* **Note 10**). Please remember to follow your institution's or company's waste disposal regulations when disposing of waste materials. Also remember to follow your institutes or company's regulations and guidelines for handling cryogens.

2.1 Sample Preparation

1. Flash freezing of plant tissue is carried out with liquid nitrogen.
2. Cryogen resistant gloves are worn during the flash freezing and pre-chilling stages of sample preparation.
3. Arabidopsis plants are grown at 22 °C in the presence and absence of light depending on the experiment.
4. Quenching is performed by flash freezing with liquid nitrogen.
5. Plant material is ground with the appropriate homogenizer (a pestle and mortar or a ball mill) (*see* **Note 11**).
6. Liquid nitrogen-resistant 2-mL centrifuge tubes are used for sample processing.
7. A liquid nitrogen resistant open centrifuge tube holder (where only the base is exposed to liquid nitrogen) is used for flash freezing.
8. The extraction buffer used is 13 mM ammonium acetate in water (adjusted to pH 5.5 with acetic acid).
9. A refrigerated microcentrifuge is used to separate the extract from plant material.
10. The extract is dried via by freeze drying.
11. The reconstitution solution is composed of the appropriate internal standards in acetonitrile–water (1:1, v/v). It is best to use structural analogs of DXP pathway intermediates as internal standards. Such internal standards accurately account for any variation in LC separation performance and MS detection.
12. DXP pathway chemical standards are stored at −20 °C in methanol.
13. Calibration curves for DXP pathway intermediates are prepared in acetonitrile–water (1:1, v/v) and the calibration range is 0.78–200 μM.

2.2 IXC-IEC-ESI-TOF MS

1. A HPLC system equipped with an autosampler, column compartment and degassing unit is used to deliver the LC eluent, which is 0.1 % formic acid in water (*see* **Note 12**).
2. Sample and standard solutions are analyzed via HPLC vials with inserts.
3. Chromatographic separations of analytes are carried out on an Aminex HPX-87H ion exclusion column (300 × 7.8 mm, Biorad) (*see* **Note 13**).
4. A TOF mass spectrometer, equipped with a heated ESI source, is used for detection (*see* **Note 13**).
5. The MS is controlled by the relevant software, which is provided by the instrument manufacturer.

2.3 HILIC-ESI-quad MS

1. A HPLC system equipped with an autosampler, column compartment and degassing unit is used to deliver the LC eluents, which are 50 mM ammonium carbonate in water and acetonitrile (*see* **Note 14**).
2. Sample and standard solutions are analyzed via HPLC vials with inserts.
3. Chromatographic separations of analytes are carried out on a ZIC-pHILIC column (150×2.1 mm, 3 μm particle size, Sequant-Nest Group).
4. A quadrupole mass spectrometer, equipped with a heated ESI source, is used for detection (*see* **Note 15**).
5. The MS is controlled by the relevant software, which is provided by the instrument manufacturer.

3 Methods

3.1 Sample Preparation (Method A)

A schematic presentation of the method is shown in Fig. 4.

1. Harvest the plant tissue and flash freeze it in liquid nitrogen.
2. Grind the plant tissue in liquid nitrogen with a pestle and mortar.

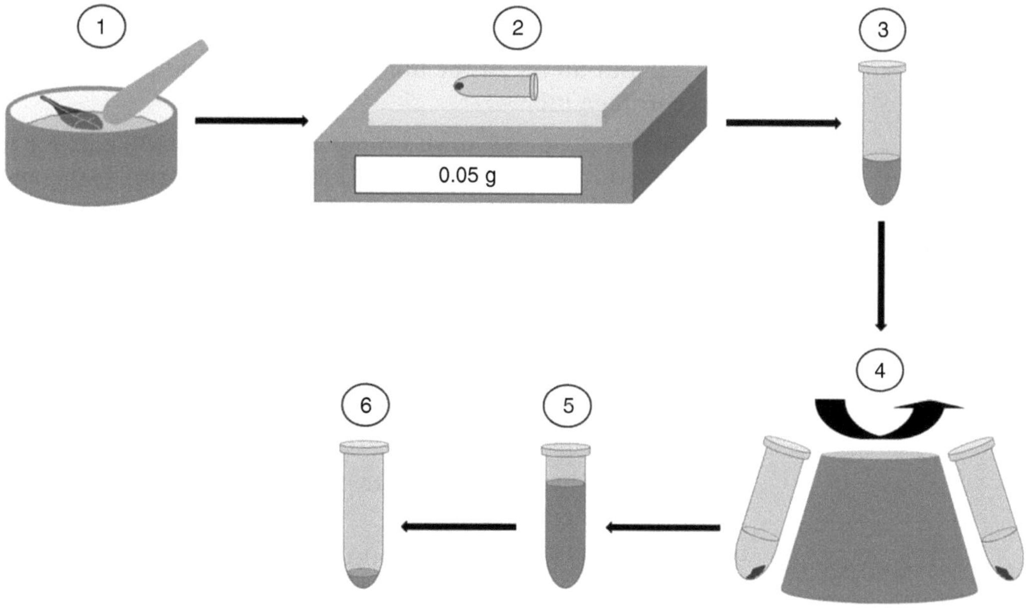

Fig. 4 Sample preparation method A. (*1*) Grind the frozen plant tissue, (*2*) weigh out 50 mg of plant tissue, (*3*) add 500 μL of 13 mM ammonium acetate solution and mix thoroughly, (*4*) centrifuge and transfer the supernatant to a separate centrifuge tube, (*5*) repeat steps 3 and 4 twice and combine the three extracts, (*6*) freeze, freeze dry, then reconstitute the dried extract in 100 μL acetonitrile–water (1:1, v/v)

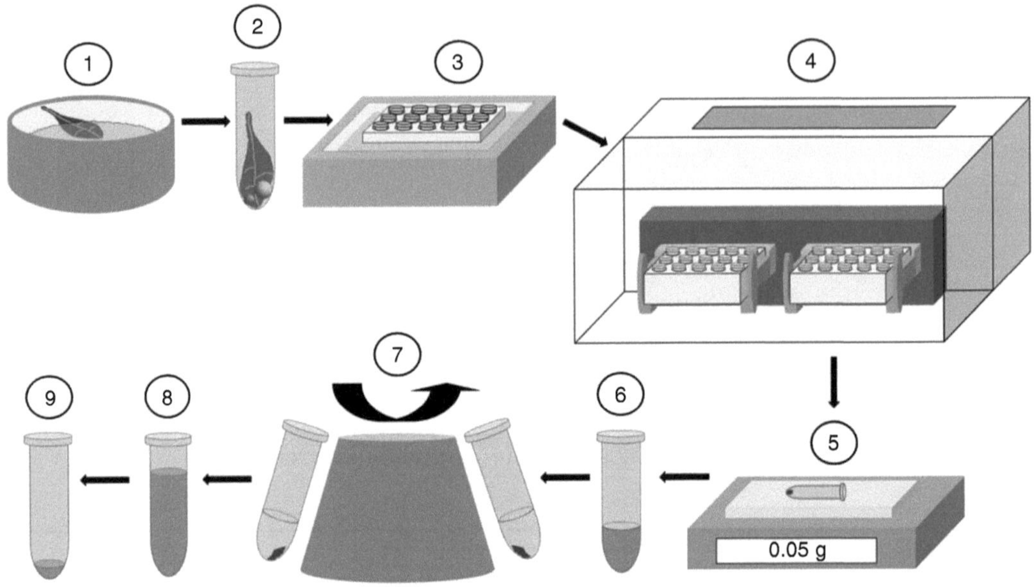

Fig. 5 Sample preparation method B. (*1*) Flash freeze the plant tissue in liquid nitrogen, (*2*) transfer the plant tissue to a centrifuge tube containing the pre-chilled grinding balls, (*3*) transfer the centrifuge tube to a pre-chilled centrifuge tube holder in a liquid nitrogen bath, (*4*) place the centrifuge tube holder in the mixing chamber of the ball mill and start the grinding cycle, (*5*) weigh out 50 mg of the plant tissue, (*6*) add 500 μL of 13 mM ammonium acetate solution and mix thoroughly, (*7*) centrifuge and transfer the supernatant to a separate centrifuge tube, (*8*) repeat steps 6 and 7 twice and combine the three extracts, (*9*) freeze, freeze dry, then reconstitute the dried extract in 100 μL acetonitrile–water (1:1, v/v)

3. Weigh out 50 mg of the homogenized/ground plant tissue in a centrifuge tube.
4. Add 500 μL of 13 mM ammonium acetate buffer in water (adjusted to pH 5.5 with acetic acid) to the plant tissue and mix thoroughly by vortexing.
5. Centrifuge at the highest speed of the bench-top microcentrifuge for 5 min at 4 °C.
6. Transfer the supernatant to a new centrifuge tube.
7. Repeat **steps 4–6** twice (these are performed on ice). Molecular weight cut-off filters may be employed to reduce the complexity of the sample.
8. Combine the three extracts together, freeze with liquid nitrogen and freeze dry.
9. Reconstitute the extract in 100 μL of the appropriate deuterated standards in acetonitrile–water (1:1, v/v).

3.2 Sample Preparation (Method B)

A schematic presentation of the method is shown in Fig. 5.

1. Pour liquid nitrogen into an appropriately sized freezer box until it is approximately a quarter full.

2. Immerse the cryogenic centrifuge tube holder (which is open at the base) in the liquid nitrogen. The holder will be pre-chilled when the nitrogen bath is no longer bubbling. Leave about a cm gap from the top of the centrifuge tube holder.
3. Pre-chill the stainless steel grinding balls (typically a 2 mm and a 4 mm ball are used) in liquid nitrogen that is housed in a clean cryogenic container.
4. Transfer a centrifuge tube to the pre-chilled centrifuge tube holder in the nitrogen bath. It should be noted that there will be liquid nitrogen in the holes of the centrifuge tube holder (as liquid nitrogen will enter from the base of the holder).
5. Wait until the nitrogen bath is no longer bubbling.
6. Quickly transfer the stainless steel grinding balls to a centrifuge tube in the pre-chilled centrifuge tube holder.
7. Harvest the plant tissue and flash freeze in liquid nitrogen.
8. Transfer the plant tissue to a centrifuge tube (containing the stainless steel grinding balls) in the pre-chilled centrifuge tube holder.
9. Place the centrifuge tube holder in the mixing chamber of the ball mill.
10. Start the grinding cycle and leave for 1 min.
11. Transfer the centrifuge tube holder to a freezer box containing dry ice.
12. Remove the steel grinding balls from the centrifuge tube.
13. Weigh out 50 mg of the homogenized/ground plant tissue in a centrifuge tube.
14. Add 500 μL of 13 mM ammonium acetate buffer in water (adjusted to pH 5.5 with acetic acid) to the plant tissue and mix thoroughly by vortexing.
15. Centrifuge at the highest speed of the bench-top microcentrifuge for 5 min at 4 °C.
16. Transfer the supernatant to a new centrifuge tube.
17. Repeat **steps 14–16** twice (**steps 14–16** are performed on ice).
18. Combine the three extracts together, freeze with liquid nitrogen and freeze dry.
19. Reconstitute the extract in 100 μL of the appropriate internal standards in acetonitrile–water (1:1, v/v).

3.3 IXC-IEC-ESI-TOF MS

A new column is flushed with 0.1 % formic acid in water overnight at a flow rate of 0.1 mL/min and at a temperature of 50 °C. This is to remove any contaminants from the previous mobile phase and to equilibrate the column(*see* **Note 15**). Before LC-ESI-TOF MS analysis, the column is flushed with 0.1 % formic acid for 30 min at a flow rate of 0.1 mL/min and at a temperature of 50 °C. The LC column compartment and autosampler thermostat temperatures are maintained at 50 °C and 4 °C, respectively. The sample injection volume used is 5–10 μL. The mobile phase is composed of 0.1 % formic acid in water and a flow rate of 0.6 mL/min is used throughout the analysis. The maximum backpressure that the LC column is able to tolerate is normally 200 bar (*see* **Note 16**). The HPLC system is coupled to a TOF MS with a 1:6 post-column split (*see* **Notes 17** and **18**). In this experiment, LC-ESI-MS coupling is achieved using an orthogonal interface. Nitrogen gas is used as both the nebulizing and drying gas to facilitate the production of gas-phase ions. The drying and nebulizing gases are set to 11 L/min and 30 psi, respectively, and a drying gas temperature of 330 °C is used throughout. Fragmentor, skimmer and ion-guide rf voltages are set to 150 V, 50 V, and 170 V, respectively. ESI is conducted in the negative-ion mode with a capillary voltage of 3.5 kV. MS experiments are carried out in the full-scan mode (m/z 145–605) at 0.86 spectra per second for the detection of $[M-H]^-$ ions. The instrument is tuned for a range of m/z 50–1,700. Prior to LC-ESI-TOF MS analysis, the TOF MS is calibrated with the manufacturers TOF tuning mix. Internal calibration of the TOF m/z axis is performed with the relevant reference masses provided by the instrument manufacturer (*see* **Notes 19–21**). A typical autosampler sequence consists of: blank → calibration curve → blank → samples → blank → calibration curve → blank. If the background noise (observed via the total ion chromatogram) is high, then more than one blank may be used at the start of the run.

1. Enter the required LC and MS parameters as well as the run sequence in the acquisition software. Normally the acquisition software will not allow the run sequence to be activated unless these parameters are saved. Once saved, the resulting method and the sequence work list can be used for future experiments.
2. Connect the mobile phase reservoir to the HPLC system.
3. Switch on the purge valve system.
4. Turn on a single HPLC pump.
5. Set the flow rate of the HPLC pump to 5 mL/min for 2 min. Purge the HPLC pump with the new mobile phase. Here the flow is directed to waste.
6. Turn off the HPLC pump.
7. Switch off the purge valve system. The flow is now directed to the column.
8. Change the flow rate to 1 or 2 mL/min depending on the HPLC system used (*see* **Note 22**).
9. Turn on the column compartment. The temperature is set to 50 °C.

10. Turn on the HPLC pump and observe the mobile phase emerging from the column compartment-to-column connection tubing.
11. Wait until the previous mobile phase has been completely replaced. One way of determining this is to test the pH of the emerging droplets (with a pH strip). If the pH is ≤3 then the previous mobile phase has been replaced. Another way is to observe the backpressure. That is, the stabilization of the backpressure is a good indication that the previous mobile phase has been replaced.
12. Connect the guard and analytical column to the column compartment.
13. Set the flow rate to 0.1 mL/min.
14. Turn on the HPLC pump and collect the column effluent for waste disposal.
15. Wait until the backpressure has stabilized.
16. Increase the flow rate to 0.2, 0.3, 0.4, 0.5, and finally to 0.6 mL/min (waiting for the backpressure to stabilize after each incremental increase in flow rate).
17. Once the back pressure has stabilized, connect the calibrant delivery system to the MS.
18. Turn on the MS and calibrate the time-of-flight mass axis.
19. Then connect the HPLC system to the ESI-MS sprayer via connection tubing.
20. Start the run sequence.
21. The total run time is 20 min.
22. Representative results are shown in Fig. 6. Once the analysis is completed flush the column with the mobile phase at 0.1 mL/min for 30 min (at 50 °C). The column can now be stored for future use. Please refer to the manufacturer's guidelines for the appropriate column cleaning and regeneration procedures.

3.4 HILIC-ESI-quad MS

LC column compartment and autosampler thermostat temperatures are maintained at 30 °C and 4 °C, respectively. The sample injection volume used is 4 μL. The mobile phase is composed of (A) 50 mM ammonium carbonate in water (*see* **Note 23**) and (B) acetonitrile. Isocratic elution is achieved at 70 % B at a flow rate of 0.2 mL/min. The maximum backpressure that the LC column is able to tolerate is 200 bar. The HPLC system is coupled to a single quadrupole mass spectrometer MS with a 1:3 post-column split. In this experiment, LC-ESI-MS coupling is achieved using an orthogonal interface. Nitrogen gas is used as both the nebulizing and drying gas to facilitate the production of gas-phase ions. The drying and nebulizing gases are set to 10 L/min and 30 psi, respectively, and a drying gas temperature of 330 °C is used throughout.

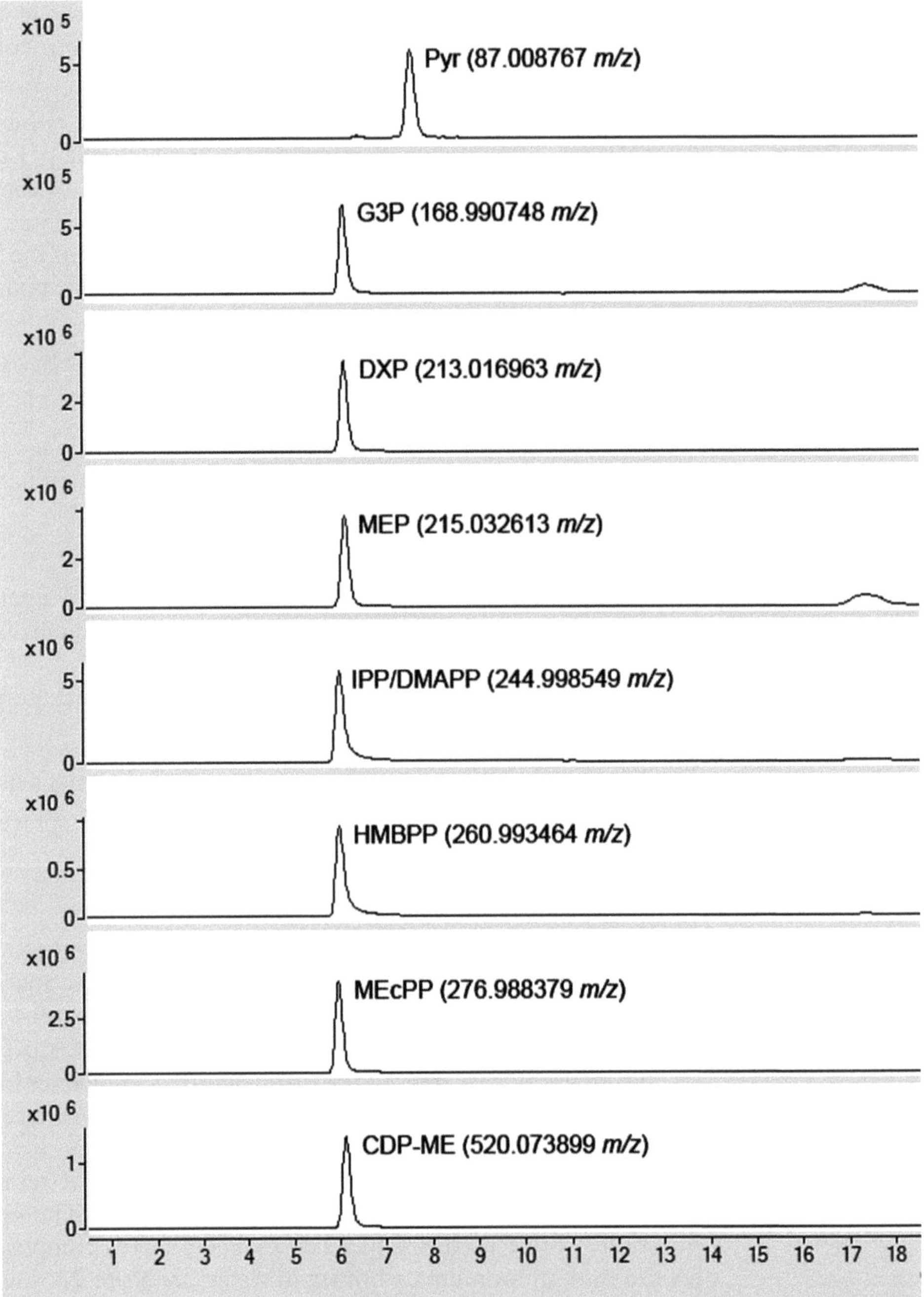

Fig. 6 IXC-IEC-ESI-TOF MS analysis of DXP pathway intermediates. Counts versus time (min). The data was acquired from 80 to 605 *m/z*

ESI is conducted in the negative-ion mode with a capillary voltage of 3.5 kV. MS experiments are carried out in the SIM mode (87, 169, 213, 215, 261, 277, and 520 *m/z*) at 3 s/cycle at a dwell time of 500 ms (or ≤1.2 s/cycle at a dwell time of 200 ms if faster scanning is required) for the detection of $[M-H]^-$ ions. The instrument is tuned for a range of m/z 50–1,700 with the appropriate tuning mix provided by the manufacturer. A typical autosampler sequence consists of: blank → calibration curve → blank → samples → blank → calibration curve → blank.

1. Enter the required LC and MS parameters as well as the run sequence in the acquisition software.
2. Connect the mobile phase reservoir to the HPLC system.
3. Switch on the purge valve system.
4. Turn on both HPLC pumps.
5. Set the flow rate of HPLC pump A to 5 mL/min for ~2 min (i.e., until the backpressure stabilizes). Purge the HPLC pump with the new mobile phase. Here the flow is directed to waste.
6. Repeat **step 5** for HPLC pump B.
7. Turn off both HPLC pumps.
8. Switch off the purge valve system. The flow is now directed to the column.
9. Reenter the appropriate mobile phase composition in the acquisition software (i.e., 30 % A and 70 % B).
10. Change the flow rate to 1 or 2 mL/min depending on the HPLC system used.
11. Turn on the column compartment. The temperature is set to 30 °C.
12. Turn on both HPLC pumps and observe the mobile phase emerging from the column compartment-to-column connection tubing.
13. Wait until the previous mobile phase has been completely replaced. To determine whether the previous mobile phase has been replaced, observe the pH of droplets (the pH should be ~9) and/or monitor the backpressure of the HPLC system.
14. Connect the guard and analytical column to the column compartment.
15. Set the flow rate to 0.1 mL/min.
16. Turn on the HPLC pump and collect the column effluent for waste disposal. The column can now be equilibrated with the starting mobile phase composition (i.e., 30 % A and 70 % B).
17. Once the backpressure has stabilized increase the flow rate to 0.2 mL/min.
18. Turn on the MS. Then connect the HPLC system to the ESI-MS sprayer via connection tubing.
19. Start the run sequence.
20. The total run time is 19 min.

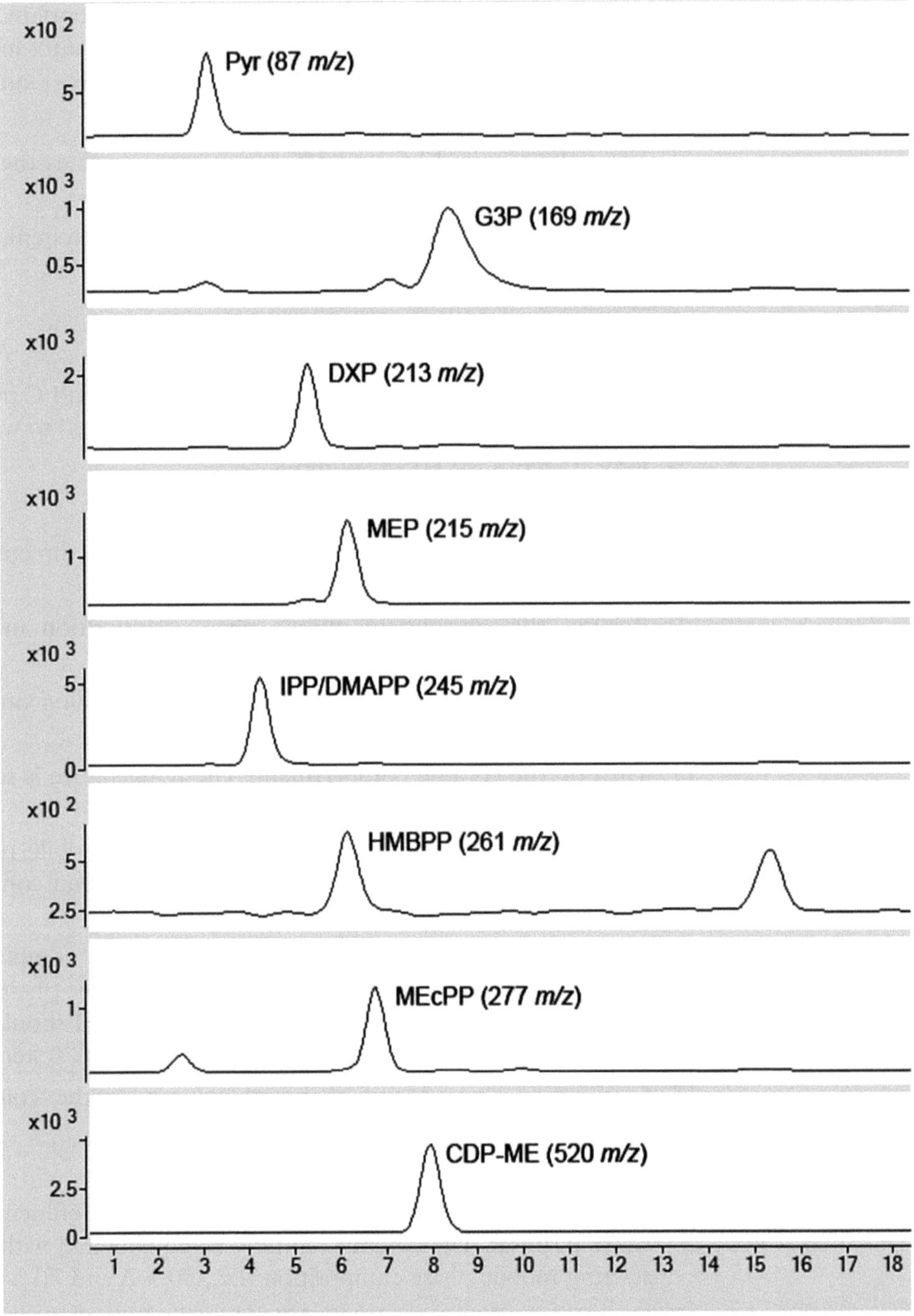

Fig. 7 HILIC-ESI-quad MS analysis of DXP pathway intermediates. Counts versus time (min). The data was acquired in the SIM mode with pre-selected ions of 87, 169, 213, 215, 261, 277, and 520 *m/z*

21. Representative results are shown in Fig. 7. Once the analysis is completed flush the column with 5 mM ammonium acetate in acetonitrile–water (8:2, v/v) at 0.1 mL/min for 20 min (at room temperature). The column can now be stored for future use. Please refer to the manufacturer's guidelines for the appropriate column cleaning and regeneration procedures.

4 Notes

1. In the SIM mode, sensitivity can be enhanced by grouping analyte ions in scheduled time segments.
2. The scan rate is the time taken for a mass analyzer to scan a given mass range (atomic mass units/s) and can significantly influence its qualitative and quantitative performance [15, 16].
3. TOF measurements are the summations of transients (the movement of ions from the pulser region to the detector) resulting from many pulses [15, 16].
4. The forward motion of the ions ensures that they do not return to the pulser region after being reflected.
5. The next set of ions is pulsed only after the previous set has reached the detector.
6. Resolution is often wrongly referred to as resolving power (Rp), which is a measure of the ability of a mass analyzer to resolve two distinct signals with a small difference and is calculated by $Rp = m/\Delta m$.
7. High mass accuracy is considered for ions whose m/z ratios are ≤2 ppm. To ensure high mass accuracy, most TOF instruments employ reference masses to recalibrate the m/z axis after a certain number of mass spectra have been averaged. Such compounds are commonly delivered to the mass spectrometer through an automated calibrant delivery system.
8. There must be enough data points across a chromatographic peak in order to accurately quantify the amount of an analyte. The acquisition rate of the mass analyzer must be high enough to achieve this.
9. Chemical standard and sample solutions are often prepared in the initial mobile phase solvent composition to improve chromatographic separation.
10. Nitrogen is generally preferred to air for drying and nebulizing ions in the ESI process as the latter can degrade compounds that are prone to oxidation.

11. The ball mill is a more precise cell disruption method than the traditionally used pestle and mortar. The former can be easily automated. The latest designs of the ball mill are comprised of 2 mL centrifuge tubes containing plant material and agitating elements (such as metallic beads), which are placed a tube holder that is situated in either a horizontal or vertical milling chamber.
12. The volatile nature of formic acid ensures that there is hardly any residue deposited on the spray shield. As a result, analysis can be conducted for an extended period of time without a significant impact on instrument performance.
13. Alternatively, the shorter Fermentation monitoring column (150×7.8 mm, Biorad) can be used for increased throughput. A flow rate of 0.5 mL/min is used with this column at a temperature of 50 °C. It should be noted that ESI is not ideal for aqueous eluents. For more efficient electrospray it is recommended that the nebulizer is orthogonal to the axis of the ESI sampling capillary (which is situated at the entrance of the MS). The orthogonal nature of the nebulizer improves the sampling of dried ions while reducing noise related to incomplete desolvation. Another benefit of orthogonal nebulization is that the sampling orifice and ion optics require only occasional cleaning.
14. The volatile nature of ammonium carbonate ensures that there is hardly any residue deposited on the spray shield. As a result, analysis can be conducted for an extended period of time without a significant impact on instrument performance. In addition to this, the use of acetonitrile as the organic mobile phase leads to increased evaporation of the nebulized droplets and, hence, a more efficient ESI process.
15. It is important that to follow the LC-MS manufacturer's instrument maintenance guidelines and cleaning practices. This will prolong the lifetime of the LC-MS system.
16. The backpressure is a good indicator of the working state of the HPLC system. An increase in pressure may suggest problems with the HPLC pump, frit, connection tubing, HPLC sample injection mechanism, guard column, or analytical column. Troubleshooting is normally conducted by the process of elimination.
17. A 1:6 post-column split is employed to ensure that ~83 % of the column effluent goes to waste in order to prolong the working performance of the LC-MS analysis.
18. The added robustness and selectivity that the IXC-IEC method provides simplifies sample preparation methods.
19. Each instrument manufacturer will have its own version of a calibrant delivery system.

20. Each instrument manufacture will have a particular time frame for an auto-tune of the ion optics (i.e., the components required to focus and transmit ions efficiently to the mass analyzer) and mass analyzer to be carried out. A successful completion of the auto-tune may be compromised by using an expired tuning mixture and/or by having a dirty ion source and ion optics.
21. It should be noted that the *m/z* axis for both the quad and TOF are calibrated during the auto-tune procedure for the mass analyzer. However, the TOF is generally calibrated daily, independent of the auto-tune.
22. HPLC systems that are used for fast separations can withstand higher backpressures. Such HPLC systems generally have smaller void volumes and so require less time to replace the previous mobile phase.
23. The HILIC separation method achieves lower LODs (due to improved separation and the basic nature of its mobile phase) and better separation than IXC-IEC, but the latter is more robust as its performance is not severely affected when analyzing extremely crude extracts. Furthermore, the Aminex HPX-87H column appears to be easier to regenerate than the ZIC-pHILIC column.

Acknowledgements

The authors would like to acknowledge that this work was part of the DOE Joint BioEnergy Institute (http://www.jbei.org) supported by the US Department of Energy, Office of Science, Office of Biological and Environmental Research, through contract DE-AC02-05CH11231 between Lawrence Berkeley National Laboratory and the US Department of Energy. The authors would also like to acknowledge Dr. Aymerick Eudes and Dr. Vladimir Tolstikov for their valuable input on sample preparation.

References

1. Xiao Y, Savchenko T, Baidoo EEK, Chehab WE, Hayden DM, Tolstikov V, Corwin JA, Kliebenstein DJ, Keasling JD, Dehesh K (2012) Retrograde signaling by the plastidial metabolite MEcPP regulates expression of nuclear stress-response genes. Cell 149:1525–1535
2. Kirby J, Keasling JD (2009) Biosynthesis of plant isoprenoids: perspectives for microbial engineering. Annu Rev Plant Biol 60:335–355
3. Li Z, Sharkey TD (2013) Metabolic profiling of the methylerythritol phosphate pathway reveals the source of post-illumination isoprene burst from leave. Plant Cell Environ 36:429–437
4. Hunter W (2007) The non-mevalonate pathway of isoprenoid precursor biosynthesis. J Biol Chem 282:21573–22177
5. Guidelines for use and care of Aminex® resin-based columns (LIT42 Rev D). Bio-Rad Laboratories, Inc. http://www.bio-rad.com/webroot/web/ pdf/lsr/literature/LIT42D.PDF
6. Harris DC (2003) Quantitative chemical analysis, 6th edn. W.H. Freeman and Company, New York
7. Gama MR, Silva RGD, Collins CH, Bottoli CBG (2012) Hydrophylic interaction chromatography. Trends Anal Chem 37:48–60

8. Buszewski B, Noga S (2012) Hydrophilic interaction liquid chromatography (HILIC)—a powerful separation technique. Anal Bioanal Chem 402:231–247
9. De Hoffman E, Stroobant V (2002) Mass spectrometry principles and applications, 2nd edn. Wiley, New York
10. Stewart II (1999) Electrospray mass spectrometry: a tool for elemental speciation. Spectrochim Acta B 54:1649–1695
11. Smyth WF (1999) The use of electrospray mass spectrometry in the detection and determination of molecules of biological significance. TrAC Trends Anal Chem 18:335–346
12. Smith JN, Flagan RC, Beauchamp JL (2002) Droplet evaporation and discharge dynamics in electrospray ionization. J Phys Chem A 106: 9957–9967
13. Wilm MS, Mann M (1994) Electrospray and Taylor-Cone theory, Dole's beam of macromolecules at last? Int J Mass Spectrom Ion Proc 136:167–180
14. Fenn JB (1993) Ion formation from charged droplets: roles of geometry, energy, and time. J Am Soc Mass Spectrom 4:524–535
15. Fjeldsted J (2003) Time-of-flight mass spectrometry. Technical overview. http://www.chem.agilent.com/Library/technicaloverviews/Public/5989-0373EN%2011-Dec-2003.pdf
16. Willoughby R, Sheehan E, Mitrovich S (1998) A global view of LC/MS: how to solve your most challenging analytical problems, 1st edn. Global View Publishing, Pittsburgh
17. Skoog DA, Holler FJ, Nieman TA (2006) Principles of instrumental analysis, 6th edn. Brooks Cole, USA
18. Jhonstone RAW, Rose ME (1996) Mass spectrometry for chemists and biochemists, 2nd edn. Cambridge University Press, Cambridge
19. Smith RM, Busch KL (1999) Understanding mass spectra—a basic approach. Wiley, New York
20. Weickhardt C, Moritz F, Grotemeyer J (1996) Time-of-flight mass spectrometry: state-of the-art in chemical analysis and molecular science. Mass Spectrom Rev 15:139–162
21. Guilhaus M (1995) Principles and instrumentation in time-of-flight mass spectrometry—physical and instrument concepts. J Mass Spectrom 30:1519–1532
22. McIntire D (2004) Effect of resolution and mass accuracy on empirical formula confirmation and identification of unknowns. Technical overview. http://www.chem.agilent.com/Library/technicaloverviews/Public/5989-1052EN%2014-May-2004.pdf

Chapter 6

Analysis of Carotenoids and Tocopherols in Plant Matrices and Assessment of Their In Vitro Antioxidant Capacity

Antonio J. Meléndez-Martínez, Carla M. Stinco, Paula Mapelli Brahm, and Isabel M. Vicario

Abstract

Carotenoids and tocopherols are lipid secondary metabolites that play essential roles in plants. They are also relevant compounds from a nutritional standpoint and attract much attention due to their proposed antioxidant properties. In this chapter, methodologies for the extraction and HPLC analysis of these compounds are described as well as a widely used protocol to assess their antioxidant capacity.

Key words Carotenoids, Citrus, Epoxycarotenoids, HPLC analysis, Oranges, Tocols, Tocotrienols, Tocopherols, Trolox equivalent antioxidant capacity (TEAC), Xanthophylls

1 Introduction

Carotenoids and tocopherols are plastidic isoprenoids involved in several functions in photosynthetic organisms. Some members of these families of plant metabolites are in addition essential components of our diets as we are unable to synthesize them de novo. In fact, some carotenoids can be converted into vitamin A and all tocopherols (α-, β-, γ-, and δ-tocopherol) exhibit vitamin E activity. Both types of isoprenoids are traditionally regarded as lipid-soluble antioxidants.

Tocopherols are only biosynthesized by photosynthetic organisms, including all plants, algae, and most cyanobacteria. The long-known role of these compounds in controlling nonenzymatic lipid peroxidation during seed storage, germination and early seedling development is well-established. Contrastingly, studies carried out on tocopherol-deficient mutant plants seem to indicate that some of the functions traditionally attributed to them need to be reevaluated [1, 2]. As far as humans are concerned, it is unarguable that tocopherols are essential nutrients and there is a large body of evidence indicating that α-tocopherol is preferentially retained and distributed.

Manuel Rodríguez-Concepción (ed.), *Plant Isoprenoids: Methods and Protocols*, Methods in Molecular Biology, vol. 1153, DOI 10.1007/978-1-4939-0606-2_6, © Springer Science+Business Media New York 2014

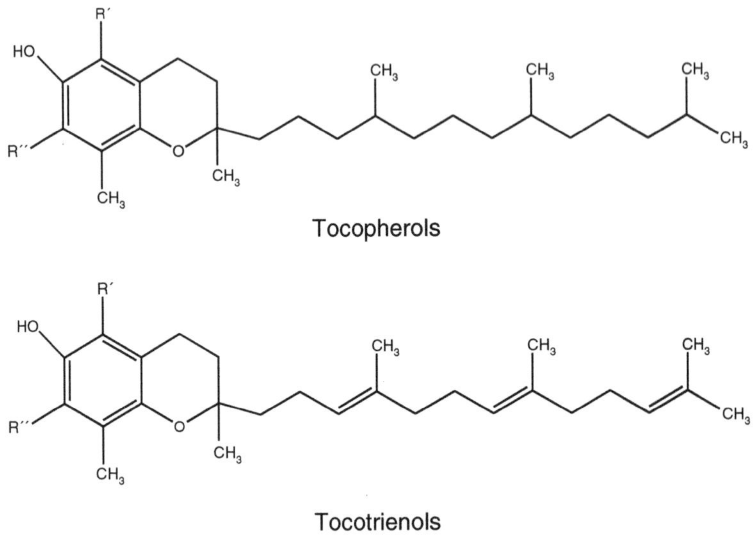

α: R´= CH_3, R´´= CH_3; β: R´= CH_3, R´´= H; γ: R´= H , R´´= CH_3; δ: R´= H , R´´= H

Fig. 1 Chemical structures of tocopherols and tocotrienols

They are thought to be potent antioxidants *in vivo* that protect membrane lipids from the oxidative damage caused by reactive oxygen species. Nonetheless in the last decade non-antioxidant biological actions of these compounds have been postulated [3, 4]. Apart from tocopherols, their related compounds tocotrienols are also attracting the interest of scientists. Tocopherols and tocotrienols are collectively known as tocols and their structures are depicted in Fig. 1 [5].

Carotenoids are synthesized by plants, algae, and some fungi and bacteria. Over 700 members of the family have been described so far, although under 100 are commonly found in our diets (structures of some of them are depicted in Fig. 2). Animals cannot synthesize carotenoids, although under certain circumstances they can modify those obtained from their diets. Virtually all carotenoids absorb visible light and therefore provide plant structures with appealing colors. As an example, many flowers and fruits owe their coloration to these pigments, hence they are important in pollination and seed dispersal. However, carotenoids are very versatile compounds involved in several key plant processes. They are present in high amounts in leaves although their color is masked by that of chlorophylls. Their functions in these structures are essential for life as we know it, since they act as accessory light-harvesting pigments and protect plants from the deleterious effects of triplet state chlorophyll and singlet oxygen, which can be formed during photosynthesis. Furthermore, some carotenoids can be cleaved to form the plant hormone abscisic acid (ABA), which plays important roles in embryo development, seed dormancy and adaptation

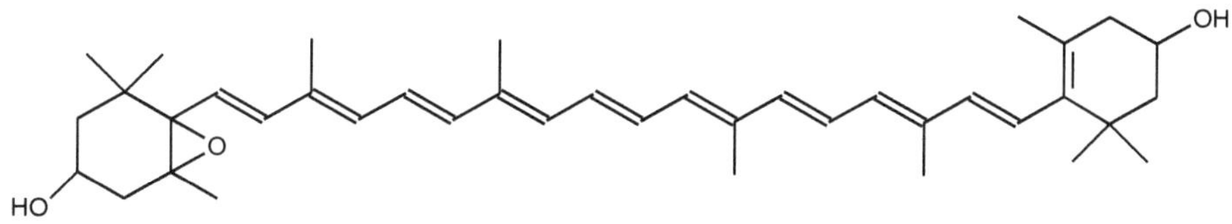

antheraxanthin

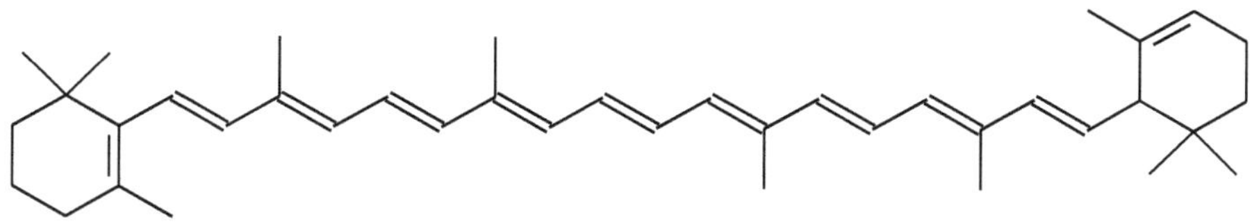

α-carotene

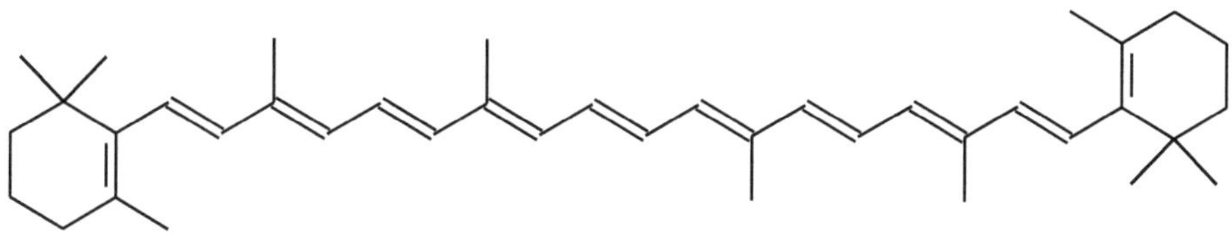

β-carotene

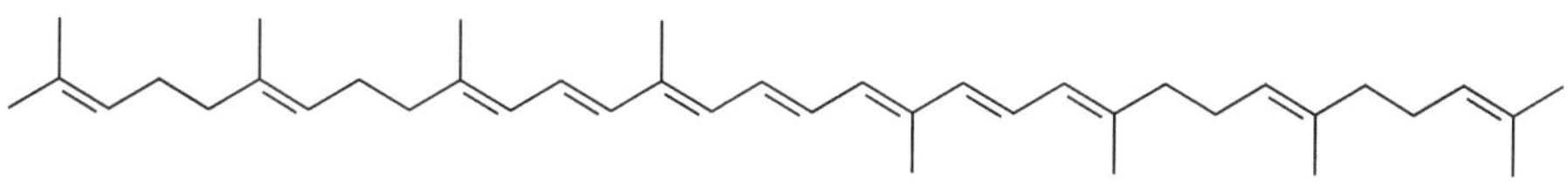

ζ-carotene

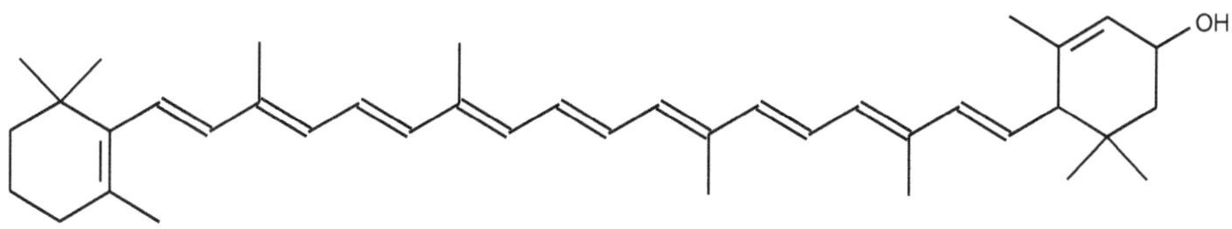

α-cryptoxanthin

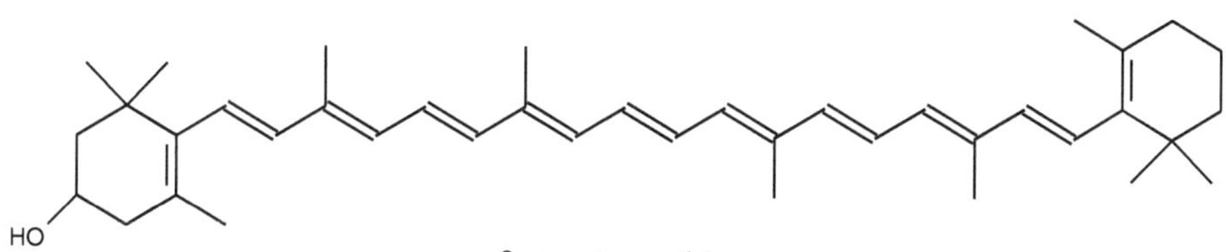

β-cryptoxanthin

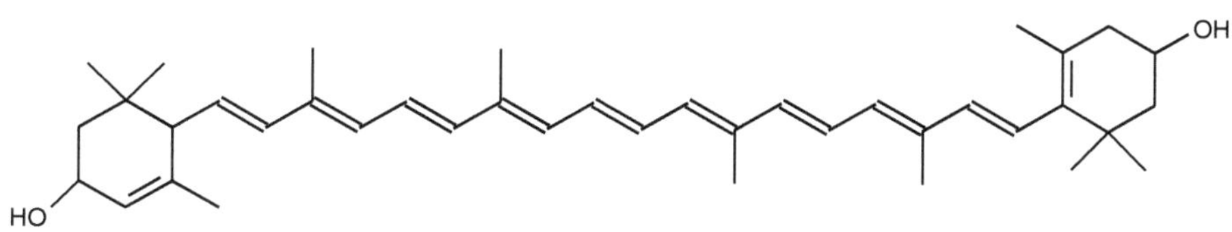

lutein

Fig. 2 Chemical structures of some plant carotenoids

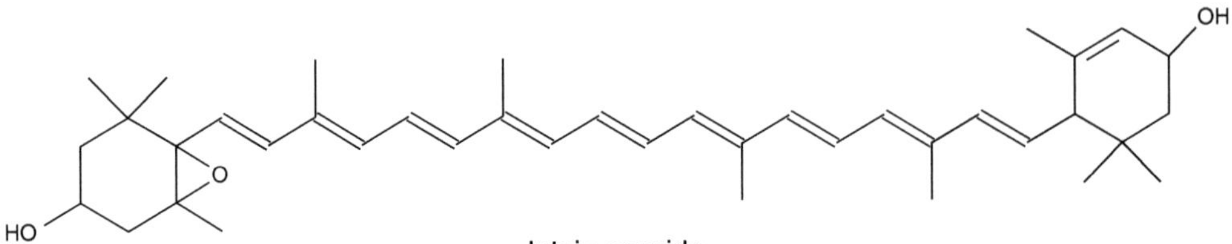

lutein epoxide

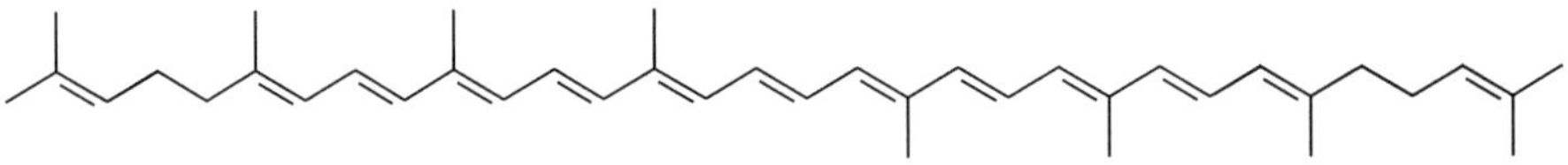

lycopene

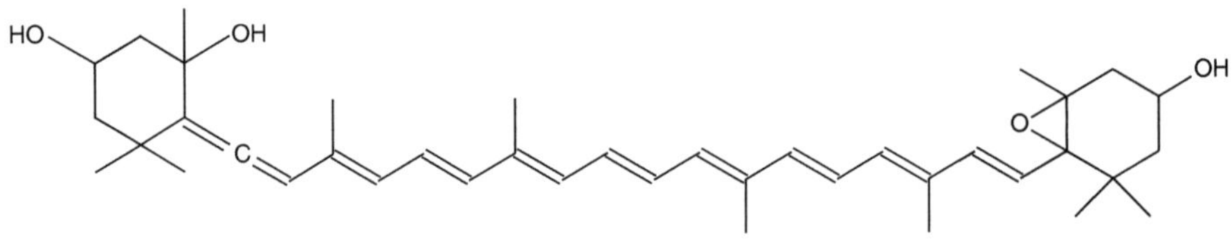

neoxanthin

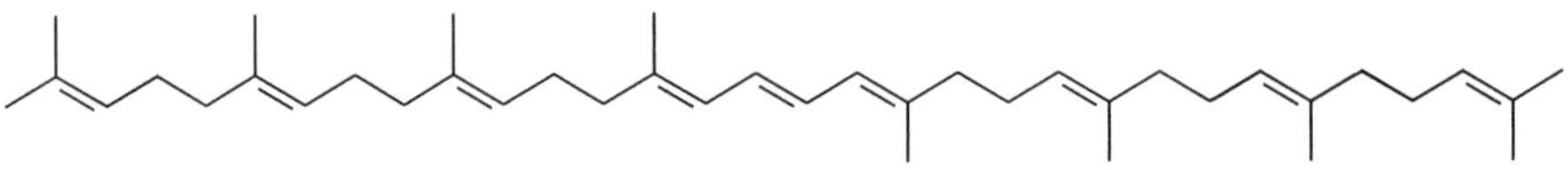

phytoene

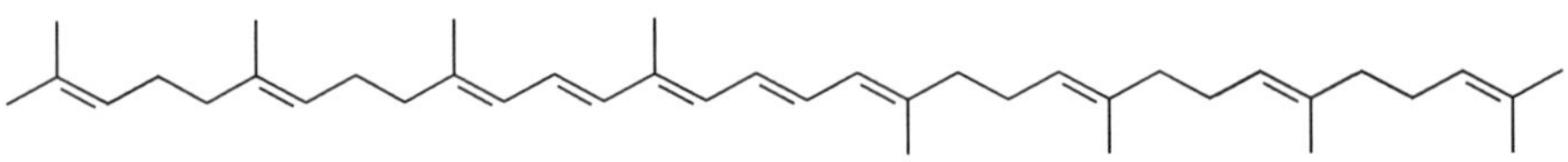

phytofluene

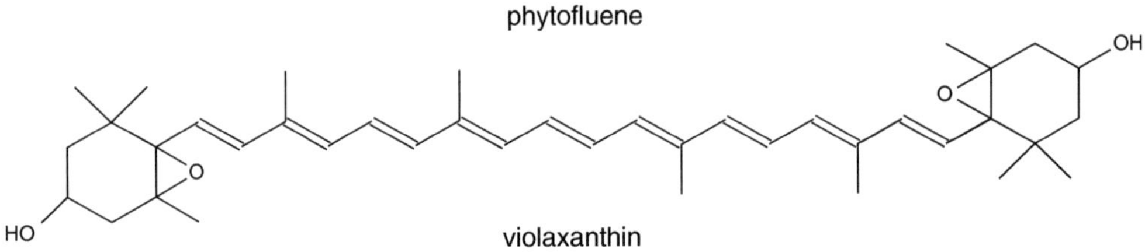

violaxanthin

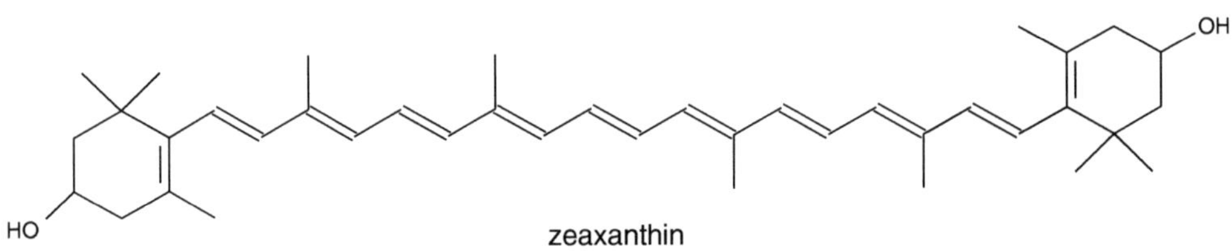

zeaxanthin

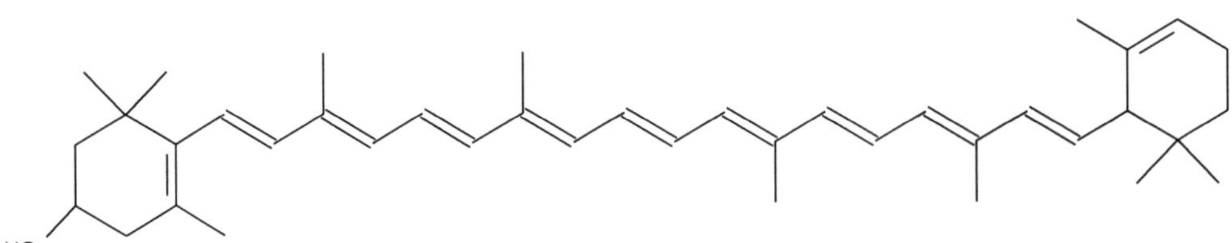

zeinoxanthin

Fig 2 (continued)

to drought and other environmental stresses. Likewise, carotenoids are precursors of aroma compounds [6, 7]. Recently, it has been reported that carotenoids are precursors of strigolactones. These compounds are considered plant hormones involved in the development of symbiotic arbuscular mycorrhizal fungi, the inhibition of plant shoot branching and the germination of parasitic plant seeds [8–10]. A few carotenoids, including in some cases diverse isomeric forms, are consistently found in humans and are thought to have nutritional relevance [11]. Apart from the already mentioned vitamin A activity of β-carotene and other carotenoids with β rings, which is well-established, carotenoids are commonly regarded as health-promoting compounds that may protect from oxidative stress and light-induced skin damage and prevent some degenerative diseases like certain types of cancer, cardiovascular disease, xerophthalmia, and age-related macular degeneration. However, in some cases there are still discrepancies on the observations and many mechanistic aspects remain obscure [12, 13].

Here we describe methodologies for the extraction and HPLC analysis of both tocopherols and carotenoids. A high-throughput microextraction protocol for the analysis of these compounds in citrus fruits and derived products with a very intricate carotenoid pattern [14] will be described stepwise. This methodology can be applied to many other plant sources and it offers several advantages. Thus, it is suitable for the analysis of the carotenoids involved in the xanthophyll cycle (violaxanthin, antheraxanthin, and zeaxanthin), an important photoprotective process [15]. Furthermore, it can be used for the analysis of 9-*cis*-violaxanthin and 9′-*cis*-neoxanthin, precursors of ABA [16], and other *cis* isomers that are thought to be the precursors of strigolactones and related compounds [17]. In fact, this methodology is appropriate for the analysis of different kinds of carotenoid isomers [18–22]. Moreover, it allows for the analysis of all the major carotenoids and tocopherols found in humans.

A widely used protocol to assess the in vitro antioxidant capacity of extracts is also described [23]. Some information to be taken into account for the sensible interpretation of these data is also provided.

2 Materials

2.1 Extraction and Analysis

1. 2 ml Eppendorf vials and HPLC amber vials and inserts.
2. Micropipettes,.
3. Microcentrifuge, vortex, shaker vortex, ultrasound bath, nitrogen gas cylinder, and vacuum concentrator.
4. HPLC system with quaternary pump and photodiode array detector.

5. C_{30} analytical HPLC column (5 μm, 4.6×250 mm).
6. HPLC grade methanol (MeOH), ethyl acetate (EtAc), dichloromethane (DCM), and methyl *tert*-butyl ether (MTBE).
7. Extraction mixture (EM): MeOH:EtAc:DCM (25:25:50) (v/v/v).
8. Methanolic KOH at 25 %.
9. High purity water for HPLC (HPW).

2.2 Assessment of Antioxidant Capacity

1. Spectrophotometer with temperature-controlled multi-cuvette device.
2. Matched disposable 1 ml plastic cuvettes.
3. Trolox (6-hydroxy-2,5,7,8-tetramethychroman-2-carboxylic acid), a water-soluble analogue of vitamin E. Prepare a 2.5 mM stock solution in ethanol. Working standards (between 0 and 15 μM) should be made fresh by dilution with ethanol.
4. Freshly-made stock solution of 7 mM ABTS diammonium salt [2,29-azinobis-(3-ethylbenzothiazoline-6-sulfonic acid)] in HPW.
5. ABTS radical cation ($ABTS^{\cdot+}$) stock solution: prepared by reacting potassium persulfate (2.45 mM, final concentration) with the aqueous 7 mM ABTS solution in the dark at room temperature over 12–16 h.
6. $ABTS^{\cdot+}$ working solutions: prepared by diluting the stock solution with ethanol up to an $A_{734nm} = 0.70 \pm 0.02$ at 30 °C.

3 Methods

Recommended procedures for the analysis of carotenoids and tocopherols can be found in reference texts [24–31]. Their reading is essential for laboratories new to the carotenoid and tocopherol fields, above all to learn factors that can affect the results of our analysis. Some of the information provided in such reference texts is summarized in this section.

Exhibiting antioxidant properties, these compounds are easily oxidized, so oxygen and other electrophiles can degrade them rapidly. The oxidation reactions are favored by light and heat, which can also cause isomerizations and other changes. Alkalis can degrade them in some cases under certain circumstances. Besides, acids can promote the isomerization of 5,6-epoxycarotenoids into their 5,8-furanoid counterparts and other changes.

Oxidation. Oxygen, especially in combination with light and heat, is a very destructive factor that can cause serious degradation of carotenoids and tocopherols, particularly once they are extracted. The presence of even traces of oxygen in stored samples (even at sub-zero temperatures) and of oxidizing agents can rapidly cause degradation.

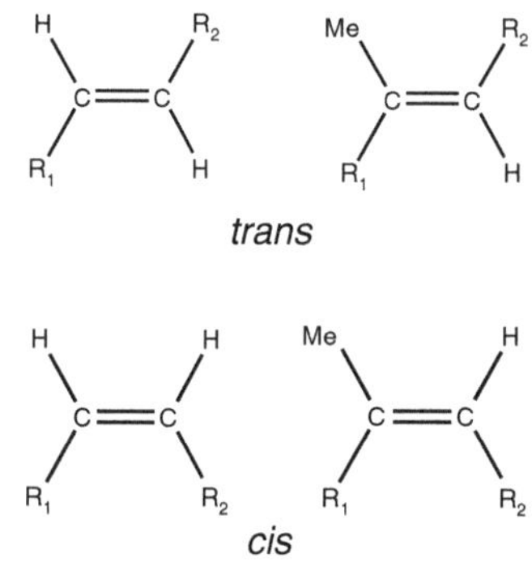

Fig. 3 *Trans* and *cis* double bonds in carotenoids

It is important to carry out the different operations in the shortest time possible such that the extracts are not exposed to air for longer than necessary. The addition of ice before homogenization and the use of cold solvents for the extractions can reduce oxidation reactions. If the extracts are not going to be analyzed immediately they must be stored in sealed containers and the oxygen must be removed by using either vacuum or argon or nitrogen atmospheres. The storage of dried extracts is preferred. If any air is present, samples can decompose in a few days, even in a freezer at −20 °C. Antioxidants like butylated hydroxytoluene (at 0.1 %), pyrogallol (at 5 %), or ascorbyl palmitate may also be used, especially when the analysis is prolonged. They can be added during sample disintegration or saponification or added to solvents, standard solutions and isolates. It is to be noted that if the extracts are going to be used to assess antioxidant activity the addition of external antioxidants can affect the results of the test.

Light. Apart from accelerating the oxidation reactions, light can cause the *cis/trans* (geometrical) isomerization of carotenoids (Fig. 3) and even the photodestruction of these compounds. Direct sunlight must be avoided in the laboratory and the work should be done in shaded daylight or subdued artificial light. Any containers with extracts must be protected from light by covering them appropriately with dark cloth or paper, aluminum foil, etc. Polycarbonate shields can be used for fluorescent lights, which emit high energy, short-wavelength radiation. The speed of manipulation and the shielding from light are especially important in extracts containing chlorophylls (e.g., extracts of green leafy or nonleafy vegetables) or other potential photosensitizers. In the presence of these, the photodegradation and isomerization of carotenoids occur very rapidly, even with brief exposure to light.

Heat. Aside from favoring the oxidation reactions of tocopherols and carotenoids, heat can lead to *cis/trans* isomerization and degradation of the latter. Thus heating should be done only when absolutely necessary. If possible, work should be done in a lab temperature of 20–22 °C. The extracts should not be left warm up too much, they must be kept cool. Excessive heating of solutions when

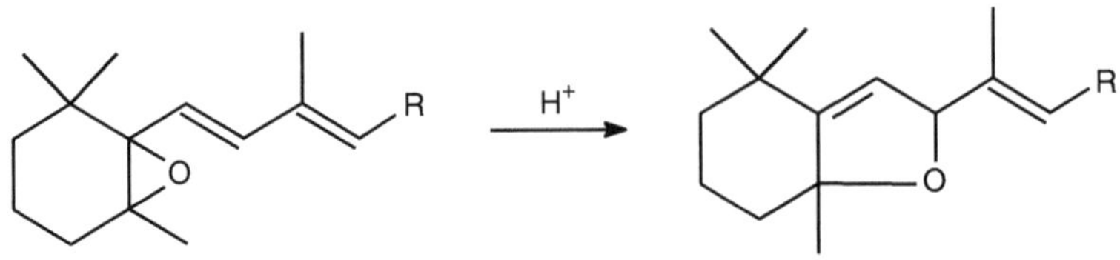

Fig. 4 Schematic representation of the isomerization of a 5,6-epoxide group (*left*) into a 5,8-furanoid group (*left*) caused by acid

concentrating must be avoided. The concentration of the extracts must be done at a temperature below 40 °C.

Acids. Carotenoids may decompose, dehydrate, or isomerize in the presence of acids. Even traces of acid favor the conversion of 5,6-epoxycarotenoids, such as violaxanthin and neoxanthin (precursors of ABA occurring in all plant photosynthetic tissues), to their 5,8-epoxides isomers (Fig. 4). Strong acids and acidic reagents should not be used near the areas where carotenoids are handled. Besides weak bases such as $NaHCO_3$, $MgCO_3$, or $CaCO_3$ (1 g/10 g sample) can be added as neutralizing agents during extraction to neutralize acids liberated from the food sample itself. If carotenoid epoxides are of special interest, all the solvents must be acid-free.

Alkalis. Saponification (i.e., the alkaline hydrolysis of lipids) is widely used for the analysis of tocols and carotenoids. In principle, the tocols naturally found in plant materials would not need to be saponified as they occur mainly as free (i.e., non-esterified) compounds. However they can be added as esters in some foods. Nonetheless, the saponification reaction can improve the extractability of tocols from some matrices. This can be owed to the softening of such materials or by weakening the interactions between the tocols and the matrices, as they can be associated with carbohydrates and proteins. To reduce the oxidation reactions during the hydrolysis, antioxidants (ascorbic acid, pyrogallol) should be added to the saponification mixture and the air should be replaced with an inert atmosphere in the container. In the case of carotenoids, the saponification reaction has several pros and cons. It can be used to remove chlorophylls and unwanted lipids from the samples which make the chromatographic separation more difficult. In samples like many fruits, xanthophylls can be found free or esterified with different fatty acids. The saponification reaction hydrolyzes the carotenol esters and simplifies the chromatographic analysis as free xanthophylls are analyzed instead of their several esters. However, the saponification reaction extends the analysis time and may lead to the formation of artifacts and the degradation of carotenoids, this being higher with higher concentration of alkali and heating.

The saponification step should be included in the analytical procedure only when it is essential. For instance, it is not indispensable in the analyses of leafy vegetables, tomato, and carrot (among other plant materials) as their lipid content is very low and they are essentially free of carotenoid esters. In the case of photosynthetic tissues, the chlorophylls that are extracted along with the carotenoids can be easily separated with most HPLC methods. Although provitamin A carotenoids (α-carotene, β-carotene, γ-carotene, and β-cryptoxanthin, among others) are very stable to saponification, lutein, violaxanthin, and other dihydroxy-, trihydroxy-, and epoxycarotenoids can be reduced considerably during the reaction and the subsequent washing steps. Other carotenoids with allylic hydroxy and keto groups (for instance astaxanthin) can undergo extensive oxidation in the presence of alkali and air. Aldol condensation is another unwanted reaction that can take place during the alkaline hydrolysis. Thus, it is well known that carotenoids with aldehydic groups (carotenals) can undergo aldol condensation in the presence of alkali and acetone, such that this solvent should be quite removed or not used when carotenal-containing extracts need to be saponified. In such samples, the extraction can be performed with methanol and ethylacetate, for example. For high-lipid samples, such as oils, lipids can be eliminated without risk of degradation of carotenoids by using a nonspecific lipase [32, 33]. A number of studies describing the simultaneous determination of free and esterified carotenoids have been published in the last decade [33–37].

Choice of solvent and other aspects related to the extraction. While there are more than 700 carotenoids, many of them differing considerably in structure and polarity, there are only a few tocols, all of them exhibiting few chemical differences that do not affect their polarity considerably (Figs. 1 and 2). Hexane, methanol, and several mixtures (hexane and ethanol, chloroform and methanol, etc.) are typically used for the extraction of tocols. If the analysis of carotenoids is also needed, the extraction solvents need to be chosen more carefully as in many cases the plant materials accumulate carotenoids with diverse structures and polarities. Normally, these extractions systems work well for tocols too.

Carotenoids are usually firstly extracted from samples containing large amounts of water, with water-miscible organic solvents like tetrahydrofuran (THF), acetone, methanol (MeOH), or ethanol (EtOH). Unlike THF and acetone, which dissolves well carotenes and oxygenated carotenoids, MeOH and EtOH dissolve xanthophylls efficiently but not the carotenes. This is why it is also common to use mixtures of solvents as extractants. The water immiscible hexane and petroleum ether (PE) are widely used in mixtures with MeOH, EtOH, and other solvents as they dissolve carotenes well. Thus, common extractant mixtures are hexane:acetone, hexane: EtOH, PE:acetone, hexane:acetone:EtOH, THF:MeOH, dichloromethane:MeOH, and so on. The approach for the extraction of hydrophilic carotenoids, like crocetin (found in the stigmas of *Crocus sativus* and the fruits of *Gardenia jasminoides*) [38] can be largely different so that more polar solvents (or mixtures of them) like H_2O, MeOH, EtOH, EtOH:H_2O (1:1), or MeOH:acetone are used. Dried or freeze-dried samples can be directly extracted with appropriate water-immiscible solvents, although rehydration and extraction with water-miscible solvents or mixtures as described above can be more efficient. Rehydration times ranging from 5 to 30 min have been reported to work well. In some samples, enzymatic digestion can improve the pigment recovery from some samples like oils [32]. Besides carotenoid and tocopherol solubility (and the miscibility of solvents), other characteristics like purity, toxicity, boiling point, and density are to be taken into account. Some solvents can contain peroxides, traces of acids, and other impurities that can bring about changes in the carotenoids. Thus, solvents of the highest quality (reagent-, UV-vis-, or HPLC-grade) must be used. If solvents of lower quality are only available they should be purified, dried, and freshly distilled to remove traces of unwanted impurities before being used for carotenoid work.

Acetone (AC) has several advantages: it is cheap, penetrates the food matrix well, dissolves carotenes and xanthophylls, and facilitates the subsequent partitioning to an apolar solvent. THF dissolves very well both carotenes and xanthophylls, although it accumulates peroxides with ease. Diethyl ether (DE) can also contain peroxides. These can be left out by distillation over reduced iron powder or calcium hydride. Although these solvents can be supplied with BHT as stabilizer it must be taken into account that there is a time limit to use it. Dichloromethane (DCM) and DE are good solvents for carotenoids and tocopherols, although they are very volatile (bp around 40 °C) and have a low flash point, so the evaporation of high quantities of these solvents should be avoided to prevent hazards. This can be done for instance by cooling the solvents before and during their use. Their volatility can also be an advantage, because if we transfer the lipophilic extract from high boiling point solvents or mixtures to DCM or DE the concentration process can be accelerated and the throughput of samples improved considerably. Hexane is thought to be neurotoxic and can be replaced with petroleum ether (bp 40–60 °C). Benzene is thought to be carcinogenic and can be replaced with toluene. Chloroform is commonly used for the extraction of carotenoids in lycopene-rich sources. It must be taken into account that it can contain traces of hydrochloric acid that can promote the isomerization of 5,6-epoxides into 5,8-furanoids. In addition, it is commonly stabilized with 1 % ethanol which can affect its chromatographic properties. It can be replaced with DCM. When extracting carotenoids the separation of a pigment-containing organic phase and an aqueous phase takes place. Depending on the density of the organic solvent the carotenoids are finally transferred to, they will be recovered from the top or the bottom of the container used for the separation of the phases.

Storage of carotenoid extracts. If they need to be stored, carotenoid extracts, fractions, or isolates should be kept dry under nitrogen or argon and kept at −20 °C or lower when not in use. Important losses of carotenoids have been observed in extracts containing other antioxidants kept at −20 °C in the dark [39]. The evaporation of the solvent should be finished under a stream of nitrogen or argon. Indeed, care should be taken to prevent the extract from going to complete dryness when they are in glass containers because it is thought that this may result in degradation of carotenoids. Additionally, the more polar carotenoids, may stick strongly to the glass walls, avoiding their complete removal from the flask. If this does occurs, redissolving the dry extract and introducing the flask for a few seconds in a ultrasound bath can facilitate the re-solubilization of the carotenoids. If the extracts, fractions, or isolates cannot be dried out for storage they should be dissolved in a petroleum ether or hexane. Leaving carotenoids in solvents such as cyclohexane, dichloromethane, diethyl ether, and acetone can lead to substantially higher degradation. It must also bear in mind that keeping carotenoids in flammable volatile solvents, such as ether, in a refrigerator is a safety hazard and should be avoided.

3.1 Extraction and Analysis of Carotenoids and Tocopherols

Initially we used in our laboratory classical low-throughput extraction methods for orange juice carotenoids involving the use of separatory funnels and other glassware. The typical sample volumes were 10–25 mL and several hundreds of milliliters of solvents were needed for each analysis in triplicate [40]. More recently we have started using a microextraction procedure that allows for the extraction of many samples in triplicate [41]. No glassware is used whatsoever, the throughput is increased considerably and the volumes of solvent needed per analysis are reduced drastically. Although oranges do not contain high amounts of tocopherols, both the extraction procedure and the HPLC method here described are suitable for the analysis of these compounds in many plant sources.

1. Take a 500 μL aliquot of the citrus juice in an eppendorf (eppendorf 1) (*see* **Note 1**).
2. Add 600 μL of the EM (*see* **Note 2**), vortex for 1 min, and place the eppendorf in an ultrasound bath for 5 min to enhance extraction.
3. Spin the mixture (18,000 × *g*, 3 min, 4 °C), collect the colored phase at the bottom and place it in a new eppendorf (eppendorf 2).
4. Add 300 μL of DCM to eppendorf 1, vortex and sonicate as in **step 2**, spin as in **step 3**, collect the colored phase, and place it in eppendorf 2.
5. Repeat **step 4** until no more color is extracted (two or three extractions in total should be enough).

6. To saponify the extract in eppendorf 2 add 600 μL of methanolic KOH at 25 %. Replace the air in the eppendorf with nitrogen and keep the reaction for 1 h in the dark with gentle shaking (*see* **Note 3**).
7. Add 700 μL of water (*see* **Note 4**), vortex for 1 min and spin the mixture (18,000×*g*, 2 min, 4 °C). Discard the aqueous upper phase.
8. Repeat **step 7** several times until the aqueous phases discarded are neutral (*see* **Note 5**).
9. Concentrate the colored saponified extract to dryness on a vacuum concentrator at temperature below 35 °C (*see* **Note 6**).
10. If the extracts are not going to be analyzed by HPLC immediately, keep them dry under an atmosphere of nitrogen at −20 °C or lower in the absence of light.
11. Redissolve the extracts in 60 μL of EtAc and spin (18,000×*g*, 2 min, 4 °C) to precipitate possible gross particles that could block the HPLC capillars. Take a 30 μL aliquot of the supernatant and transfer it to an insert placed inside an amber glass HPLC vial (*see* **Notes 7** and **8**).
12. Inject 20 μL on the HPLC system. The following conditions, based on a method described elsewhere [42], can be used:
    (a) Mobile phase: methanol containing 0.1 % of ammonium acetate (MeOH), methyl *tert*-butyl ether (MTBE), and water (H_2O), according to the following gradient:

Time (min)	% MeOH	% MTBE	% H_2O	Curve
0	90	5	5	Linear
12	95	5	0	Linear
25	89	11	0	Linear
40	75	25	0	Linear
62	50	50	0	Linear
65	90	5	5	Linear
70	90	5	5	Linear

    (b) Flow: 1 mL/min.
    (c) Column: YMC C30 column (5 μm, 250×4.6 mm)
    (d) Column temperature: 20 °C.
    (e) Wavelengths: select them according to the spectroscopic characteristics of the compounds of interest in the mobile phase. In this HPLC method phytoene can be monitored at 285 nm, the tocopherols at 290–300 nm, phytofluene at 350 nm, and the most common plant carotenoids at 440–470 nm (*see* **Note 9**).

The chromatograms corresponding to the separation of saponified orange juice carotenoids and a mixture of tocopherol standards using this gradient is shown in Figs. 5 and 6.

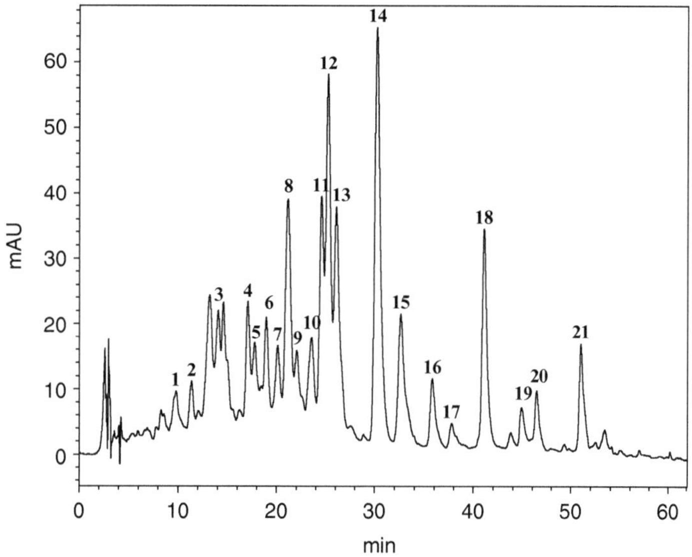

Fig. 5 Chromatogram at 450 nm of a saponified carotenoid extract from a retail orange juice. For peak identification refer to Table 1

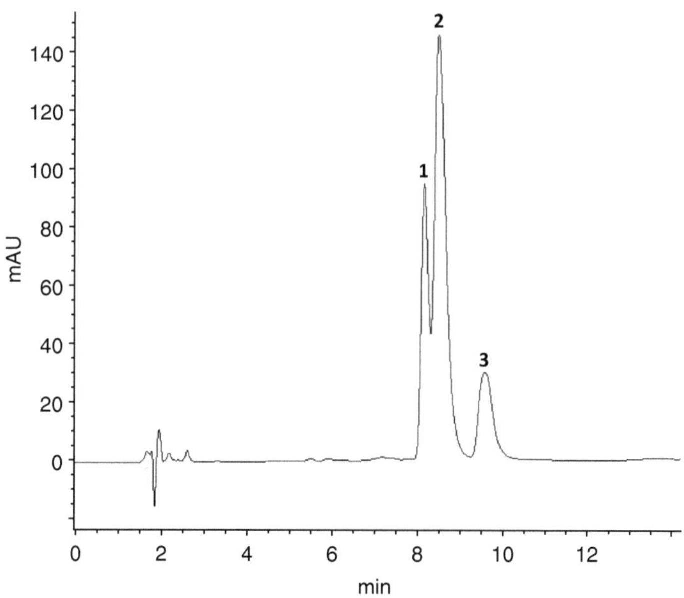

Fig. 6 Chromatogram at 290 nm of a mixture of tocopherol standards. For peak *identification refer to Table 2

Table 1
Peak identification

Peak	Rt (min)	Carotenoid
1	9.41	Unidentified. Maxima at 400, 422, and 448 nm
2	11.22	Unidentified. Maxima at 400, 422, and 448 nm
3	13.2-14.6	Violaxanthin isomers
4	17.05	*Trans*-luteoxanthin
5	17.75	*Cis*-antheraxanthin
6	18.93	*Trans*-auroxanthin
7	20.09	*Trans*-auroxanthin
8	21.12	Antheraxanthin
9	22.02	*Trans*-auroxanthin
10	23.54	*Cis*-luteoxanthin
11	24.56	Mutatoxanthin
12	25.23	Lutein
13	26.07	Mutatoxanthin
14	30.21	Zeaxanthin
15	32.62	9- or 9′-*Cis*-antheraxanthin
16	35.79	Zeinoxanthin
17	37.75	9-*Cis*-zeaxanthin
18	41.07	β-Cryptoxanthin
19	44.86	*Cis*-ζ-carotene
20	46.44	α-Carotene
21	51.01	β-Carotene

Saponified orange juice carotenoid extract

Table 2
Peak identification

Peak	Rt (min)	Tocopherol
1	8.32	γ-Tocopherol
2	8.47	β-Tocopherol
3	9.57	α-Tocopherol

Mixture of tocopherol standards

3.2 Assessment of Trolox Equivalent Antioxidant Capacity (TEAC)

The method is based on the decolorization of the preformed radical cation $ABTS^{\cdot+}$ by antioxidants, which can be monitored by the decrease in absorbance at 734 nm [23].

1. Prepare the $ABTS^{\cdot+}$ solution by reacting ABTS and potassium persulfate the day before (*see* **Note 10**).
2. Turn on the spectrophotometer and the temperature control module. Set the former at 734 nm and the latter at 30 °C (*see* **Note 11**).
3. Prepare the working Trolox solutions and keep them in the fridge until their use (*see* **Note 12**).
4. Make a blank measurement of the spectrophotometer with ethanol.
5. Prepare an $ABTS^{\cdot+}$ working solution with an absorbance of 0.70 at 734 nm (*see* **Note 13**).
6. Make a blank reading with 1.0 mL of the $ABTS^{\cdot+}$ solution and 10 μL of ethanol. Record the absorbance at 8 min (*see* **Note 14**).
7. Measure the absorbance drop caused by the Trolox solutions after 8 min of incubation.
8. Dilute the test extracts so that the percentage of inhibition ranges from 20 to 80 % when a 10-μL aliquot is added.
9. Run a blank with 1.0 mL of the $ABTS^{\cdot+}$ solution and V_1 (μL) of the solvent of the extracts in triplicate (*see* **Note 15**).
10. Measure the absorbance drop caused by V_1 (μL) of the test extracts after 8 min of incubation.
11. Run a blank with 1.0 mL of the $ABTS^{\cdot+}$ solution and V_2 (μL) of the solvent of the extracts.
12. Measure the absorbance drop caused by V_2 (μL) of the extracts after 8 min of incubation.
13. The percentage of inhibition of A_{734nm} is calculated and plotted as a function of the concentration of the test extracts and the Trolox solutions. To calculate the TEAC, the gradient of the plot for the samples is divided by the gradient of the plot for Trolox (Fig. 7). The TEAC value is an assessment of the μmol of Trolox necessary to provide the same antioxidant capacity as one gram of the test extract (*see* **Note 16**).

4 Notes

1. The use of freeze-dried material in the form of fine powder can facilitate the extraction. In this case, add water (500 μL), vortex (1 min), leave the mixture stand for 10–20 min, spin (18,000×*g*, 2 min, 4 °C), and discard the aqueous extract before starting the extraction. This will help remove sugars.

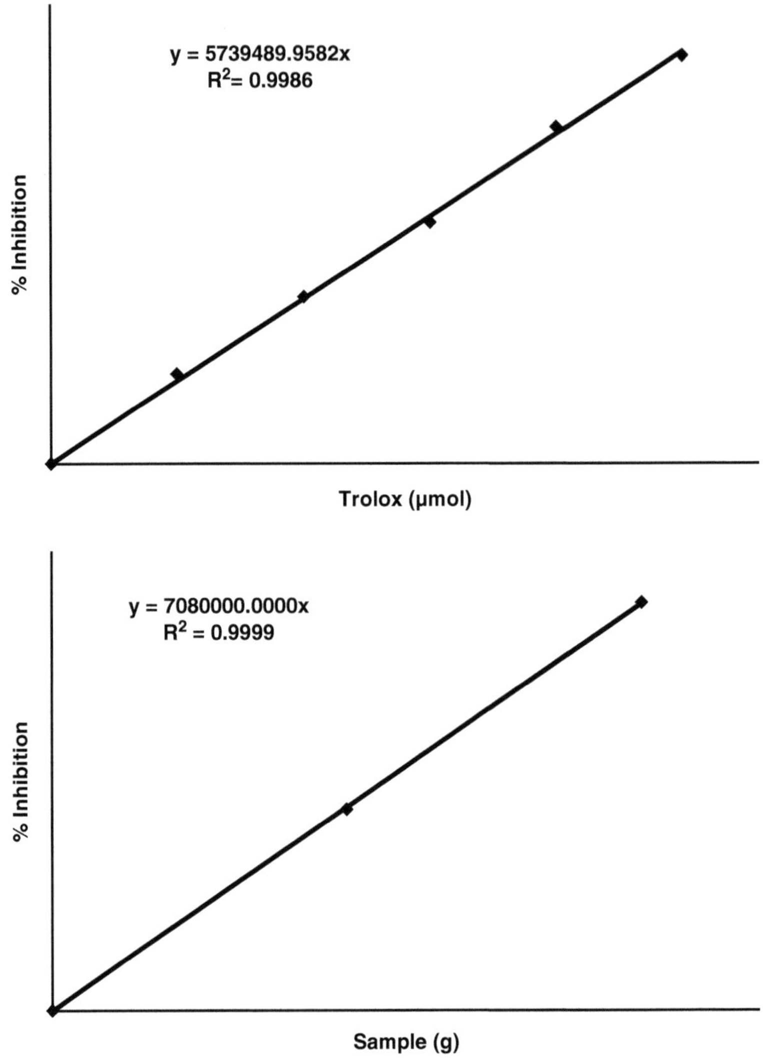

Fig. 7 Plots of the percentage of inhibition of A_{734nm} corresponding to Trolox solutions and orange juice lipophilic extracts

2. We have noticed that if the orange juices are squeezed and analyzed fast, there is little formation of 5,8-epoxides, even if weak bases are not added for the extractions. On the contrary, we have noticed that orange juices frozen for several months and, above all, those available in the market do contain large quantities of 5,8-epoxycarotenoids. In some cases, no 5,6-epoxycarotenoids are found at all. In any case the presence of 5,8-furanoids (also commonly referred to as 5,8-epoxides) in the chromatograms can denote that acid-driven isomerizations have taken place, either during the steps need to analyze carotenoids or during the processing, storage or manipulation of the samples. In general, the orange juices purchased from

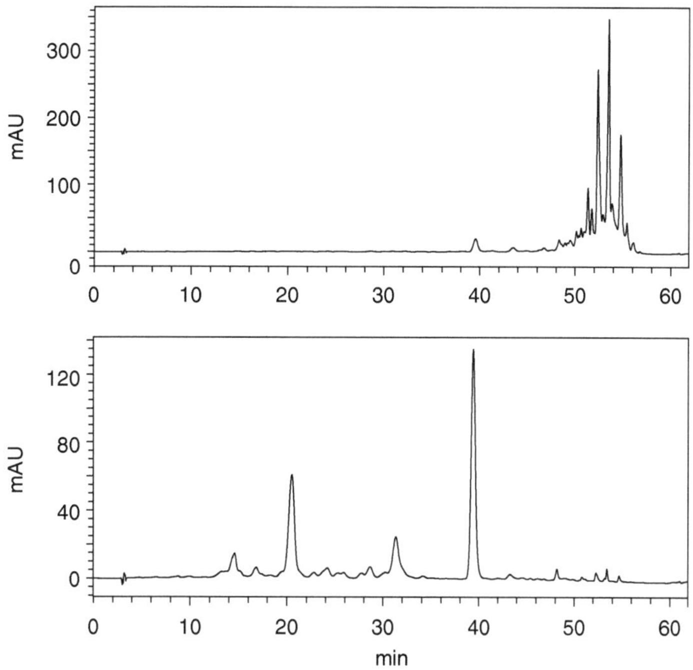

Fig. 8 Chromatograms at 450 nm of a native (unsaponified) and a saponified carotenoid extract from mandarins. The HPLC conditions were those described in Subheading 3

retailers contain high amounts of 5,8-epoxycarotenoids, as it can be observed in Fig. 5.

3. Alternatively, the alkaline hydrolysis can be carried out by using a lower concentration of alkali and a longer reaction time. Some authors use 5–10 % of alkali and keep the saponification overnight. In any case it is important to check that the esters are completely hydrolyzed whatever the conditions used are. Chromatograms of totally unesterified carotenoid extracts have fewer peaks than those of the same extract containing carotenoid esters. Esters of carotenoids are less polar than the corresponding carotenoids so the presence of several late eluting peaks (very often poorly resolved) in reverse phase HPLC methods can denote that the saponification has not been complete (Fig. 8).
4. An aqueous solution of NaCl (5–10 %) can also be used. These saline solutions can be useful to reduce the risk of formation of emulsions and enhance the transfer of the compounds of interest to the lipophilic phase.
5. An indicator paper can help determine how many washes are needed. This will depend on the protocol used for the saponification. Generally, the lower the concentration of the KOH solution the fewer washes will be necessary.

6. Some authors recommend to finishing the concentration under a stream of nitrogen.
7. These volumes are orientative and should be adjusted depending on the samples to be analyzed and the compounds of interest.
8. Necessary precautions to avoid evaporations (use of refrigerated HPLC trays, proper capping of the vials, etc.) should be taken as with such small volumes the loss of solvent can lead to important errors in the quantification.
9. More specific data on the spectroscopic features of carotenoids can be found in reference carotenoid texts [30, 43]. The detection of phytofluene and tocopherols can be enhanced considerably by using fluorescence detection. Phytofluene is known to fluoresce at around 510 nm when excited with near-UV light [44]. Excitation wavelengths in the interval 290–296 nm and emission wavelengths in the range of 325–330 nm have been used to detect tocopherols [26].
10. This stock solution is stable in the dark at room temperature for 2 days. Apart from being stable, the radical cation is easy to handle, absorbs strongly in the visible region of the spectrum and has a high molar extinction coefficient, hence the reaction with antioxidant can be monitored with conventional UV/visible absorption spectrophotometers. Furthermore the experiments are easy to repeat so this methodology is widely used to compare the *in vitro* antioxidant capacity of extracts from different sources. However, one major criticism to this protocol is that the $ABTS^{\cdot+}$ is a nonphysiological radical, that is, a radical not found in biological systems.
11. By choosing 734 nm the interference of other absorbing compounds and turbidity is minimized.
12. The Trolox stock solution has been shown to be stable for several months when stored in a freezer. The Trolox working standards must be freshly prepared from the stock on a daily basis.
13. The absorbance of the $ABTS^{\cdot+}$ working solution may need to be readjusted to 0.7 with certain frequency.
14. It has been shown that Trolox, carotenoids and tocopherols react with the radical cation quite rapidly under these conditions, hence 8 min is considered more than enough time for the reaction to take place. However, the fact that some antioxidants can react more slowly has to be considered. Thus, in certain extracts a longer endpoint may be necessary since if the absorbance readings are taken before the reaction is finished the TEAC values would not be estimated properly.
15. The solvent used must dissolve properly the compounds of interest without causing changes in them (please, refer to the section

dedicated to the solvents used for the analysis of carotenoids and tocopherols) and be miscible with the ABTS$^{\cdot+}$ working solution in ethanol. In any case, the mixing of the reaction mixture is recommended.

16. The TEAC is an assessment of the relative ability of hydrogen or electron-donating antioxidants to scavenge the ABTS$^{\cdot+}$ compared with that of Trolox [45]. It is very useful to establish standardized comparisons in the scavenging capacities of extracts from different samples or developmental stages. In any case is of paramount importance to understand that the extrapolation of the *in vitro* results obtained to biological systems is not possible for a number of reasons. One of them is that ABTS$^{\cdot+}$ is a nonphysiological radical, as stated before. Furthermore, the different antioxidants in the extracts have been removed from their biological milieus, where they actually exert their actions. As an example, different antioxidant species that may not be in the same location *in vivo* can be together and interact in the extract used for the *in vitro* measurements. When it comes to try to extrapolate the *in vitro* observations to likely health benefits in humans, other factor like absorption, metabolism, excretion, subcellular location, interactions with other food components, etc. have to be taken into account [46, 47]. For example, it is well known that not all the carotenoids and tocopherols are absorbed with the same efficiency. Also, some carotenoids can be cleaved to give retinoids or apocarotenoids, as a result of which molecules with antioxidant properties different from those of the parent compound are formed.

Acknowledgements

A.J.M.M. acknowledges funding from the European Union (FP6-24671-QUALITOM and FP7-224789-INTERCAROT) and the Spanish programs Juan de la Cierva (JCI-2008-1767) and Ramón y Cajal (RYC-2010-07115). The authors are members of the IBERCAROT network, funded by CYTED (ref. 112RT0445).

References

1. DellaPenna D, Pogson BJ (2006) Vitamin synthesis in plants: tocopherols and carotenoids. Annu Rev Plant Biol 57:711–738
2. Maeda H, DellaPenna D (2007) Tocopherol functions in photosynthetic organisms. Curr Opin Plant Biol 10:260–265
3. Atkinson J, Epand RF, Epand RM (2008) Tocopherols and tocotrienols in membranes: a critical review. Free Rad Biol Med 44: 739–764
4. Azzi A (2007) Molecular mechanism of [alpha]-tocopherol action. Free Rad Biol Med 43:16–21
5. AOCS (2013) AOCS lipid library. http://lipidlibrary. aocs.org/. Accessed 12 Dec 2012
6. Britton G (1995) Structure and properties of carotenoids in relation to function. FASEB J 9: 1551–1558
7. Britton G, Liaaen-Jensen S, Pfander H (2008) Carotenoids. Volume 4: Natural functions. Birkhäuser, Basel

8. Akiyama K, Matsuzaki K, Hayashi H (2005) Plant sesquiterpenes induce hyphal branching in arbuscular mycorrhizal fungi. Nature 435: 824–827
9. Akiyama K, Hayashi H (2006) Strigolactones: chemical signals for fungal symbionts and parasitic weeds in plant roots. Ann Bot 97: 925–931
10. Gomez-Roldan V, Fermas S, Brewer PB et al (2008) Strigolactone inhibition of shoot branching. Nature 455:189–194
11. Melendez-Martinez AJ, Stinco CM, Liu C et al (2013) A simple HPLC method for the comprehensive analysis of cis/trans (Z/E) geometrical isomers of carotenoids for nutritional studies. Food Chem 138:1350
12. Krinsky N, Mayne ST, Sies H (2004) Carotenoids in health and disease. Marcel Dekker, New York
13. Britton G, Liaaen-Jensen S, Pfander H (2009) Carotenoids. Volume 5: Nutrition and Health. Birkhäuser, Basel
14. Meléndez-Martínez AJ, Britton G, Vicario IM et al (2008) The complex carotenoid pattern of orange juices from concentrate. Food Chem 109:546–553
15. Demmig-Adams B, Adams WW III (1996) The role of xanthophyll cycle carotenoids in the protection of photosynthesis. Trends Plant Sci 1:21–26
16. Qin X, Zeevaart JAD (1999) The 9-*cis*-epoxycarotenoid cleavage reaction is the key regulatory step of abscisic acid biosynthesis in water-stressed bean. Proc Natl Acad Sci U S A 96:15354–15361
17. Alder A, Jamil M, Marzorati M et al (2012) The path from β-carotene to carlactone, a strigolactone-like plant hormone. Science 335: 1348–1351
18. Meléndez-Martínez AJ, Britton G, Vicario IM et al (2006) HPLC analysis of geometrical isomers of lutein epoxide isolated from dandelion (*Taraxacum officinale* F. Weber ex Wiggers). Phytochemistry 67:771–777
19. Meléndez-Martínez AJ, Vicario IM, Heredia FJ (2007) Geometrical isomers of violaxanthin in orange juice. Food Chem 104:169–175
20. Meléndez-Martínez AJ, Escudero-Gilete ML, Vicario IM et al (2009) Separation of structural, geometrical and optical isomers of epoxycarotenoids using triacontyl-bonded stationary phases. J Sep Sci 32:1838–1848
21. Meléndez-Martínez AJ, Britton G, Vicario IM et al (2005) Identification of isolutein (lutein epoxide) as *cis*-antheraxanthin in orange juice. J Agric Food Chem 53:9369–9373
22. Meléndez-Martínez AJ, Britton G, Vicario IM et al (2008) Does the carotenoid neoxanthin occur in orange juice? Food Chem 107:49–54
23. Re R, Pellegrini N, Proteggente A et al (1999) Antioxidant activity applying an improved ABTS radical cation decolorization assay. Free Radic Biol Med 26:1231–1237
24. Britton G, Khachik F (2009) Carotenoids in food. In: Britton G, Liaaen-Jensen S, Pfander H (eds) Carotenoids. Volume 5: Nutrition and health. Birkhäuser, Basel, pp 45–66
25. Britton G, Liaaen-Jensen S, Pfander H (1995) Carotenoids Volume 1A: Isolation and analysis. Birkhäuser, Basel
26. Lampi A-M (2011) Analysis of tocopherols and tocotrienols by HPLC. http://lipidlibrary.aocs.org/topics/tocopherols/index.htm. Accessed 14 Dec 2012
27. Lee J (2004) Analysis of tocopherols and tocotrienols in foods. In: Eitenmiller R, Lee J (eds) Vitamin E. Food chemistry, composition, and analysis. CRC, Boca Raton, FL, pp 323–424
28. Meléndez-Martínez AJ, Vicario IM, Heredia FJ (2007) Review: analysis of carotenoids in orange juice. J Food Comp Anal 20:638–649
29. Mercadante AZ (2007) Analysis of carotenoids. In: Socaciu C (ed) Food colorants: chemical and functional properties. CRC, Boca Raton, FL, pp 447–478
30. Rodriguez-Amaya DB (2001) A guide to carotenoid analysis in foods. ILSI Press, Washington, DC
31. Rodriguez-Amaya DB (2010) Quantitative analysis, in vitro assessment of bioavailability and antioxidant activity of food carotenoids—a review. J Food Comp Anal 23:726–740
32. Lietz G, Henry CJK (1997) A modified method to minimise losses of carotenoids and tocopherols during HPLC analysis of red palm oil. Food Chem 60:109–117
33. Delgado-Pelayo R, Hornero-Méndez D (2012) Identification and quantitative analysis of carotenoids and their esters from sarsaparilla (*Smilax aspera* L.) berries. J Agric Food Chem 60:8225–8232
34. Breithaupt DE, Schwack W (2000) Determination of free and bound carotenoids in paprika (Capsicum annuum L.) by LC/MS. Eur Food Res Technol 211:52–55
35. Breithaupt DE, Wirt U, Bamedi A (2002) Differentiation between lutein monoester regioisomers and detection of lutein diesters from marigold flowers (Tagetes erecta L.) and Several fruits by liquid chromatography-mass spectrometry. J Agric Food Chem 50: 66–70

36. Giuffrida D, La Torre L, Manuela S et al (2006) Application of HPLC–APCI–MS with a C-30 reversed phase column for the characterization of carotenoid esters in mandarin essential oil. Flavour Frag J 21: 319–323
37. Murillo E, Giuffrida D, Menchaca D et al (2013) Native carotenoids composition of some tropical fruits. Food Chem. http://dx.doi.org/10.1016/j.foodchem.2012.11.014
38. Pfister S, Meyer P, Steck A et al (1996) Isolation and structure elucidation of carotenoid-glycosyl esters in gardenia fruits (*Gardenia jasminoides* Ellis) and Saffron (*Crocus sativus* Linne). J Agric Food Chem 44:2612–2615
39. Davey MW, Mellidou I, Keulemans W (2009) Considerations to prevent the breakdown and loss of fruit carotenoids during extraction and analysis in Musa. J Chromatogr A 1216: 5759–5762
40. Meléndez-Martínez AJ, Britton G, Vicario IM et al (2005) Color and carotenoid profile of Spanish Valencia late ultrafrozen orange juices. Food Res Int 38:931–936
41. Stinco CM, Fernández-Vázquez R, Escudero-Gilete ML et al (2012) Effect of orange juice processing on the color, particle size and bioaccessibility of carotenoids. J Agric Food Chem 60:1447–1455
42. Mouly PP, Gaydou EM, Corsetti J (1999) Determination of the geographical origin of Valencia orange juice using carotenoid liquid chromatographic profiles. J Chromatogr A 844:149–159
43. Britton G, Liaaen-Jensen S, Pfander H (2004) Carotenoids handbook. Birkhäuser, Basel
44. Davies BH (1976) Carotenoids. In: Goodwin TW (ed) Chemistry and biochemistry of plant pigments. Academic, London, pp 38–165
45. Antolovich M, Prenzler PD, Patsalides E et al (2002) Methods for testing antioxidant activity. Analyst 127:183–198
46. Sies H (2007) Total antioxidant capacity: appraisal of a concept. J Nutr 137:1493–1495
47. Frankel EN, Finley JW (2008) How to standardize the multiplicity of methods to evaluate natural antioxidants. J Agric Food Chem 56:4901–4908

Chapter 7

Simultaneous Analyses of Oxidized and Reduced Forms of Photosynthetic Quinones by High-Performance Liquid Chromatography

Masaru Shibata and Hiroshi Shimada

Abstract

Plastoquinones and phylloquinones are the major plant quinones localized in chloroplasts, and they act as photosynthetic electron redox mediators in thylakoid membranes. These quinones are analyzed by two processes: extraction with organic solvents and quinone assay by high-performance liquid chromatography (HPLC) analysis. Solvent choice is very important from the viewpoint of stability of the redox state in the extraction processes and during storage of plant quinones. We introduce procedures and solvents to avoid changes in the redox state of quinones, in addition to achieving high extraction efficiency. Traditional methods have problems of low sensitivity and require preparation steps to remove interfering substances, such as plant pigments. HPLC systems have been developed utilizing the fluorescent properties of quinols (reduced forms) to measure quinones. Plastoquinones were detected by reversed-phase HPLC with dual detectors (ultra-violet and fluorescence detection). However, the peak of phylloquinone and plastoquinone isomers with shorter side chains often overlaps with a large peak of fast-eluting pigments. To address these issues, HPLC with fluorescence detection after post-column reduction to convert quinones to fluorescent quinol was applied for measurement of fast-eluting quinones (low hydrophobicity quinones and quinols) such as phylloquinone. Using post-column reduction methods with sodium borohydride or platinum black, not only the reduced forms (fluorescent) but also the oxidized forms (non-fluorescent) could be clearly measured by HPLC with a fluorescence detector.

Key words Plastoquinone, Phylloquinone, High-performance liquid chromatography (HPLC), Post-column reduction, Quinone redox

1 Introduction

The cells of higher plants contain various prenylquinones, which are closely involved in photosynthesis and respiration [1, 2]. Plant quinones act as electron carriers, triggers of gene expression, and quench reactive oxygen species. While quinone contents are dependent on leaf age and growth conditions, most abundant quinone in green leaves is plastoquinone (PQ) [3]. Plastoquinones consist of a redox quinone ring (1,4-benzoquinone) and an isoprenoid side

Manuel Rodríguez-Concepción (ed.), *Plant Isoprenoids: Methods and Protocols*, Methods in Molecular Biology, vol. 1153, DOI 10.1007/978-1-4939-0606-2_7, © Springer Science+Business Media New York 2014

chain and are classified on the length of the side chain. PQ with a side chain of nine isoprene units as a lipid tail, PQ-9, is accumulated preferentially in green leaves, and PQs with shorter side chains of 8, 4, or 3 isoprene units are known as PQ-8, PQ-4, and PQ-3, respectively [4, 5]. Furthermore, PQs with side chain modifications can be divided mainly into three types (PQ-A, PQ-B, and PQ-C), and PQ-A is most abundant in leaves [1, 4–7]. Plastoquinones are synthesized in chloroplasts and are distributed in three chloroplast compartments, i.e., the thylakoid membranes, envelopes, and plastoglobuli [3, 8, 9]. Plastoquinone molecules in chloroplasts are the electron carriers from photosystem II to cytochrome b_6/f complex and translocate protons across the thylakoid membranes. Furthermore, many cellular responses to variable environments are stimulated *via* the PQ redox state and regulate chloroplast gene expression of proteins of light-harvesting complex and the photosystem reaction center [10–13].

On the other hand, naphthoquinones are another class of quinones involved in the photosynthetic electron transport chain. They are divided into two main types, phylloquinone (vitamin K_1, VK) and menaquinones (vitamin K_2), on the basis of the structure of the polyprenyl side chain. Phylloquinone is found only in chloroplasts of plants and cyanobacteria. Functional VK as an electron acceptor is localized in thylakoid membranes, and two molecules of VK are bound to the reaction center of photosystem I (electron acceptor A_1 site) [14, 15]. To study the physiological responses and regulations of gene expression *via* quinone redox in photosynthetic electron transport, the oxidized and reduced forms of plant quinones should be measured accurately.

In currently available analytical methods for prenylquinone derivatives, plant quinones are extracted by organic solvents and subsequently determined by high-performance liquid chromatography (HPLC) with suitable detectors. The quinones are liable to undergo a change in redox state due to reaction with co-extracted cellular components and atmospheric oxygen. The properties of the solvent used for quinone extraction are thought to be major limiting factors for the extraction and chromatography efficiency. Therefore, extraction solvents must be carefully selected to achieve high extraction efficiency and to prevent redox change of quinones.

As the redox state of PQ can change during extraction and storage, the effects of various solvents on the redox state of PQ have been examined in detail [16–18]. The oxidation of PQH_2 to PQ was promoted by oxygen in air at room temperature, regardless of the kind of solvent used. While antioxidants such as butylated hydroxytoluene (BHT) and ascorbic acid are often applied to plant extracts to prevent peroxidation of plant materials, addition of some antioxidants does not appear useful for determination of the redox state of PQ, as reduction of the oxidized form may occur in protic solvents. Interestingly, Kruk et al. reported that reduction of a portion of PQ occurred in water/acetone solvent containing a high concentration of ascorbic acid [16]. Therefore, a good solvent must be selected taking into account both extraction efficiently and lack of alternation in the redox state in the extraction process and during storage of plant quinones. To prevent undesirable changes in the quinone redox state, ethylacetate [16], acetone [17], and tetrahydrofuran (THF) [18] have been used as good solvents for plant quinones.

In HPLC analyses, small amounts of reduced forms of plant quinones can be detected with high sensitivity by an ultraviolet (UV) detector, because the molar extinction coefficient of quinol at a maximum absorption of UV is smaller than that of the oxidized form. For determination of plant quinones, the Okayama group introduced highly sensitive and selective analysis of PQs and VK by HPLC with an electrochemical detector [5]. However, this detection required specialized apparatus, such as an electrochemical detector that is able to detect electroactive or redox substances [5, 19]. High sensitivity measurement of the reduced form of quinone can be performed by utilizing the fluorescence properties of quinols (reduced forms).

Recent HPLC methods are based on normal or reversed-phase HPLC coupled with dual detectors, which connected in series UV (for oxidized form) and fluorescence detectors (for reduced form) [12, 16, 17, 20, 21]. However, the quinones with shorter side chains, such as VK with four isoprene units, are insufficiently separated from the pigments with strong absorption in the UV range by HPLC. Using HPLC with fluorescence detection after post-column reduction to convert quinones to fluorescent quinol, plant quinones can be measured both selectively and sensitively. There are three post-column reduction methods: (1) chemical reduction by reductant, (2) catalytic reduction by catalyzer, and (3) electrochemical reduction by a potentiostat [11, 15, 22–27]. Recently, a novel method for plant quinones was developed that was as rapid, highly sensitive, and highly selective HPLC analysis. Ultra-high pressure liquid chromatography with quadrupole time of flight mass spectrometer as a detector was reported as a more sensitive and powerful method in wide quinone profiles of plants [18].

In this chapter, we introduce quinone extraction procedures and analytical methods by HPLCs with dual detectors as normal method and with post-column using chemical and catalytic reduction methods as highly sensitive and selective methods.

2 Materials

2.1 Equipment

1. Common HPLC system (such as Shimadzu LC-10 system; Shimadzu) with fluorescence (RF-10Av) and/or UV-Vis (SPD-10A) detectors. A HPLC pump to flow chemical reductant.

2. Spectrophotometer (UV-2450; Shimadzu) to conduct spectrophotometric assay of PQ and VK.
3. Centrifugal concentrator (CC-105; Tomy) to evaporate the solvents.
4. Centrifuge (MX-305; Tomy).

2.2 Reagents and Supplies

1. Ethanol, spectrochemical analysis grade.
2. $NaBH_4$, sodium borohydride, EP grade.
3. PQ oxidized form. PQ can be purified from isolated chloroplasts by acidic aluminum oxide column and HPLC preparation [5].
4. PQ reduced form (PQH_2). Hydrogenate PQ to PQH_2 with $NaBH_4$ in ethanol. Estimate the concentration of PQH_2 in ethanol from $\varepsilon = 3.39$ mM^{-1} cm^{-1} at 290 nm in spectrum of PQH_2 or from $\varepsilon = 14.8$ mM^{-1} cm^{-1} at 255 nm in redox difference spectrum of PQ [16].
5. 100 mM $FeCl_3$. Ferric chloride solution is freshly prepared.
6. Hexane, acetone, isopropyl alcohol (IPA), and THF, GR grade.
7. Internal standard solution for PQ analysis. Dissolve a small amount of decylplastoquinone in ethanol. Calculate a concentration of decylplastoquinone from $\varepsilon = 17.9$ mM^{-1} cm^{-1} at 255 nm in ethanol [18].
8. Internal standard solution for VK analysis. Dissolve a small amount of menaquinone-4 in ethanol. Estimate the concentration of menaquinone-4 from $\varepsilon = 18.6$ mM^{-1} cm^{-1} at 248 nm in ethanol [24].
9. Methanol and ethanol as HPLC eluent, HPLC grade.
10. HPLC eluent. Mix 500 mL of methanol and 500 mL of ethanol, and degas under reduced pressure.
11. 0.035 % (w/v) $NaBH_4$ solution for post-column reduction. Weigh 105 mg of $NaBH_4$ and dissolve in 300 mL of ethanol. It takes some time until $NaBH_4$ dissolves completely (*see* **Note 1**).
12. Platinum black cartridge (10×4 mm i.d.) as a catalytic reduction column (*see* **Note 2**).
13. Analytical columns. Luna 5 μC18(2) VP-ODS (4.6 mm i.d. × 150 mm; Phenomenex, Inc) and Shimpack ODS (4.6 mm i.d. × 250 mm; Shimadzu).
14. Tee connector and coil for HPLC with post-column chemical reduction. Reaction coil made of stainless steel (1,500 × 0.5 mm i.d.) and a three-way tee to connect between lines of reductant solution and separated quinones.

3 Methods

3.1 Spectrophotometric Assay of Plastoquinones

1. To assay PQ in organic solvent, dilute with ethanol to approximate PQ concentration (ca. 0.02 mM) (*see* **Note 3**).
2. Measure the spectrum from 200 to 340 nm against an ethanol blank. The UV spectrum of PQ has two peaks at 255 and 262 nm. If the absorbance is too high at 255 nm, dilution of sample with ethanol would be necessary.
3. Add a small amount of $NaBH_4$ to PQ solution in a cuvette, and gently mix to reduce PQ to PQH_2 with a micropipette. After 20 s, take the spectrum in the range between 200 and 340 nm. At this time, absorption of PQ decreases at 255 nm but that of PQH_2 increases at 290 nm depending on reduction of PQ (*see* **Note 4**).
4. Again, read the spectrum from 200 to 340 nm to check whether PQ has been completely reduced. If the second PQH_2 line coincides or runs close to the first PQH_2 line, the reduction of PQ has been carried out completely (*see* **Notes 5** and **6**).
5. To calculate PQ concentration, draw the redox difference spectrum obtained by subtracting the reduced spectrum from the oxidized spectrum.
6. Check three isosbestic points of PQs, 232, 276, and 308 nm, where the reduced line crosses the oxidized line on the quinone spectrum in ethanol. These points are indicators of the purity of PQ. [4, 17]
7. Calculate PQ concentration from the following equation [5, 6, 17, 28]:

$$\text{PQ (mM)} = \Delta A \text{ (oxidized}-\text{reduced spectrum at 255 nm)}/\Delta\varepsilon.$$

The $\Delta\varepsilon$ (255 nm, oxidized minus reduced) is the millimolar difference extinction coefficient at maximum absorption for difference spectrum of PQ, which is 14.8 (mM^{-1} cm^{-1}) for any PQs in ethanol solvent (*see* **Note 7**).

3.2 Plastoquinone Extraction with THF

1. Grind approximately 100 mg of leaf material in a mortar with liquid nitrogen. Extract quinones from leaves with 500 μL of THF containing 10 μM decylplastoquinone as an internal standard (*see* **Notes 8–11**).
2. After vortexing, centrifuge at 14,000 × *g* for 2 min, and transfer 400 μL of green supernatant to a new microtube.
3. Immediately subject to HPLC analysis to avoid undesirable redox changes and degradation of quinones (*see* **Note 12**).

3.3 Phylloquinone Extraction with IPA/Hexane

1. Grind approximately 100 mg of leaf material in liquid nitrogen (*see* **Notes 10**, **11**, and **13**).
2. Extract quinones from leaves with 0.8 mL of IPA/hexane (3:1, v/v) after addition of 50 μL of menaquinone-4 (10 μM in ethanol, corresponding to 0.5 nmol) as an internal standard.

3. After vortexing, centrifuge at 14,000 × *g* for 2 min, and transfer the green phase to a 2.0-mL microtube.
4. Re-extract the remaining pellet with 0.5 mL of hexane. Add 0.5 mL of methanol/water (9:1) to the combined organic phase, and after vortexing and centrifugation, transfer the upper hexane phase to a new microtube.
5. Evaporate the solvent with nitrogen gas under reduced pressure and dissolve the residue in 100 μL of ethanol.
6. Immediately subject to HPLC analysis to avoid undesirable redox changes and degradation of quinones (*see* **Note 12**).

3.4 Simple Extraction of Quinones with Acetone

1. Grind approximately 100 mg of leaf material in liquid nitrogen (*see* **Notes 10, 11, 14**, and **15**).
2. Extract quinones from plant material with 800 μL of freezer-cooled acetone in a mortar chilled with liquid nitrogen. To prevent undesirable changes in the quinone redox state, this process should be conducted as quickly as possible.
3. Transfer the homogenate to a microtube and vortex. Centrifuge at 10,000 × *g* for 5 min, and transfer the green supernatant to a new 2.0-mL microtube.
4. Again, extract the remaining pellet with 400 μL of acetone, and finally 200 μL of hexane is used to retrieve the residual quinones.
5. Combine organic solvents, weigh the acetone/hexane solvent, and calculate the volume of extract (specific gravity = 0.77 g/mL). The total extract volume is approximately 1.5 mL. The quinones in leaf segments are mostly extracted by these procedures (>98 % of total PQs).
6. Immediately subject to HPLC analysis to avoid undesirable redox changes and degradation of quinones (*see* **Note 12**).

3.5 HPLC Analysis of Plastoquinones

1. Connect a fluorescence detector to the outlet of a spectrophotometric detector such as a photodiode array detector (PDA) and a UV detector (Fig. 1a) (*see* **Note 16**).
2. Subject 1–20 μL of standard PQ solution or leaf extracts to reversed-phase HPLC system. Plastoquinones can be separated by the reversed-phase column (Luna 5 μC18(2) VP-ODS, 150 × 4.6 mm i.d.) at 40 °C with isocratic elution of a mixture of an equivalent volume of methanol and ethanol as the mobile phase at a flow rate of 1.0 mL/min. Detection of PQ and PQH_2 is performed by UV absorbance at 254 nm and fluorescence at 330 nm, respectively, with excitation at 290 nm.
3. Calculate the amounts of PQ from the peak area by comparison with standard curves obtained for authentic PQ under the

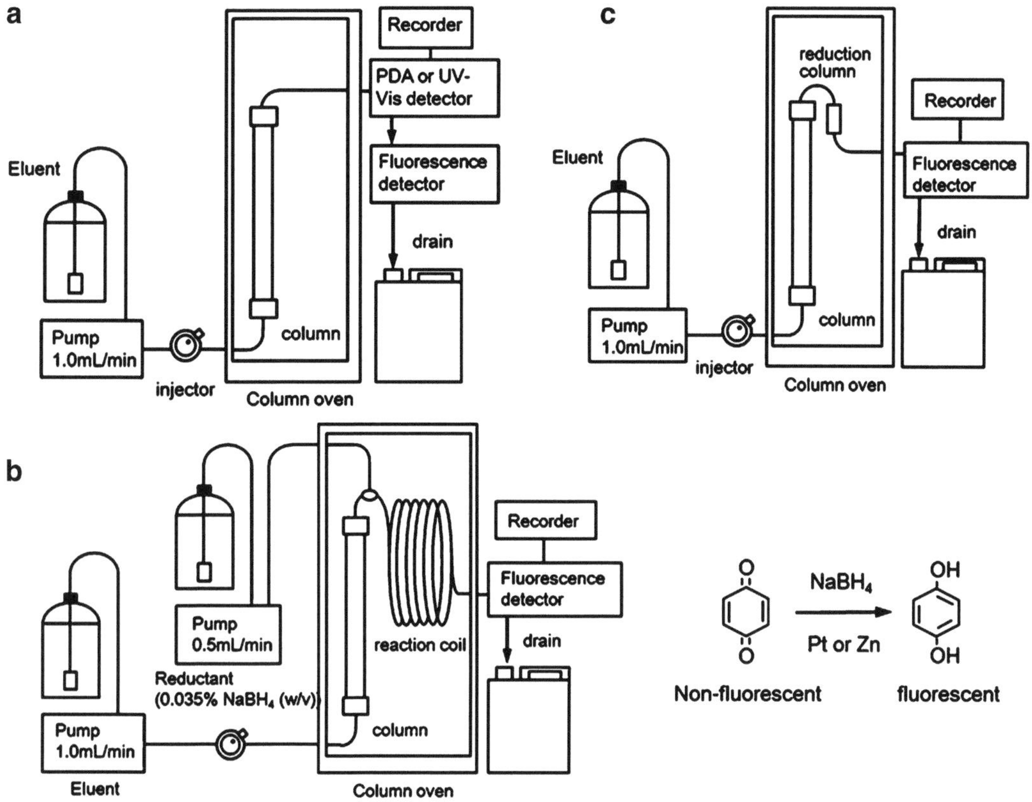

Fig. 1 Diagram of HPLC to detect quinones. (**a**) Dual detector. (**b**) Post-column chemical reduction. (**c**) Post-column catalytic reduction

same conditions. The calibration curves constructed by plotting the peak area against the amount of authentic PQ were linear over the range tested from 2.5 to 1,200 pmol.

3.6 HPLC Analysis of Phylloquinones by Post-column Chemical Reduction to Fluorescent Quinol (see Note 17)

1. Connect the reaction coil (1,500 × 0.5 mm i.d.) to the outlet of a separation column using a three-way tee. Lead a line tube from another outlet of the tee to an HPLC fluorescence detector (Fig. 1b).
2. Put the column with the coil in a column-oven, to maintain the reaction coil at a constant temperature for quinone analysis.
3. To reduce non-fluorescent quinones to fluorescent quinols in the reaction coil, flow ethanol solution containing 0.035 % (w/v) $NaBH_4$ at a flow rate of 0.5 mL/min[1] (*see* **Notes 1, 18,** and **19**).
4. Subject 1–20 μL of standard VK solution or leaf extracts to the reversed-phase HPLC system. Phylloquinone can be separated by the reversed-phase column (Shimpack ODS, 250 × 4.6 mm i.d.) at 40 °C with isocratic elution of a mixture of an equivalent volume of methanol and ethanol as the mobile phase at a

flow rate of 1.0 mL/min. Reduced form of VK in the eluent is detected by monitoring emission at 430 nm with excitation at 320 nm (*see* **Note 17**) [29].

5. Calculate the amounts of VK from the peak area by comparison with standard curves obtained for authentic VK under the same conditions. The calibration curves constructed by plotting the peak area against the amount of authentic VK.

3.7 HPLC Analysis of Phylloquinones by Post-column Platinum Reduction to Fluorescent Quinol

1. Connect the Pt-reduction column (10×4 mm i.d.) to the outlet of a separation column (*see* **Notes 20–22**) (Fig. 1c).
2. Place the analytical column with the Pt-column in a column-oven to maintain a constant temperature of Pt catalyzing reaction for quinone analysis.
3. Lead a line tube from another outlet of the Pt-column to the HPLC fluorescence detector.
4. Subject 1–20 μL of standard VK solution or leaf extracts to the reversed-phase HPLC system. Phylloquinone can be separated by the reversed-phase column (Shimpack ODS, 250×4.6 mm i.d.) at 40 °C with isocratic elution of a mixture of methanol/ethanol (1/1, v/v) as the mobile phase at a flow rate of 1.0 mL/min. Detection of VK is performed by fluorescence of 430 nm with excitation at 320 nm.
5. Calculate the amounts of VK from the peak area by comparison with standard curves obtained for authentic VK under the same conditions.

3.8 Identification of Plant Quinones

Conventional HPLC with spectrophotometric detection lacks both selectivity and sensitivity, and measurement of quinones requires concentration of the extract and removal of substances that strongly absorb UV light. To address these issues, HPLC with fluorescence detection after post-column reduction to convert quinones to fluorescent quinol was applied for measurement of PQ and VK contents without pretreatment of crude extracts.

3.8.1 Peak Identification and Quantification of Plastoquinones

Plastoquinones can be identified by their retention times corresponding to oxidized and reduced forms. A representative chromatogram obtained by HPLC with dual detectors is shown in Fig. 2a, b. The retention times of each quinone are listed in Table 1. PQ was detected with high sensitivity by monitoring changes in absorbance at 254 nm. On the other hand, the peak corresponding to PQH_2 at 290 nm, the maximum absorption of PQH_2, was very small and was approximately at the limit of quantification, due to the smaller molar extinction coefficient of quinol relative to that of quinone. Using the fluorescence properties of quinol (ex. 290 nm, em. 330 nm), PQH_2 could be clearly determined. Separation of both reduced forms of the PQ-9 and ubiquinone-9 could not be

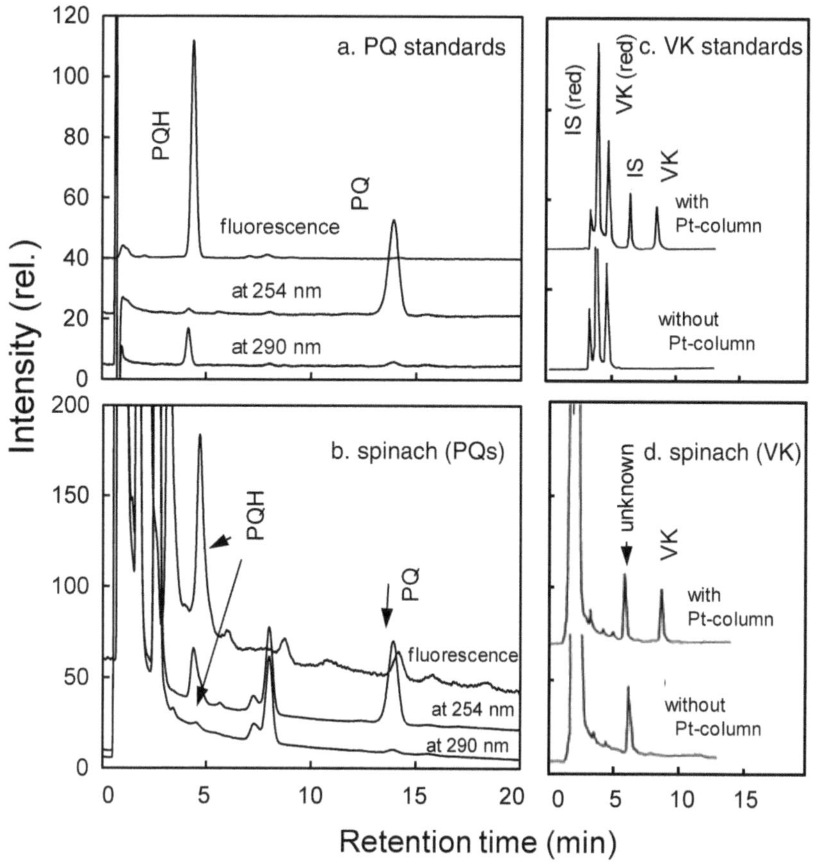

Fig. 2 HPLC profiles of plastoquinones and phylloquinones. Standard samples (**a** and **c**) and quinones from spinach leaves (**b** and **d**) were analyzed by HPLC with dual detectors (**a** and **b**) for PQs and with fluorometric detector after post-column reduction (**c** and **d**) for VKs. PQ and PQH_2 were detected with UV at 254 nm and fluorescence at 320 nm with excitation at 290 nm, respectively. Oxidized and reduced forms of VK measured by fluorescence of quinol HPLC after post-column Pt-reduction and no reduction

achieved under the same conditions [17]. Although ubiquinol and PQH_2 have the same maximum absorption wavelength of 290 nm, determination of PQH_2 by fluorescence detector was not interfered with by reduced ubiquinone contamination due to the differences in fluorescence wavelength and fluorescence quantum yield of both quinols [28, 29].

3.8.2 Peak Identification and Quantification of Phylloquinones

As the peak of VK often overlaps on the shoulder of a large peak of fast-eluting pigments, the baseline is unstable, and the quantification limit of VK is relatively high. Sensitive and selective assay of VKs in leaves could be accomplished by HPLC with post-column reduction methods (Fig. 2d). The HPLC profile by post-column chemical reduction, as well as Pt-reduction, showed that HPLC with post-column reduction is very useful for such analyses (data not shown). In standard naphthoquinones, reduced VK was observed

Table 1
Retention times of prenylquinones and detection conditions

Quinone (redox form)	Retention time (min)	Conditions: Detection	Conditions: Analytical column, post-column
Plastoquinones		Dual detectors (Fluorescence, UV)	
Reduced	4.38	em. at 330 nm ex. at 290 nm	Luna VP-ODS
Oxidized	13.79	UV at 254 nm	–
Phylloquinones		Fluorescence	
Reduced	6.92	em. at 430 nm ex. at 320 nm	Reduction by $NaBH_4$
Oxidized	8.65		Shimpack ODS
Phylloquinones		Fluorescence	Reduction by platinum black
Reduced	7.18	em. at 430 nm ex. at 320 nm	Shimpack ODS
Oxidized	8.95		

by fluorescence detection without using reductant, and oxidized VK was elucidated using reductant as shown in Fig. 2c, d. In spinach leaves, only VK was detected (Fig. 2d). These results were consistent with the report that VK is the only naphthoquinone in plant leaves. Using these methods, the quantification limit of VK is less than 1.0 pmol, corresponding to about 15 mg leaves, depending on plant species and growth conditions.

In addition to the Pt-reduction column, a Zn-column was also applied to the post-reduction methods [24, 25]. Furthermore, PQ isomers can be measured accurately by these post-column methods as well as VKs. Especially, these HPLC methods are suitable for the measurement of PQ species with shorter side chains and hydroxyl group modifications. A part of PQs and VK are known to occur in other compartments not involved in photosynthetic electron transfer, such as plastoglobuli and chloroplast envelopes [16]. The total amounts of quinones by the present HPLC methods indicate quinone contents in the whole compartments, i.e., the contents and redox state of quinones do not represent the PQ pool and photosystem I content in the photosynthetic electron transfer chain.

4 Notes

1. $NaBH_4$ solution can be used at room temperature for 2 days. When the $NaBH_4$ solution becomes cloudy, new reductant ethanolic solution is freshly prepared.

2. A Pt-column for practical use can be prepared in the laboratory by filling a column with Pt black powder reagent as follows.
   (a) Add Pt black powder into an empty column (10×4 mm i.d., or 30×4 mm i.d.).
   (b) Wash the filled column with ethanol at a flow rate of 0.5 mL/min using an HPLC pump or a syringe.
   (c) Open the top nut of the column to check for the void in the column. Refill with Pt black when voids are formed in the column.

   Instead of Pt-reduction column, zinc material can be also used for post-column reduction of quinones [24, 25]. Zn-reduction columns are commercially available.
3. Oxidation of PQH_2 in the sample can be performed by the following procedures:
   (a) Add 10 μL of 100 mM $FeCl_3$ to a test tube containing 1.0 mL of reduced PQ (or a mixture of reduced and oxidized PQs) ethanol solution and bring the final $FeCl_3$ concentration to 1.0 mM. Gently swirl the reaction solution for 1 min to oxidize PQH_2.
   (b) Add 1.0 mL of hexane and distilled water, and vortex. Centrifuge at 5,000×*g* for 5 min and collect the upper layer. The remaining aqueous phase is extracted with 1.0 mL of hexane. Combine hexane phases to obtain PQ solution. Wash twice with 2.0 mL of distilled water to remove ferric chloride in hexane.
   (c) Evaporate hexane phase containing PQ to dryness under reduced pressure, and dissolve the dry film in a small amount of ethanol. By this procedure, most PQH_2 is oxidized to PQ.
   (d) Determine the amounts of total PQs (sum of oxidized and reduced form) from the PQ redox difference spectrum according to Subheading 3.1.
4. Upon reduction with $NaBH_4$, all PQs develop a new absorption maximum at 290 nm. The residual PQ after reduction is estimated from the spectrum in ethanol, because the peak at 255 nm depending on the remaining oxidized form does not disappear due to incomplete reduction of PQ. When PQ reduction is partial, the residual PQ in ethanol can be discriminated by the spectrum with a peak at 255 nm depending on the oxidized form. While further reduction is required, the over-reduction by excess $NaBH_4$ causes degradation of PQ.
5. Hydrogen bubbles are likely to form due to excess amounts of $NaBH_4$ in ethanol. Therefore, the PQH_2 spectrum is disturbed by adhesion of hydrogen bubbles on the cuvette. Instead of $NaBH_4$ particles, 0.02 % $NaBH_4$ ethanolic solution (addition

of 10–50 μL to 1.0 mL of PQ solution) can be used as the reducing agent.

6. PQH_2 formed by $NaBH_4$ can be used as standard PQH_2 to plot the calibration curve. The redox difference spectrum indicates concentration of PQH_2 from hydrogenation of PQ.
7. The spectrum of PQ is independent of prenyl side chain modifications. Therefore, PQs such as PQ-A, PQ-B, and PQ-C give the same absorption spectrum with maximum absorption at 255 nm in ethanol (molar absorption coefficient, $\varepsilon = 17.9\ mM^{-1}\ cm^{-1}$) [6].
8. In comparison with other organic solvents, THF is a very useful solvent for extraction of various plant quinones with high efficiency.
9. Although antioxidants such as BHT and ascorbate are often added to the plant extracts to prevent peroxidation of plant materials, some antioxidants can stimulate hydrogenation of quinone to quinol in organic solvents containing a small amount of water [16]. Purified PQ reduction occurs in water/organic solvent with high ascorbic acid contents, but hydrogenation of PQ can be substantially suppressed by using a large amount of extraction solvent relative to the sample [17, 18].
10. If PQ oxidation occurs in course of quinone extraction, add glutathione (GSH) to the extract solvent to a final concentration of 1.0 mM. While ascorbate and BHT are capable of reducing plant quinones to quinols in aqueous–organic solvents, quinone reduction by GSH is not observed. For these reasons, ascorbate and BHT at least are undesirable as antioxidants, to prevent quinol oxidation, but GSH could be used as an antioxidant for extracts [16–18].
11. For extraction of plant quinones with a bead crusher, the extraction procedure is carried out according to the following steps.
    (a) Grind 100–150 mg of leaf segments to a powder in a mortar with liquid nitrogen, accurately weigh approximately 100 mg of leaf powder, and transfer to a new microtube. At this time, take care to prevent thawing of frozen leaf samples.
    (b) Promptly suspend in 500 μL of suitable solvent containing 10 μM decylplastoquinone as an internal standard, and add glass beads.
    (c) Homogenize leaf samples for 2 min at 2,000 rpm with a bead crusher, and centrifuge (10,000 × *g* for 3 min at 4 °C).
    (d) Transfer supernatant of sample solution to a new microtube, store at below −60 °C until analysis of quinones.

12. The stabilities of quinones are strongly influenced by temperature and light conditions. When quinones are not used immediately after extraction, they should be stored below –60 °C.
13. Although IPA, as well as acetone, is a better extraction solvent for VK, the oxidation from PQH_2 to PQ occurs in the process of PQ extract from leaves of plants under high light conditions. Therefore, IPA cannot be used as the extraction solvent depending on the growth conditions of plants [18, 24].
14. In contrast to the good properties for the stability of quinone redox state, ethyl acetate shows lower extraction efficiency of PQH_2 and VK compared to THF. When quinones are extracted with ethyl acetate, the extraction is performed according to the following steps.
    (a) Homogenize leaves for 30 s in 1 mL of freezer-cooled ethyl acetate in a refrigerator pre-cooled mortar.
    (b) Add 1 mL of ethyl acetate and homogenize for 30 s.
    (c) Transfer 400 μL of the extract to a microtube and immediately evaporate to dryness under a stream of nitrogen gas.
    (d) Dissolve the dried extract in 200 μL of ethanol, briefly centrifuge, and analyze immediately by HPLC.
15. Solvent selection is very important from the viewpoints of the stability of redox state in the extraction process and during storage of plant quinones. Acetone has the ability to efficiently extract most quinones with different side chain lengths, i.e., with a wide range of hydrophobicity. As the water in leaves is co-extracted by acetone, quinones may be reduced to quinols by endogenous antioxidants. However, hydrogenation of PQ can be substantially suppressed by using a solvent volume five-fold greater than the sample, because the reducing agents present in plant tissues are not concentrated enough to promote reduction of PQ [16–18].
16. The pressure resistance of a flow cell in PDA (or UV-Vis) detector is higher than that in a fluorescence detector. Therefore, the detector would be serially arrayed in order to a PDA and a fluorescence detector to avoid damage to the flow cell. The length of the line connecting between detectors should be as short as possible to suppress the broad peak depending on internal diffusion of the separated quinones.
17. It is possible to simultaneously measure oxidized and reduced forms of PQ by a post-column HPLC method with slight modifications. Detection of PQs is carried out by monitoring fluorescence at 320 nm with excitation at 290 nm.
18. Under the same conditions, reduction time of quinone to quinol in the reaction coil is approximately 12 s, and 1.0 nmol of quinone can be fully reduced with 0.01 % of $NaBH_4$.

19. If hydrogen bubbles appear in 0.035 % $NaBH_4$ ethanol solution, 0.025 % solution is used as an alternative reductant to suppress bubble formation.
20. Regularly check the reducing power of the Pt-column to gradually decrease the reducing capacity of Pt black by repeated measurements and long-term storage.
21. Zinc can be used as an alternative to platinum in the reduction column. Phylloquinone analysis by the post-column reduction HPLC method with a Zn-reduction column is performed as follows. The solvent is composed of 10 mM $ZnCl_2$, 5 mM acetic acid, and 5 mM sodium acetate in a mixture of methanol and dichloromethane (9/1) (v/v). Phylloquinones are separated by isocratic chromatography (flow rate: 1.0 mL/min) on a reversed-phase RP18 column equipped with a post-column Zn cartridge (30 × 4 mm, filled with zinc powder, 63-μm particle size) to reduce all quinones [24, 25].
22. The length of the line between the analytical column and Pt-column should be as short as possible to suppress the broad peak depending on internal diffusion of the separated quinones.

References

1. Nowicka B, Kruk J (2010) Occurrence, biosynthesis and function of isoprenoid quinones. Biochim Biophys Acta 1797:1587–1605
2. Pulido P, Perello C, Rodriguez-Concepcion M (2012) New insights into plant isoprenoid metabolism. Mol Plant 5:964–967
3. Lichtenthaler HK (2007) Biosynthesis, accumulation and emission of carotenoids, α-tocopherol, plastoquinone, and isoprene in leaves under high photosynthetic irradiance. Photosynth Res 92:163–179
4. Barr R, Crane FL (1967) Comparative studies on plastoquinones. III. Distribution of plastoquinones in higher plants. Plant Physiol 42:1255–1263
5. Okayama S (1984) Reversed-phase high-performance liquid chromatography of prenylquinones in green leaves using an electrochemical detector. Plant Cell Physiol 25:1445–1449
6. Barr R, Crane FL (1971) Quinones in algae and higher plants. Methods Enzymol 23:372–408
7. Kruk J, Burda K, Schmid GH, Radunz A, Strzalka K (1998) Function of plastoquinones B and C as electron acceptors in photosystem II. Fatty acids analysis of plastoquinone. Photosynth Res 58:203–209
8. Lichtenthaler HK, Prenzel U, Douce D, Joyard J (1981) Localization of prenylquinones in the envelope of spinach chloroplasts. Biochim Biophys Acta 641:99–105
9. Lundquist PK, Poliakov A, Bhuiyan NH, Zybailov B, Sun Q, van Wijk KJ (2012) The functional network of the arabidopsis plastoglobule proteome based on quantitative proteomics and genome-wide coexpression analysis. Plant Physiol 158:117–1192
10. Kruk J, Trebst A (2008) Plastoquinol as a singlet oxygen scavenger in photosystem II. Biochim Biophys Acta 1777:154–162
11. Nowicka B, Kruk J (2012) Plastoquinol is more active than α-tocopherol in singlet oxygen scavenging during high light stress of *Chlamydomonas reinhardtii*. Biochim Biophys Acta 1817:389–394
12. Pshybytko NL, Kruk J, Kabashnikova LF, Strzalka K (2008) Function of plastoquinone in heat stress reactions of plants. Biochem Biophys Acta 1777:1393–1399
13. Zer H, Ohad I (2003) Light, redox state, thylakoid-protein phosphorylation and signaling gene expression. Trends Biochem Sci 28:467–470
14. Johnson TW, Zybailov B, Jones AD, Bittl R, Zech S, Stehlik D, Golbeck JH, Chitnis PR (2001) Recruitment of a foreign quinone into the A_1 site of photosystem I. *In vivo* replacement

of plastoquinone-9 by media-supplemented naphthoquinones in phylloquinone biosynthetic pathway mutants of *Synechocystis* sp. PCC 6803. J Biol Chem 276:39512–39521

15. Shimada H, Ohno R, Shibata M, Ikegami I, Onai K, Ohto M, Takamiya K (2005) Inactivation and deficiency of core proteins of photosystems I and II caused by genetical phylloquinone and plastoquinone deficiency but retained lamellar structure in a T-DNA mutant of Arabidopsis. Plant J 41:627–637
16. Kruk J, Karpinski S (2006) An HPLC-based method of estimation of the total redox state of plastoquinone in chloroplasts, the size of the photochemically active plastoquinone-pool and its redox state in thylakoids of *Arabidopsis*. Biochem Biophys Acta 1757: 1669–1675
17. Yoshida K, Shibata M, Terashima I, Noguchi K (2010) Simultaneous determination of *in vivo* plastoquinone and ubiquinone redox states by HPLC-based analysis. Plant Cell Physiol 51:836–841
18. Martinis J, Kessler F, Glauser G (2011) A novel method for prenylquinone profiling in plant tissues by ultra-high pressure liquid chromatography-mass spectrometry. Plant Methods 7:23, doi: 2310.1186/1746-4811-7-23
19. Isogai Y, Mishimura M, Okayama S (1987) Simultaneous determination of oxidized and reduced forms of plastoquinone A in spinach chloroplasts by high-performance liquid chromatography with an electrochemical detector. Plant Cell Physiol 28:1301–1306
20. Yoshida K, Watababe CK, Hachiya T, Tholen D, Shibata M, Terashima I, Noguchi K (2011) Distinct responses of the mitochondrial respiratory chain to long- and short-term high-light environments in *Arabidopsis thaliana*. Plant Cell Physiol 34:618–628
21. Cheng Z, Sattler S, Maeda H, Sakuragi Y, Bryant DA, Penna DD (2003) Highly divergent methyltransferases catalyze a conserved reaction in tocopherol and plastoquinone synthesis in cyanobacteria and photosynthetic eukaryotes. Plant Cell 15:2343–2356
22. Hiraike H, Kimura M, Itokawa Y (1986) Measurement of K vitamins in human placenta by high-performance liquid chromatography with fluorometric detection. Jpn J Hyg 41:764–768
23. Hiraike H, Kimura M, Itokawa Y (1988) Determination of K vitamins (phylloquinone and menaquinones) in umbilical cord plasma by a platinum-reduction column. J Chromatogr 430:143–148
24. Lohmann A, Schöttler M, Bréhélin C, Kessler F, Bock R, Cahoon EB, Doėrmann P (2006) Deficiency in phylloquinone (Vitamin K_1) methylation affects prenyl quinone distribution, photosystem I abundance, and anthocyanin accumulation in the *Arabidopsis AtmenG* mutant. J Biol Chem 281:40461–40472
25. Jakob E, Elmadfa I (1996) Application of a simplified HPLC assay for the determination of phylloquinone (vitamin K_1) in animal and plant food items. Food Chem 56:87–91
26. Shibata M, Tsuyama M, Takami T, Shimizu H, Kobayashi Y (2004) Accumulation of menaquinones with incompletely reduced side chains and loss of α-tocopherol in rice mutants with alternations in the chlorophyll moiety. J Exp Bot 55:1989–1996
27. Kusube K, Abe K, Hiroshima O, Ishiguro Y, Ishikawa S, Hoshida H (1984) Determination of vitamin K analogues by high performance liquid chromatography with electrochemical derivatization. Chem Pharm Bull 32:179–184
28. De Vitry C, Carles C, Diner BA (1986) Quantitation of plastquinone-9 in photosystem II reaction center particles. FEBS Lett 196:203–206
29. Kruk J, Strzalka K, Leblanc RM (1993) Fluorescence properties of plastoquinol, ubiquinol and α-tocopherol quinol in solution and liposome membranes. J Photochem Photobiol B Biol 19:33–38

Chapter 8

Determination of Sterol Lipids in Plant Tissues by Gas Chromatography and Q-TOF Mass Spectrometry

Vera Wewer and Peter Dörmann

Abstract

Sterols are an abundant lipid class in the extraplastidic membranes of plant cells. In addition to free sterols, plants contain different conjugated sterols, i.e. sterol esters, sterol glucosides, and acylated sterol glucosides. Sterol lipids can be measured by gas chromatography after separation via thin-layer chromatography. Here, we describe a comprehensive technique based on the quantification of all four sterol classes by direct infusion quadrupole time-of-flight (Q-TOF) mass spectrometry.

Key words Sitosterol, Stigmasterol, Mass spectrometry, Arabidopsis, Derivatization, Hydrolysis

1 Introduction

Sterols represent one of the three major membrane lipid classes in plants, next to glycerolipids and sphingolipids. The availability of suitable methods for sterol quantification is the prerequisite for understanding the functions of sterol lipids in plants. In contrast to animals and fungi, where cholesterol or ergosterol, respectively, are predominant, plants contain a larger variety of sterols. In *Arabidopsis thaliana* and most other plant species, the main sterols are the Δ^5-sterols sitosterol, campesterol and stigmasterol, with lower amounts of cholesterol. In contrast, some plants such as the legume species *Medicago sativa* contain Δ^7-sterols, such as spinasterol, which are isomers of the more common Δ^5-sterols [1]. Sterols can occur in their free form, or as conjugated sterol lipids such as sterol glucosides (SGs), acylated sterol glucosides (ASGs), and sterol esters (SEs). In *Arabidopsis thaliana* free sterols (FS) are predominant, whereas the conjugated sterol lipid classes are less abundant [2, 3]. In contrast, other plant species, e.g., members of the *Solanaceae* family, can contain considerable amounts of glucosylated sterol lipids [4]. While FSs, SGs, and

Manuel Rodríguez-Concepción (ed.), *Plant Isoprenoids: Methods and Protocols*, Methods in Molecular Biology, vol. 1153, DOI 10.1007/978-1-4939-0606-2_8, © Springer Science+Business Media New York 2014

ASGs are constituents of all extraplastidic membranes, the nonpolar SEs localize to the oil bodies in the cytosol. They are believed to be involved in regulating the free sterol homeostasis in plant membranes. Thus, SE contents increase during senescence or under stress conditions [5–7].

Free sterols, and with limited sensitivity SEs, can be directly quantified by gas chromatography with flame ionization detector (GC-FID) or gas chromatography-mass spectrometry (GC-MS) analysis. However, the separation and detection by GC of free sterols and SEs is strongly improved by converting the sterol moieties into their respective trimethylsilyl ethers after derivatization with silylating agents such as *N*-methyl-*N*-(trimethylsilyl)-trifluoroacetamide (MSTFA). Due to the high polarity of the sugar moiety, SGs and ASGs are not volatile and cannot be directly measured by GC. Therefore, the GC-based analysis of SGs, ASGs, and SEs requires the liberation of the sterol residue from the conjugated sterol lipid by acidic or alkaline cleavage of the fatty acid or sugar residues. Quantification by GC analysis can subsequently be carried out by measuring the fatty acid (for SEs) or the sterol residue (for SGs, ASGs, and SEs). This procedure requires careful separation of the different sterol lipid classes prior to analysis, which can be achieved by a combination of solid phase extraction (SPE) and thin layer chromatography (TLC) (Fig. 1).

An alternative, comprehensive method for the quantification of all four sterol lipid classes in plants by direct infusion Q-TOF MS analysis has recently been developed [8] (Fig. 2). This LC-MS-based technology allows the direct measurement of intact conjugated sterol molecules after extraction of lipids from plant tissue. The method requires only low amounts of plant material due to the high sensitivity of the Q-TOF mass spectrometer and is characterized by an easy sample preparation. While this method was established on a Q-TOF mass spectrometer coupled to a direct infusion nanospray ion source, the technique can also be adapted to other tandem mass spectrometers such as triple quadrupole instruments equipped with a nanospray or ESI source. The use of internal standards for each sterol lipid class allows for very precise and reliable lipid quantification. Furthermore, this method yields additional information about the molecular species composition of the different sterol lipid classes that cannot be obtained by GC analysis of the sterol residue alone.

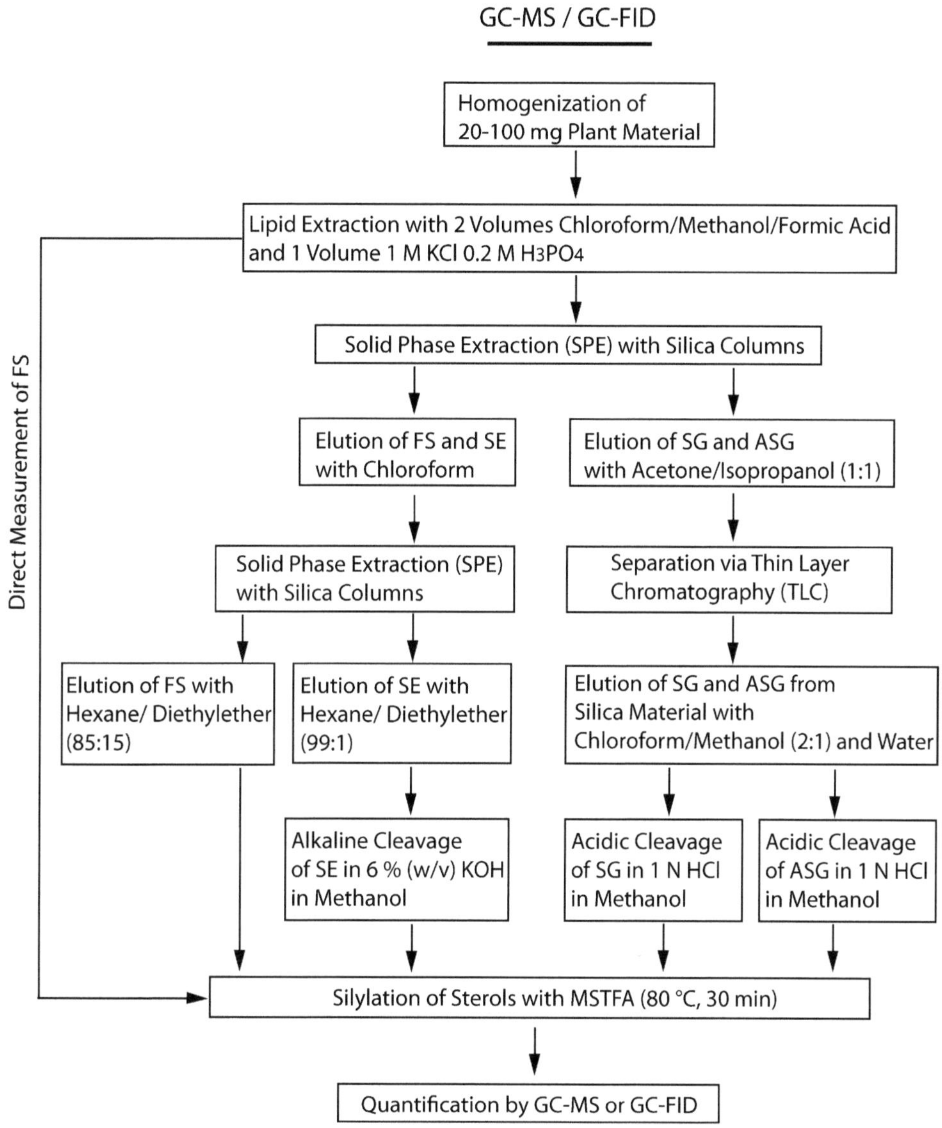

Fig. 1 Sample preparation for sterol lipid analysis by gas chromatography. Lipids are extracted from plant material with chloroform/methanol in the presence of internal standards. The crude lipid extract is purified by SPE, resulting in a nonpolar fraction containing FSs and SEs and a polar lipid fraction containing SGs and ASGs. These fractions are further processed and submitted to a second SPE step or to TLC to separate all sterol lipid classes. Conjugated sterols are then submitted to alkaline or acidic hydrolysis to liberate the sterol residues. Sterols are subsequently analyzed by gas chromatography after silylation with *N*-methyl-*N*-(trimethylsilyl)-trifluoroacetamide (MSTFA)

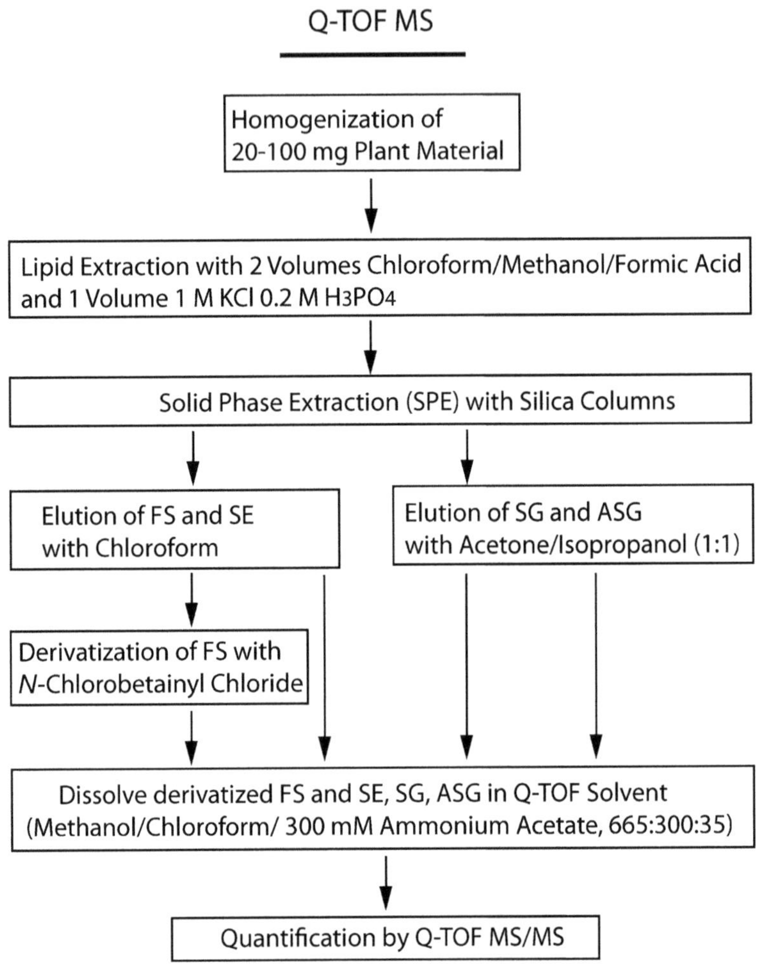

Fig. 2 Sample preparation for sterol lipid analysis by Q-TOF mass spectrometry. Lipids are extracted from plant material with chloroform/methanol in the presence of internal standards. The crude lipid extract is purified by SPE, resulting in a nonpolar fraction containing FSs and SEs and a polar lipid fraction containing SGs and ASGs. FSs are derivatized with *N*-chlorobetainyl chloride prior to analysis by Q-TOF MS, while SEs, SGs, and ASGs are measured directly

2 Materials

Use only high-purity grade solvents (HPLC-grade or better) (*see* **Note 1**). Store solvents at room temperature and lipids at −20 °C. Close caps tightly to prevent solvent evaporation upon storage. Use glass ware whenever possible, avoid plastic ware. Use of 2 mL reaction tubes (Eppendorf) for lipid extraction is acceptable. Rinse glass ware with chloroform/methanol (2:1) prior to use. Use glass pipettes (Hamilton) for pipetting exact amounts of lipid standards (*see* **Note 2**) or Pasteur pipettes for larger volumes.

2.1 Internal Standards

2.1.1 FS (Free Sterols)

Cholesterol, Cholestanol, Stigmasterol, Stigmastanol (Sigma).

2.1.2 SE (Sterol Esters)

16:0-Cholesterol, 16:1-Cholesterol, 18:0-Cholesterol, 18:1-Cholesterol (Sigma).

2.1.3 Synthesis of SG (Sterol Glucosides)

1. Glass tubes with screw caps and teflon septa.
2. Pasteur pipettes.
3. SPE silica columns, Strata Si-1 1 mL (Phenomenex).
4. Heating block (20–130 °C).
5. Glass funnel.
6. Filter paper, 8 cm diameter.
7. pH test strips.
8. Cholestanol, Stigmastanol (Sigma).
9. Glucopyranosyl-bromide-tetrabenzoate (Sigma).
10. Drierite (Sigma).
11. Celite (Sigma).
12. Cadmium carbonate (Sigma).
13. Solvents: toluene, chloroform, methanol.
14. 0.15 M Sodium methylate (methanolic solution) (Merck).
15. 1 N HCl in methanol: 100 mL 3 N methanolic HCl (Sigma) + 200 mL methanol (Store at 4 °C).

2.1.4 Hydrogenation of ASGs (Acylated Sterol Glucosides)

1. Esterified (acylated) sterol glucoside mix, from soybean (Matreya LLC).
2. Pasteur pipettes.
3. Glass tubes with screw caps and teflon septa.
4. Platinum(IV)oxide (Sigma).
5. Chloroform.
6. Hydrogen gas.

2.2 Quantification of Sterol Lipids

1. Pentadecanoic acid (15:0) (Sigma).
2. Chloroform/methanol (2:1).
3. Pasteur pipettes.
4. Glass tubes with screw caps and teflon septa.
5. Heating block or water bath (20–100 °C).
6. 6 % (w/v) KOH in methanol.
7. 1 N HCl in methanol: 100 mL 3 N methanolic HCl (Sigma) + 200 mL methanol. Store at 4 °C.

8. NaCl solution: 0.9 % NaCl (w/v) in H_2O.
9. Hexane.
10. *N*-Methyl-*N*-(trimethylsilyl)-trifluoroacetamide (MSTFA).
11. Gas chromatograph: Agilent 7890A Plus with flame ionization detector.
12. Gas chromatograph: Agilent 7890A Plus with mass spectrometer.
13. GC column: 30 m HP5-MS (Agilent). (Temperature gradient for sterol analysis: start at 150 °C increase to 280 °C at 10 °C min, hold for 10.5 min, and decrease to 150 °C at 20 °C min; Temperature gradient for analysis of FAMEs: start at 100 °C increase to 160 °C with 25 °C/min, then to 220 °C with 10 °C/min, and decrease to 100 °C with 25 °C/min).

2.3 Extraction of Lipids from Plant Material

1. Tissue homogenizer Precellys® (PeQlab).
2. 2 mL reaction tubes (Eppendorf).
3. Glass syringe (100 μL) (Hamilton).
4. Pasteur pipettes.
5. Chloroform/methanol/formic acid (1:1:0.1, v/v/v).
6. 1 M KCl /0.2 M H_3PO_4.
7. Chloroform.
8. Centrifuge 5417 R (Eppendorf).
9. Glass tubes with screw caps and teflon inserts.

2.4 Fractionation of Lipid Extracts

1. Glass tubes with screw caps and teflon inserts.
2. Pasteur pipettes.
3. SPE silica columns, Strata Si-1 1 mL (Phenomenex).
4. Chloroform.
5. Acetone/isopropanol (1:1).
6. Sample concentrator with nitrogen gas stream (Techne).
7. Methanol/chloroform/300 mM ammonium acetate (665:300:35, v/v/v).

2.5 Separation of Sterol Lipid Classes

1. Glass tubes with screw caps and teflon inserts.
2. Pasteur pipettes.
3. SPE silica columns, Strata Si-1 1 mL (Phenomenex).
4. Hexane.
5. Hexane/diethylether (99:1).
6. Hexane/diethylether (85:15).
7. Sample concentrator with nitrogen gas stream (Techne).
8. Baker Si 250 TLC plates with concentration zone (J.T. Baker).
9. Glass tank with lid for TLC plate development.

10. Chloroform/methanol (2:1).
11. Iodine.
12. Centrifuge 5810 R (Eppendorf).

2.6 Derivatization of FS

1. Oven (20–110 °C).
2. Mortar and pestle (porcelain).
3. Heating block (20–80 °C).
4. Glass tubes with screw caps and teflon septa and inserts.
5. Pasteur pipettes.
6. Vortexer.
7. Betaine hydrochloride.
8. Thionyl chloride (Fluka).
9. Toluene.
10. Anhydrous methylene chloride.
11. *N*-Chlorobetainyl chloride.
12. Anhydrous pyridine.
13. Chloroform.
14. Methanol.
15. Water.
16. Centrifuge 5810 R (Eppendorf).
17. Sample concentrator with nitrogen gas stream (Techne).
18. Methanol/chloroform/300 mM ammonium acetate (665:300:35, v/v/v).

2.7 Q-TOF MS/MS Analysis

Q-TOF mass spectrometer: Agilent 6530 Accurate Mass Q-TOF LC-MS unit equipped with direct infusion nanospray source: MS 1200 with infusion chip (Agilent).

3 Methods

3.1 Preparation of Internal Standards

The choice of suitable internal standards is critical for reliable lipid quantification by mass spectrometry. Internal standards should be similar in size and molecular structure to the naturally occurring plant lipids for whose quantification they are employed. If possible, these standards should be added to the samples early during lipid extraction. This way, the internal standards are subject to the exact same treatment as the authentic plant sterols. Plant sterols such as cholesterol and sitosterol contain a double bond at position C5 of their ring structure. Saturation of this double bond with hydrogen leads to the formation of cholestanol and stigmastanol, which are adequate standards for sterol lipid quantification. For the quantification of sterol esters, cholesterol esters containing a saturated or monounsaturated fatty acid may be used, as the predominant sterol esters in plants are sitosterol, stigmasterol, and campesterol esterified to 18:2 or 18:3 fatty acids. While the internal standards for the quantification of FSs and SEs are commercially available, those for the quantification of SG and ASG have to be synthesized or modified (*see* Subheadings 3.1.3 and 3.1.4).

The sterol moieties of all sterol lipids can be quantified by GC analysis, after hydrolysis of the conjugated sterols and subsequent derivatization with MSTFA. The trimethylsilyl ethers of sterols can be quantified in the presence of an internal sterol standard of a known concentration. The fatty acid residues of sterol esters and ASGs can be quantified by GC-FID analysis after formation of fatty acid methyl esters (FAMEs) in the presence of pentadecanoic acid (15:0) as internal standard. Both the trimethylsilyl ethers of sterols and the FAMEs can be identified according to their retention times in comparison to those of standard lipids (*see* **Note 3**).

Prepare a master mix containing internal standards for FSs, SEs, SGs, and ASGs in chloroform/methanol (2:1). Final amounts of internal standards should be between 1 and 5 nmol in 100 μL for tissue samples of up to 200 mg fresh weight.

3.1.1 FS

Use cholestanol and stigmastanol as internal standards for the quantification of free sterols by Q-TOF MS. Weigh the sterols and prepare 1 nmol/μL stock solutions in chloroform.

3.1.2 SE

Use 16:0-cholesterol, 16:1-cholesterol, 18:0-cholesterol, and 18:1-cholesterol as internal standards for sterol ester quantification (*see* **Note 4**).

1. Weigh the sterol esters and dissolve in chloroform to prepare an equimolar 1 nmol/μL stock solution.
2. Confirm concentrations by GC-FID analysis of FAMEs. To this end, dry aliquots of SE stock solutions under a stream of air and add 1 mL of 6 % (w/v) KOH in methanol [9] and 100 μL of a pentadecanoic acid standard (50 μg/mL 15:0 in methanol).
3. Incubate at 90 °C for 1 h in a glass tube with closed cap and teflon septum.
4. Extract FAMEs with 1 mL of hexane and 1 mL of 0.9 % (w/v) NaCl. Shake vigorously and centrifuge (3 min, 2,000 × *g*) for phase separation.
5. Harvest the upper phase, transfer to a fresh glass tube and dry the lipids under a stream of air.
6. Add 1 mL of 1 N methanolic HCl and incubate at 80 °C for 30 min to methylate the internal standard pentadecanoic acid. Repeat **steps 4** and **5**.
7. Dissolve in 200 μL of hexane and transfer into GC glass vials for GC-FID analysis (*see* **Note 5**).

3.1.3 Synthesis of SG

1. Dissolve 1.25 g of glucopyranosyl-bromide-tetrabenzoate in 10 mL of toluene [10].
2. To a solution of 0.61 g of sterol (cholestanol and/or stigmastanol) in 12 mL of toluene in a glass tube (with teflon-septum), add 2 g of drierite and 2 g of cadmium carbonate.
3. Let the mixture boil for 7 h at more than 110 °C.
4. Cool down to room temperature, then dilute with 24 mL of chloroform, add one spatula of celite and filter the mixture through a small round filter paper in a glass funnel. Collect the lipids in the filtrate in a glass vial.
5. Dry the lipids under a stream of air.
6. Redissolve the dried lipids in 40 mL of 0.15 M sodium methylate (methanolic solution).
7. Let the mixture stir over night at room temperature (in glass tube on a shaker or with a magnetic stir bar).
8. Neutralize the pH with 1 N methanolic HCl. Use pH strips to monitor the pH after adding a few drops of methanolic HCl to the solution. Do not place pH strip into the solution, but take a Pasteur pipette to apply single drops onto the pH strip.
9. If necessary, divide the solution into aliquots in several glass tubes for the following steps to allow for the addition of larger quantities of solvent.
10. For the extraction of sterol lipids add 2 volumes (80 mL) of chloroform/methanol (2:1) and 1 volume (40 mL) of water. Shake the mixture vigorously, then centrifuge briefly for phase separation (e.g., 3 min, 2,000 × *g*).
11. Harvest lower organic phase, evaporate the solvent under a stream of air and redissolve the dried lipids in ca. 5 mL of chloroform.
12. Apply the sterol glucosides to a 500 mg SPE silica column (equilibrated with chloroform). Wash with 10 mL of chloroform to elute free sterols and discard this fraction. Elute sterol glucosides with methanol (5 mL). Make sure no free sterol is present in the methanol fraction. Concentrate the SG fraction under air flow. Store at –20 °C.
13. Quantify the sterol glucoside content by GC analysis:
    (a) Transfer an aliquot of the SG into a glass vial and add a different sterol (e.g., cholesterol) of a defined amount as internal standard for GC analysis.
    (b) Dry the samples under a stream of air and add 1 mL of 1 N methanolic HCl.
    (c) Incubate at 90 °C for 1 h with closed cap. Extract liberated sterols with 1 mL of hexane and 1 mL of 0.9 % (w/v)

NaCl. Shake vigorously and centrifuge (3 min, 2,000 × *g*) for phase separation.

(d) Harvest the upper hexane phase and dry the free sterols under a stream of air.

(e) Add 50 μL of MSTFA and incubate at 80 °C for 30 min for derivatization of the sterols.

(f) Evaporate the MSTFA under a stream of air and dissolve the lipids in 100–200 μL of hexane.

(g) Transfer the samples into GC glass vials for subsequent GC analysis.

14. Adjust concentrations of SGs to 1 nmol/μL of cholestanol-glc and stigmastanol-glc.

3.1.4 Hydrogenation of ASG

The commercially available esterified ASG mixture from soybean (Matreya) contains mainly sitosterol-glc, stigmasterol-glc, and campesterol-glc esterified to 16:0, 18:2, and 18:3 fatty acids, respectively. To obtain a suitable standard for ASG quantification by Q-TOF MS, saturation of the double bonds in the fatty acids and the sterol residues is required. To this end, ASGs are hydrogenated [11].

1. Dissolve soybean ASGs (Matreya) in 5 mL of chloroform in a 40 mL glass tube and add a small amount (5–10 mg) of platinum(IV)oxide.
2. Hydrogenation takes place at room temperature by application of a stream of hydrogen gas to the lipid solution. The process may take several hours or overnight to complete. *Attention*: Hydrogen gas is highly flammable. Therefore, perform this experiment in a fume hood in the absence of any ignition sources!
3. Centrifuge the suspension (5 min, 2,000 × *g*) to sediment the platinum(IV)oxide and harvest the chloroform containing the hydrogenated ASGs.
4. Apply the hydrogenated ASGs onto a silica SPE column (equilibrated with chloroform), wash with 5 mL of chloroform and then elute ASGs with methanol (5 mL). Check complete saturation of the sterol and the fatty acid moieties by Q-TOF MS/MS analysis. Store the saturated ASGs in chloroform/methanol (2:1) (*see* **Note 6**).
5. To determine the concentration of the ASG solution, quantify the fatty acid residue after methylation via GC-FID analysis, and the sterol moiety after derivatization with MSTFA by GC-MS. Acidic hydrolysis of ASGs and subsequent analysis of the sterol residue by GC-MS is performed as described for sterol glucosides (see above). Methylation of the fatty acid can be done by acidic methanolysis of ASGs and FAMEs are quantified by GC-FID as described for the acyl groups derived from SEs (see above).

3.2 Extraction of Lipids from Plant Material

1. Harvest 25–100 mg of plant material and determine the exact fresh weight quickly to minimize the risk of lipid degradation by lipases. Quickly freeze the samples in liquid nitrogen and immediately proceed with extraction or store the material at −80 °C.
2. Grind the samples to a fine powder in 2 mL reaction tubes using a ball mill (Precellys®) at 5,000 Hz for 10 s (*see* **Note 7**). Grind the samples a second time in the presence of chloroform/methanol/formic acid (1:1:0.1) to ensure complete homogenization of the plant material. The acidic extraction buffer inactivates lipid degrading lipases [12, 13] (*see* **Note 8**).
3. Add 100 μL of the master mix of internal standards.
4. Extract lipids with 2 volumes of chloroform/methanol/formic acid (1:1:0.1) and 1 volume of 1 M KCl/0.2 M H_3PO_4. Vortex and centrifuge at 5,000 × *g* for 3–5 min for phase separation. Harvest the lower organic phase with a Pasteur pipette.
5. Re-extract the upper phase twice with chloroform.
6. Combine the organic phases in a fresh glass tube and dry the lipids under an air stream.
7. Dissolve the dried lipids in 500 μL of chloroform for fractionation by SPE (*see* **Note 9**).

3.3 Fractionation of Lipid Extracts

The reliable quantification of conjugated sterols requires the purification of the crude lipid extracts by SPE on silica columns prior to Q-TOF MS/MS analysis.

1. Apply crude lipid extracts in chloroform to a 100 mg SPE silica column equilibrated with chloroform. Collect flow-through.
2. Elute free sterols and sterol esters with approx. 5 mL of chloroform and combine with flow-through (*see* **Note 10**). Divide the chloroform fraction into two aliquots. Use one aliquot for the analysis of free sterols after derivatization with *N*-chlorobetainyl chloride (*see* Subheading 3.5) and the other one for the analysis of sterol esters by Q-TOF MS/MS.
3. Elute sterol glucosides and ASGs with approx. 3 mL of acetone/isopropanol (1:1). Collect eluate in fresh glass tube.
4. For Q-TOF MS/MS analysis of sterol lipids, dry the lipids of the chloroform fraction and the acetone/isopropanol fraction, respectively, under a stream of air.
5. Dissolve the samples in 200–500 μL of Q-TOF solvent: methanol/chloroform/300 mM ammonium acetate (665:300:35, v/v/v) [14] and transfer the samples into Q-TOF autosampler vials (*see* **Note 11**).
6. Proceed with Q-TOF MS/MS analysis (*see* Subheading 3.6).

3.4 Separation of Sterol Lipid Classes

In an additional purification step for subsequent analysis of sterol lipids by GC-MS, the lipid fractions obtained by a first purification step via SPE (*see* Subheading 3.3) containing FS/SE or SG/ASG can be further separated. For the separation of FSs from SEs, an additional SPE step with a different solvent system can be employed (see http://www.cyberlipid.org). SGs needs to be separated from ASGs by TLC.

3.4.1 Separation of FS and SE by SPE

1. Take the chloroform fraction obtained by SPE (*see* Subheading 3.3) containing FSs and SEs and dry the lipids under a stream of air.
2. Dissolve the lipids in hexane and apply to a silica column (equilibrated in hexane).
3. Elute SEs with 5 mL of hexane/diethylether (99:1).
4. Transfer the silica column to a fresh glass tube and elute FSs with 5 mL of hexane/diethylether (85:15).
5. Dry the lipids under a stream of air and prepare the samples for GC analysis, as described in Subheading 3.1 (**step 3**).

3.4.2 Separation of SG and ASG by TLC

1. Take the acetone/isopropanol (1:1) fraction obtained by SPE (*see* Subheading 3.3) containing SGs and ASGs and dry the lipids under a stream of air.
2. Dissolve the lipids in 20–50 μL of chloroform/methanol (2:1) and load onto the TLC plate.
3. Apply co-migrating standards (*see* **Note 12**).
4. Develop the plate for 1 h with acetone/toluene/water (91:30:8).
5. Remove the TLC plate from the tank and dry.
6. Stain the part of the plate containing the co-migrating standards with iodine vapor. To this end, place iodine crystals into a Pasteur pipette, which is connected to airflow and let the iodine vapor stain the standards. To avoid staining of the sample lanes, cover the rest of the TLC plate with paper.
7. Identify the sterol lipids from plant extracts by comparison with the stained standards. Scrape the silica material off the glass plate with a razor blade and extract the lipids with 2 volumes of chloroform/methanol (2:1) and 1 volume of water for further analysis by GC-FID (*see* **Note 13**). Centrifuge the mixture for 5 min at 2,000 × *g* to achieve phase separation.
8. Harvest the lower organic phase with a Pasteur pipette and transfer lipids to a fresh glass tube. Continue with acidic hydrolysis of SGs or ASGs (*see* Subheading 3.1.3).

3.5 Derivatization of FS with N-Chlorobetainyl Chloride for Q-TOF MS/MS Analysis

N-Chlorobetainyl chloride is required for the derivatization of FSs prior to Q-TOF MS/MS analysis [15].

3.5.1 Synthesis of N-Chlorobetainyl Chloride

1. Grind 1 g of betaine hydrochloride crystals (Sigma) in a mortar and incubate the powder at 105 °C for 1 h to remove residual moisture.
2. Add the betaine hydrochloride powder to 0.93 g of thionyl chloride (Sigma) in a glass tube and close lightly with a teflon-lined cap.
3. In the fume hood, heat the mixture slowly to 70 °C and vortex regularly until SO_2 and HCl develop. Open the cap occasionally to release pressure. Keep at 70 °C for at least another 1.5 h and vortex regularly until all betaine hydrochloride has dissolved. If necessary incubate overnight (*see* **Note 14**).
4. Add 1 mL of toluene (prewarmed to 70 °C).
5. Cool the mixture to room temperature while vortexing.
6. Add warm toluene and heat the mixture until *N*-chlorobetainyl chloride has dissolved (*see* **Note 15**).
7. Cool the mixture to room temperature while vortexing (*see* **Note 16**).
8. Wash the crystals several times with methylene chloride at room temperature. Add methylene chloride to the crystals and vortex (*see* **Note 17**).
9. Dry the crystals under a stream of air. A white powder will become visible (*see* **Note 18**).

3.5.2 Derivatization of FS with N-Chlorobetainyl Chloride.

FSs are quantified by Q-TOF MS/MS after derivatization with *N*-chlorobetainyl chloride. This technique has been described before for diacylglycerol derivatization [16] and leads to a high signal intensity and allows quantification of the betaine group after fragmentation.

1. Take an aliquot of the chloroform fraction obtained by SPE purification of a crude lipid extract (*see* Subheading 3.4). This fraction contains FS (*see* **Note 19**).
2. Dry the lipids under a stream of air and redissolve in 0.5 mL of anhydrous methylene chloride. Use a glass tube and screw caps with Teflon septum.
3. Add about 5 mg of *N*-chlorobetainyl chloride and 50 μL of anhydrous pyridine (fume hood).
4. Incubate the mixture at room temperature overnight or for 4 h at 42 °C. Keep the caps firmly closed.
5. To extract derivatized sterols, add 2 volumes of chloroform/ methanol (1:1) and 1 volume of water. Shake vigorously.

6. Centrifuge the samples briefly to achieve phase separation (e.g., 3 min, 2,000 × *g*). Harvest the lower organic phase with a Pasteur pipette and transfer into a fresh glass tube.
7. Dry the solvent under a stream of air and dissolve the derivatized sterols in 200–500 μL of Q-TOF solvent: methanol/chloroform/300 mM ammonium acetate (665:300:35, v/v/v) and transfer the samples into Q-TOF autosampler vials.
8. Proceed with Q-TOF MS/MS analysis (*see* **Note 20**).

3.6 Q-TOF MS/MS Analysis

Fractionation of the crude lipid extract by SPE is used to enrich the sterol lipid classes and reduce the amounts of contaminants which might lead to ion suppression (*see* Subheading 3.3). All experiments are carried out on an Agilent 6530 Q-TOF LC-MS instrument equipped with an Agilent HPLC chip/MS 1200 ion source with infusion chip. The instrument is operated in positive mode and lipids are injected via nanospray infusion. The solvent system consisting of methanol/chloroform/300 mM ammonium acetate (665:300:35, v/v/v) has previously been described for the analysis of phospho- and galactolipids [14] and is also suitable for the analysis of sterol lipids. When this solvent is employed, SGs, ASGs, and SEs can be detected as ammonium adducts. Betainylated sterols carry a positive charge and do not form adducts with other ions.

3.6.1 Q-TOF Parameters

1. Direct injection: Injection volume: 5–10 μL, Flow rate: 1 μL/min.
2. Ion Source: Capillary voltage (V_{Cap}): 1,700 V, Fragmentor Voltage : 200 V.
3. Quadrupole: Operation Mode (MS/MS): Narrow Range (1.2 m/z).
4. Collision Cell: Nitrogen Gas temperature: 300 °C, Nitrogen Gas Flow Rate: 8 L/min, Quadrupole Operation Mode (MS/MS): Narrow Range (1.2 *m*/*z*).
5. Collision Energies: FS (betainylated): 35 V, SE: 13 V, SG: 10 V, ASG: 15 V.
6. Data Recording: 1 spectrum/second, MS spectra recorded every fifth MS/MS spectrum.

Targeted analysis of sterol lipids is performed after fragmentation of the selected ions in the collision cell at specific collision energies (MS/MS mode). For each lipid class, a characteristic fragment can be observed which can be used for identification and quantification. Fragmentation of betainylated FSs leads to the formation of the betaine group, while fragmentation of SEs, SGs, and ASGs gives rise to the corresponding sterol backbones. Saturated ASGs show a different fragmentation pattern leading to a high signal peak for the acylated glucose fragment (*see* **Notes 21** and **22**) [8]. MS/MS spectra are recorded at least five times for each molecular ion and the average of the detected signal intensities was calculated for quantification. Exact molecular masses of the ammonium adducts of SEs, SGs, and ASGs as well as of the betainylated FSs can be calculated using specific software tools such as the mass calculator of the Agilent MassHunter software [8].

3.6.2 Data Analysis

Data can be analyzed using the Agilent MassHunter qualitative analysis software and further processed by Microsoft Excel 2007. The peak height of the selected fragment ion derived from of the respective sterol lipid can be used for quantification using the extracted ion chromatogram function of the MassHunter software. This is comparable to a precursor ion scan function of triple quadrupole instruments.

3.6.3 Isotopic Overlap

The phenomenon of isotopic overlap due to the presence of one or more ^{13}C atoms is inherent to all fatty acid containing lipids and to sterols carrying a variable number of double bonds. For example, in sterol ester molecules containing 45 carbon atoms, the statistical probability for the presence of one ^{13}C atom is 48.67 % and for two ^{13}C atoms it is 11.58 %. The presence of a sterol ion containing two ^{13}C atoms leads to an overlap with ions that differ in mass to charge ratio (m/z) by two, equivalent to two hydrogen atoms. For instance, the presence of two ^{13}C atoms in a monounsaturated sterol results in an m/z of (M+2) almost identical to the m/z of a saturated sterol. Therefore, the signal of the saturated sterol has to be corrected, by subtracting from the total peak height the calculated contribution of the monounsaturated sterol containing two ^{13}C [17]. These values can be calculated with the help of an Isotope Distribution Calculator taking into consideration the fragmentation patterns of the ions. Lipid classes such as SEs and ASGs can carry double bonds in both the sterol and the fatty acid moiety, which renders the calculation of the isotopic overlay more complex. For these lipid classes a stepwise calculation can be performed, considering both possible scenarios, i.e., a sterol moiety containing two ^{13}C or an acyl moiety containing two ^{13}C.

3.6.4 Correction Factors

Correction factors can be employed when the analysis of an equimolar lipid mixture results in lower signal intensities for certain lipid molecules than for others. This is the case when analyzing unsaturated plant SGs in the presence of saturated SGs as internal standards. Fragmentation of saturated SGs results in a number of fragments with low signal intensities as opposed to the fragmentation of unsaturated plant SGs. This can be compensated by the use of a correction factor. This factor was 0.17 as determined by Q-TOF MS/MS on the Agilent 6530 instrument. Similarly the quantification of free sterols after derivatization resulted in lower signals for unsaturated sterols than for saturated sterols. Therefore, a correction factor of 1.61 was employed during data analysis (*see* **Note 23**).

3.6.5 Size-Dependent Correction

Some sterol lipids also show size-dependent differences in signal intensity, similar to phospholipids [14, 18]. Therefore, two internal standards with different sizes were used for all sterol lipid classes. The sterol lipids were then quantified after correction for size dependence based on a linear regression curve of the two internal standards.

4 Notes

1. Q-TOF grade ultrapure water should be used for all buffers. In our hands, water that was double deionized and filtrated by Millipore purifiers is of a very high purity.
2. Rinse glass pipettes with chloroform/methanol (2:1) (I) three times prior to and after use. Repeat these steps with chloroform/methanol (2:1) (II) from a second bottle, to prevent carry-over of lipid contaminations. Regular cleaning of the glass pipettes with acetone and hexane is recommended.
3. GC-MS analysis can also be used to identify the sterols and fatty acids present in a sample.
4. SEs carrying a saturated fatty acid residue are very prone to ion suppression when measured in complex lipid extracts by Q-TOF MS/MS. In contrast, SEs with fatty acid residues carrying at least one double bond rarely suffer from ion suppression. Therefore 16:0 and 18:0 cholesterol esters are used for the quantification of plant SEs with saturated fatty acids, while plant SEs with unsaturated fatty acids are quantified in relation to 16:1 and 18:1 cholesterol esters.
5. SEs used as internal standards are quantified via their fatty acid residues by measuring the FAMEs obtained after alkaline cleavage. At the same time, the sterol residue is liberated and can be quantified by GC-MS after derivatization with MSTFA. This second procedure is recommended for the analysis of SEs from plants after fractionation of the crude lipid extract via SPE (*see* Subheadings 3.3 and 3.4).
6. To achieve a ratio of chloroform/methanol of 2:1 simply add chloroform to the lipids dissolved in methanol. Do not evaporate solvents unless necessary, to avoid accumulation of contaminants.
7. Do not allow plant material to thaw. Alternatively you can use mortar and pestle for manual tissue homogenization.
8. Do not exceed 50 % of the maximal volume of the tube to avoid leakage or breaking of the tube.
9. Low abundant lipids may suffer from ion suppression during direct infusion Q-TOF MS/MS analysis. Therefore, SPE is

performed to enrich sterol lipids and reduce ion suppression resulting from the presence of more abundant polar lipids. GC-MS analysis of free sterols can be performed at this step without additional purification.

10. The chloroform fraction of leaf lipid extracts is green due to the presence of chlorophyll. Ensure that the chloroform eluate is colorless before proceeding to the next elution step.
11. Ion suppression is a common problem if lipid concentration in the sample is too high. Dilute the samples with Q-TOF solvent if necessary.
12. Co-migrating standards are used to help identifying the sterol lipids after separation by TLC. Use commercial standards for SGs and ASGs.
13. For an improved extraction of the lipids from the silica material, add the chloroform/methanol first, then incubate the mixture for several minutes with occasional shaking. Subsequently add the water, shake the mixture well and proceed with centrifugation.
14. The solution will turn yellow over time. Overnight incubation might even lead to a dark brown color. This does not affect derivatization efficiency of *N*-chlorobetainyl chloride.
15. *N*-chlorobetainyl chloride might not dissolve completely. In this case, heat for 10 min to 70 °C, then continue with the protocol.
16. Vigorous vortexing during cooling can prevent the formation of crystals, which are hard to break up later.
17. If the solution is yellow or brown, the resulting powder will eventually become white after washing the crystals repeatedly. However, with each washing step some material is lost and the yield is reduced. Therefore, the wash steps should be limited (about three times).
18. When drying the powder, take care to use gentle air flow to avoid loss of the material. Transfer the powder to several small glass tubes, close them tightly and store at 4 °C. Keep away from moisture.
19. If only FSs are measured, the crude lipid extract can be used without fractionation by SPE.
20. Derivatized sterols should be measured immediately to avoid degradation.
21. Parameters such as fragmentor voltage and collision energies may differ when the experiments are carried out on a different instrument. Higher fragmentor voltages may lead to premature fragmentation of lipids before reaching the collision cell. In this case signal peaks for the respective fragment ions can be

detected in MS only mode. Use different internal standards and natural plant sterols to optimize the MS/MS parameters.

22. Fragmentation of lipids results in characteristic fragment ions which allow for a very reliable identification of the selected ions. However, rarely other molecules with a similar molecular mass might show similar fragmentation patterns. Therefore, when detecting unusual molecular species during Q-TOF MS/MS analysis, it is recommended to compare the results with data obtained by GC-MS analysis of the same sample.
23. Correction factors can be determined by analyzing equimolar mixtures of saturated and unsaturated sterols by Q-TOF MS/MS. If differences in signal intensity are observed, compare the results with data observed in MS mode without fragmentation. Thus, it can be determined whether the deviations result from differential fragmentation. To determine correction factors, prepare the same samples for GC analysis and compare the results with the quantification by Q-TOF MS/MS analysis.

Acknowledgements

This work was supported by an Instrument grant (Forschungsgrossgeräte-Antrag) and by the Forschungsschwerpunkt 1212 of Deutsche Forschungsgemeinschaft (grant Do520/9). We would like to thank Isabel Dombrink, Helga Peisker, and Katharina vom Dorp for their support during method development.

References

1. Huang L, Grunwald C (1988) Effect of light on sterol changes in *Medicago sativa*. Plant Physiol 88:1403–1406
2. Nomura T, Kitasaka Y, Yakatsuto S et al (1999) Brassinosteroid/sterol synthesis and plant growth as affected by lka and lkb mutations of pea. Plant Physiol 119:1517–1526
3. Schaeffer A, Bronner R, Benveniste P, Schaller H (2001) The ratio of campesterol to sitosterol that modulates growth in Arabidopsis is controlled by Sterol Methyltransferase 2;1. Plant J 25:605–615
4. Duperon R, Thiersault M, Duperon P (2001) High level of glycosylated sterols in species of Solanum and sterol changes during the development of the tomato. Phytochemistry 23: 743–746
5. Dyas L, Goad L (1993) Steryl fatty acyl esters in plants. Phytochemistry 34:17–29
6. Sturley SL (1997) Molecular aspects of intracellular sterol esterification: the acyl coenzyme A: cholesterol acyltransferase reaction. Curr Opin Lipidol 8:167–173
7. Schaller H (2004) New aspects of sterol biosynthesis in growth and development of higher plants. Plant Physiol Biochem 42:465–476
8. Wewer V, Dombrink I, Vom Dorp K, Dörmann P (2011) Quantification of sterol lipids in plants by quadrupole time-of-flight mass spectrometry. J Lipid Res 52:1039–1054
9. Schaller H, Grausem B, Benveniste P et al (1995) Expression of the Hevea brasiliensis (H.B.K.) Müll. Arg. 3-hydroxy-3-methylglutaryl-coenzyme A reductase 1-in tobacco results in sterol overproduction. Plant Physiol 109: 761–770
10. Iga S, Schmidt RR, Buzasc MC (2005) Chemical synthesis of cholesteryl β-D-galactofuranoside and -pyranoside. Carbohydr Res 340:2052–2054
11. Buseman CM, Tamura P, Sparks AA et al (2006) Wounding stimulates the accumulation

of glycerolipids containing oxophytodienoic acid and dinor-oxophytodienoic acid in Arabidopsis leaves. Plant Physiol 142: 28–39

12. Bligh EG, Dyer WJ (1959) A rapid method of total lipid extraction and purification. Can J Biochem Physiol 37:911–917
13. Browse J, Warwick N, Somerville CR, Slack CR (1986) Fluxes through the prokaryotic and eukaryotic pathways of lipid synthesis in the '16:3' plant *Arabidopsis thaliana*. Biochem J 235:25–31
14. Welti R, Li W, Li M et al (2002) Profiling membrane lipids in plant stress responses. Role of phospholipase Dα in freezing-induced lipid changes in Arabidopsis. J Biol Chem 277: 31994–32002
15. Vassel B, Skelly WG (1963) *N*-Chlorobetainyl chloride. Org Synth 4:154
16. Li YL, Su X, Stahl PD, Gross ML (2007) Quantification of diacylglycerol molecular species in biological samples by electrospray ionization mass spectrometry after one-step derivatization. Anal Chem 79:1569–1574
17. Ejsing CS, Duchoslav E, Sampaio JL et al (2006) Automated identification and quantification of glycerophospholipid molecular species by multiple precursor ion scanning. Anal Chem 78:6202–6214
18. Brügger B, Erben G, Sandhoff R et al (1997) Quantitative analysis of biological membrane lipids at the low picomole level by nanoelectrospray ionization tandem mass spectrometry. Proc Natl Acad Sci U S A 94:2339–2344

Chapter 9

Analysis of Plant Polyisoprenoids

Katarzyna Gawarecka and Ewa Swiezewska

Abstract

Polyisoprenoid alcohols are representatives of high-molecular terpenoids. Their hydrocarbon chains are built of 5 to more than 100 isoprene units giving rise to polymer molecules that differ in chain-length and/or geometrical configuration. Plants have been shown to accumulate diverse polyisoprenoid mixtures with tissue-specific composition. In this chapter, methods of analysis of polyisoprenoid alcohols in plant material are described, including isolation and purification of polyisoprenoids from plant tissue, fast semi-quantitative analysis of the polyisoprenoid profile by thin-layer chromatography (straight phase adsorption and reversed phase partition techniques), and quantification of polyisoprenoids with the aid of high performance liquid chromatography. This approach results in full characterization of complex polyisoprenoid mixtures accumulated in various plant tissues and other matrixes.

Key words Polyisoprenoids, Polyprenol, Dolichol, TLC, HPLC

1 Introduction

Polyisoprenoid alcohols are highly hydrophobic isoprenoid products synthesized in all living cells. Despite their regular structure composed of numerous ($5 > n > 100$ and more) isoprene units, several structural variants have been described [1]. Firstly, hydrogenation status of the α-terminal double bond in the molecule decides whether a polyisoprenoid alcohol belongs to the subgroup of polyprenols (α-unsaturated) or dolichols (α-saturated) type. Polyprenols are considered as typical components of plant photosynthetic tissues and bacterial cells, while dolichols are found in plant roots and all animal and yeast cells [1]. Secondly, the diversity of the chain-length resulting from the number of isoprene units polymerized in the molecule of natural polyisoprenoid alcohols, which might be a few and reach more than 100. Interestingly, this variation is much more highly pronounced in the case of plant polyprenols than mammalian dolichols [1]. In all eukaryotic cells polyisoprenoids are always accumulated as a mixture of homologues commonly called a "family," with one component being most abundant and a

Manuel Rodríguez-Concepción (ed.), *Plant Isoprenoids: Methods and Protocols*, Methods in Molecular Biology, vol. 1153, DOI 10.1007/978-1-4939-0606-2_9, © Springer Science+Business Media New York 2014

Gaussian-like distribution of homologues. Interestingly, in some plant tissues a two- or even three-family pattern of polyisoprenoid alcohols has been noted [2]. Thirdly, the occurrence of the double bond in the isoprene units gives rise to geometrical *trans* (*E*) or *cis* (*Z*) isomers. The most typical polyisoprenoid alcohols are either of the di-*trans* (ω-t_2-$c_{n\text{-}3}$) or the tri-*trans* (ω-t_3-c_{n-3}) type. However, all-*trans* (ω-t_{n-1}) and α-*trans* (ω-t_3-c_{n-4}-t) variants have also been reported [3]. Thus, due to the natural heterogeneity, polyisoprenoids with their modest molecules composed of the repetitive, head-to-tail condensed, isoprene units require attention during the procedure of isolation and quantitative analysis.

A main enzyme synthesizing the polyisoprenoid hydrocarbon skeleton is *cis*-prenyl transferase (CPT) [4]. It is also worth mentioning that plant dolichols, as shown for those accumulated in the hairy roots, are "mosaic" isoprenoid compounds built from the isopentenyl diphosphate molecules derived from both the methylerythritol phosphate (MEP) and the mevalonate (MVA) pathways [5]. The vital biological function of dolichyl phosphate found in a minute amount in eukaryotic cells comes from its involvement as a cofactor in the biosynthesis of *N*-, *O*-, and *C*-glycosylated and GPI-anchored proteins [6]. The function of the cellular predominant chemical forms of polyisoprenoids, polyprenols and dolichols accumulated as free alcohols and carboxylic esters remains elusive. Since their accumulation is considerably increased in senescing eukaryotic cells and upon pathological and adverse environmental conditions [7–9], their role in cell defense against stress has been suggested. Moreover, polyisoprenoids being the structural components of the biological membranes affect their chemico-physical properties. Extrapolating the results of the biophysical experiments (phospho)polyisoprenoids have been suggested as membrane fluidizers, permeabilizers, and vesicle fusion inducers [10–12]. Additionally, the observed species-specific composition of the polyprenol mixture in plant photosynthetic tissues gives rise to the possibility of their exploitation as chemotaxonomic markers [13, 14].

Analyses of the structure and content of polyisoprenoids have been performed in various biological matrixes. Due to their specific properties, e.g., high hydrophobicity and complex tissue composition, characterization of the polyisoprenoid profile requires specific analytical approaches and various chromatographic methods have been introduced in this field. For qualitative and semi-quantitative analysis, fast and simple thin layer chromatography (TLC) seems the method of choice although one should remember that two complementary TLC systems have to be used in parallel, i.e., adsorption and partition chromatography. Adsorption chromatography (straight phase system on silica gel plates is most commonly used) provides information on the general structural features of polyisoprenoids such as the esterification status of the hydroxyl group and hydrogenation status of the α-terminal

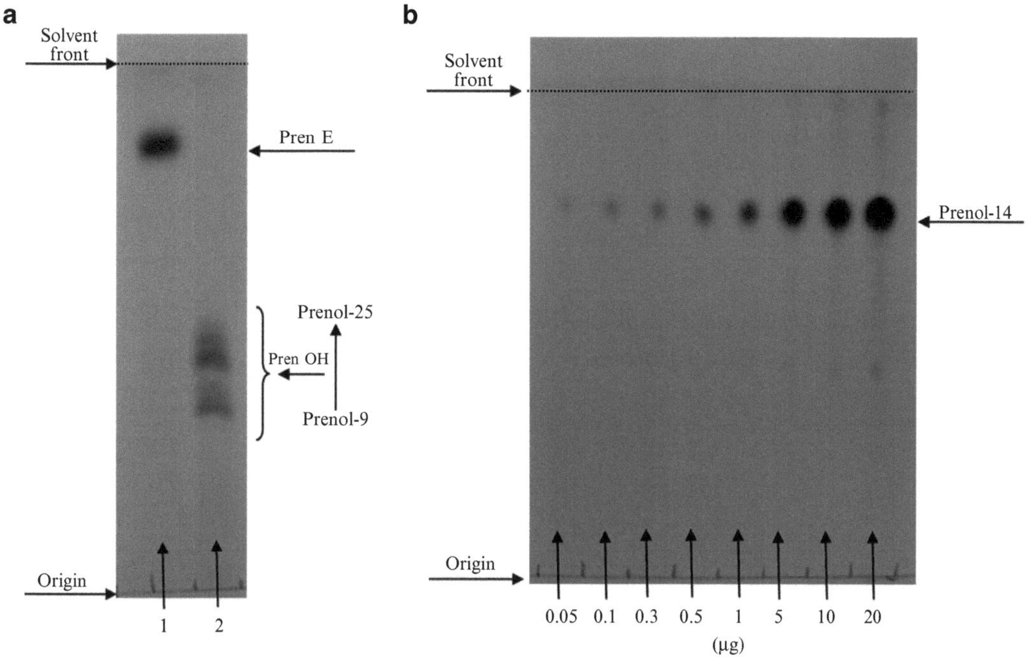

Fig. 1 TLC chromatograms of polyisoprenoid standards. (**a**) Mixture of native polyprenyl esters (PrenE) isolated from *Picea abies* (lane 1) and a qualitative polyprenol standard (polyprenol mixture composed of Pren-9, -11, up to -23 and Pren-25; lane 2). (**b**) Concentration curve of Pren-14. The indicated amount of Pren-14 dissolved in 10 μL of hexane was spotted. Lipids were resolved on TLC Silica gel plate using: (a) solvent mixture recommended in Materials (*see* Subheading 2.3, **item 1**) or (b) modified solvent mixture composed of toluene/ethyl acetate 4/1, by vol

isoprene residue. Due to their extreme hydrophobicity, esters of polyprenols and dolichols show significantly faster mobility than free alcohols upon analysis on the silica gel TLC plates (Fig. 1a). Modification of the polarity of the developing solvent makes more detailed analysis possible for free alcohols (Fig. 1). Highly unpolar solvents (mixtures of hexane and toluene) are recommended for an analysis of esters, while more polar solvents (e.g., toluene supplemented with 5–10 % of ethyl acetate) are preferred for alcohols. Interestingly, the latter chromatographic system appears useful to distinguish in addition structural modifications within the polyisoprenoid hydrocarbon skeleton [3], e.g., *cis* vs. *trans* configuration of the α-terminal double bond in the polyprenol molecule (higher chromatographic mobility of an α-*cis* than α-*trans* polyprenol of the same chain-length) or hydrogenation status of the α-terminal isoprenoid residue (higher chromatographic mobility of polyprenol than dolichol of the same chain-length). However, this system permits only rough identification of the chain-length of analyzed polyisoprenoids with no clear information on the ratio of the particular homologues comprising the polyisoprenoid family. In contrast, partition chromatography in a reversed phase system is used for identification of the composition of the polyisoprenoid

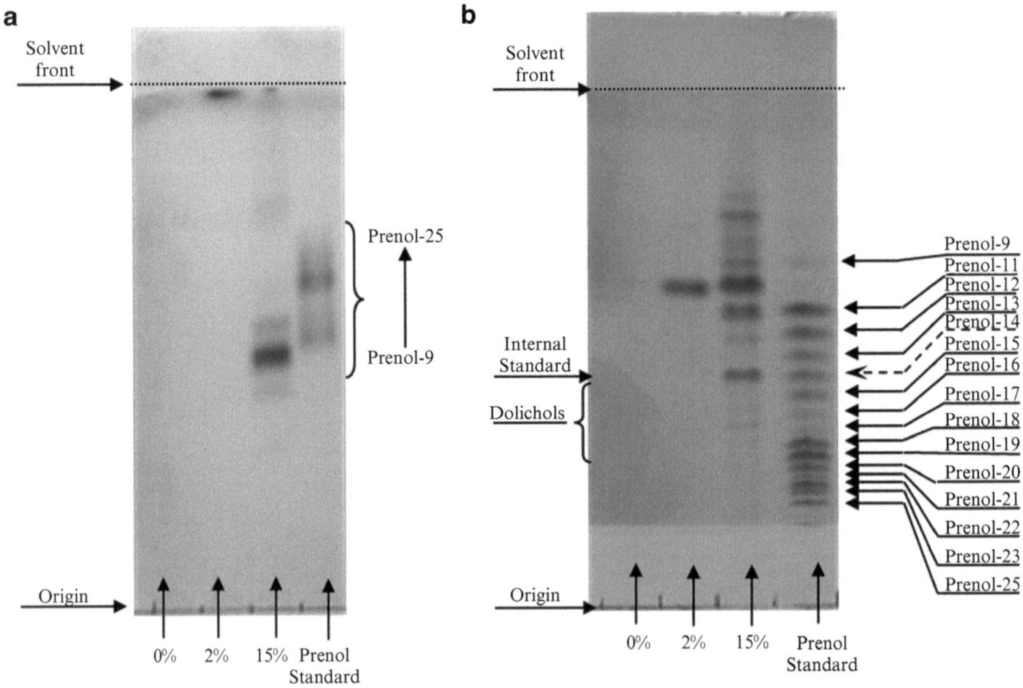

Fig. 2 TLC chromatograms of lipids extracted from 3-week-old *Arabidopsis thaliana* seedlings. Three fractions (0, 2, and 15 %) obtained after separation of a unsaponifiable lipids on silica gel column and a qualitative polyprenol standard (polyprenol mixture composed of Pren-9, -11, up to -23 and Pren-25) were resolved on TLC (**a**) Silica gel and (**b**) RP-18 plates (*see* Subheading 2.3). Please note that lipid(s) appearing as a strong band in fraction 2 % do(es) not correspond to polyisoprenoid alcohol since R_f of this compound(s) on silica gel plate (**a**) precludes this possibility

family (the "pattern" of polyprenols or dolichols). Modulation of the polarity of the developing solvent, e.g., by changing the methanol:acetone ratio, permits analysis of polyisoprenoid mixtures composed of long- as well as short-chain length alcohols. One should remember, however, that separation of polyprenols and dolichols is difficult to obtain on the RP-HPTLC chromatogram (Fig. 2).

In general, TLC is a valuable screening method convenient for tentative identification of the polyisoprenoid content and profile. The detection limit for the TLC methods for polyisoprenoids is approx. 0.1–0.5 μg of the single homologue which makes the analysis of 0.5–1 g of fresh plant tissue plausible. For semiquantitative analysis, the calibration curve has to be analyzed in parallel (Fig. 1b). To visualize the spots, TLC chromatograms are most often stained by iodine vapors—unfortunately this convenient method results in transient, easily destaining chromatograms which have to be immediately scanned. Alternatively, TLC plates might be permanently visualized by spraying with, e.g., sulfuric acid (10 % sulfuric acid in ethanol) or anisaldehyde (5 % anisaldehyde in ethanol) followed by heating (approx. 100 °C). Densitometric analysis of permanently stained chromatograms provides semi-quantitative data on the polyisoprenoid content.

A quantitative analysis of polyisoprenoids requires application of high performance liquid chromatography (HPLC) methods. Reversed phase partition chromatography columns (e.g., C-18 modified silica gel) and UV detector fixed at 208–210 nm are used. Application of solvent gradient is recommended and optimization of the gradient program usually results in good quality spectra of complex polyisoprenoid mixtures, otherwise quite difficult to resolve. The detection limit of the HPLC/UV analysis of polyisoprenoids practically achieved in our laboratory is approx. 1 ng. Careful selection of the appropriate internal standard is recommended after identification of the polyisoprenoid profile in the tissue of interest. Since all eukaryotic organisms accumulate polyisoprenoids as mixtures of homologues, qualitative identification of their signals by comparison with standards is usually quite easy.

Finally, it should be kept in mind that the unequivocal identification of polyisoprenoid alcohols of an unknown, not yet analyzed tissue, requires confirmation by mass spectrometry and/or nuclear magnetic resonance analysis. HPLC/MS also permits the dolichol/polyprenol ratio to be established, even without full chromatographic separation of these species. Additionally, HPLC/MS with Selected Ion Monitoring (SIM) scanning gives significantly enhanced sensitivity (approx. 50 pg) of polyisoprenoid alcohols [15]. Analysis of polyisoprenoids by the HPLC/MS method with various ionization variants has been applied in the literature, while in our hands Electrospray Ionization (ESI) and Atmospheric Pressure Photoionization (APPI) seem to give the best results [16]. In this chapter, the reader finds a detailed description of the analytical method currently used in our laboratory to analyze plant polyisoprenoids. The procedure of polyisoprenoid analysis described below might be used for all types of tissues and the protocol has been developed for 3-week-old Arabidopsis seedlings.

2 Materials

Use analytical or, when indicated, HPLC grade reagents and ultrapure water (MilliQ, prepared by purifying deionized water to attain a resistivity of 18 MΩ cm at 25 °C). Work inside the ventilation hood while using organic solvents. Use only glass laboratory equipment (tubes, pipettes, flasks, bottles, etc.), plastic materials must not be used. Tubes for storage of lipids and organic solvent solutions should be equipped with teflon seals. Carefully follow all waste disposal regulations when disposing of waste materials.

2.1 Hydrolyzing Mixture

1. Weigh 1.2 g of NaOH and transfer to a screw-capped tube. Add 0.6 mL of water (*see* **Note 1**) and vortex. Add 3.3 mL of ethanol, 3.95 mL of toluene, and 0.005 g of pyrogallol (*see* **Note 2**). This amount is sufficient for ten samples, each of 3 g of fresh weight.

2.2 HPLC Mobile Phases

1. Solvent A: add 900 mL of methanol to a 1 L screw-capped bottle. Add 100 mL of water and mix carefully. Use HPLC grade methanol and MilliQ water (*see* **Note 3**).
2. Solvent B: add 500 mL of methanol, 250 mL of 2-propanol, and 250 mL of hexane to a 1 L screw-capped bottle and mix carefully. Use HPLC grade solvents (*see* **Note 3**).

2.3 TLC Solvents

1. S-TLC: mix 90 mL of toluene with 10 mL of ethyl acetate in a TLC tank (*see* **Note 4**).
2. RP-TLC: add a desired volume of acetone to a TLC tank (*see* **Note 4**).

2.4 Silica Gel for Column Chromatography

1. Load 40 g of silica gel powder (Silica gel 60, 0.040–0.063 mm) into a 250 mL screw-capped bottle. Add 100 mL of hexane and mix (*see* **Note 5**).

2.5 Solvents for Column Chromatography

1. 2 % E/H: add 98 mL of hexane to a 100 mL screw-capped bottle. Add 2 mL of diethyl ether and mix (*see* **Note 6**).
2. 15 % E/H: add 170 mL of hexane to a 250 mL screw-capped bottle. Add 30 mL of diethyl ether and mix (*see* **Note 6**).

2.6 Internal Standard and Extraction Solvent

1. Load 5 mg of Prenol-14 into a screw-capped tube. Add 5 mL of hexane and mix carefully. Store the solution at −20 °C (*see* **Note 7**).
2. C/M: add 100 mL of methanol and 100 mL of chloroform to a 250 mL bottle and mix. This amount is sufficient for 10 samples (3 g of fresh weight each). Use methanol and chloroform of p.a. grade (*see* **Note 8**).

3 Methods

Carry out all the procedures at room temperature unless specified otherwise. All the manipulations with solvents should be performed inside the ventilation hood.

3.1 Polyisoprenoid Isolation

1. Cut fresh plant tissue into pieces with scissors, weigh approx. 3 g, and place in a homogenization screw-capped tube (*see* **Note 9**).
2. Add 10 μL of the internal standard and 20 mL of CM extraction solvent and homogenize the sample with a mechanical homogenizer (we use Ultra-Turrax homogenizer for approx. 5 min) (*see* **Note 10**).
3. Put the sample into a dark place for 24 h at room temperature, shaking the tube occasionally (*see* **Note 11**).

4. After that time, add 5 mL of chloroform and 2 mL of water (*see* **Note 12**) and wait 15 min until phases are separated.
5. Transfer the lower (chloroform) phase to the next screw-capped tube using a Pasteur pipette (*see* **Note 13**).
6. Re-extract the tissue by adding 12 mL of chloroform to a homogenization tube containing the upper methanol–water phase and vortex. Wait 15 min until phases are separated.
7. Collect the lower (chloroform) phase and pool both lower phases.
8. Evaporate this crude extract to dryness (oily residue) under a stream of nitrogen or by using an evaporator (*see* **Note 14**).

3.2 Hydrolysis

1. Add 0.8 mL of the hydrolyzing mixture to the sample (*see* **Note 15**).
2. Close the tube tightly and incubate in a pre-warmed water bath at 97 °C for 1 h.
3. Let the tube cool (10–15 min) and add 2 mL of water and 2 mL of hexane to the cooled hydrolyzed mixture. Vortex carefully and then wait 10 min until the two phases are separated (*see* **Note 16**).
4. Collect the top organic (hexane) phase and put into a new previously weighed tube.
5. Re-extract the hydrolyzed mixture three times by adding 2 mL of hexane each time. Pool all four hexane phases (*see* **Note 17**).
6. Evaporate hexane under a stream of nitrogen and dissolve the remaining (unsaponifiable) lipids in 0.4 mL of hexane.

3.3 Polyisoprenoid Purification

1. Place three tubes in a rack (*see* **Note 18**).
2. Into the first tube insert a standard glass Pasteur pipette plugged with a small piece of cotton wool (*see* **Note 19**).
3. Fill this column with a prepared silica gel suspension up to the final column volume of 2.5 mL (*see* **Note 20**).
4. Check whether the column is correctly packed (*see* **Note 21**).
5. Wash the column with 3 mL of hexane.
6. Load unsaponifiable lipids on the top of the column (*see* **Note 22**).
7. Rinse the sample tube three times with 0.4 mL of hexane and load these solutions on the top of the column too.
8. When the entire volume of hexane is down in the tube, rinse the column with an additional 1 mL of pure hexane. When almost all hexane is down in the tube transfer the column to a second tube.
9. Elute the column with 6–8 mL of 2 % E/H (*see* **Note 23**) and transfer the column to a third tube.

10. Continue the elution using 14–15 mL of 15 % E/H.
11. Evaporate all three fractions under a stream of nitrogen and dissolve each remaining residual material in 0.2 mL of hexane.

3.4 Polyisoprenoid Analysis

3.4.1 TLC Analysis

1. Use both the silica gel and RP HPTLC plates. Spot 5 μL of the sample from each tube on both TLC plates (*see* **Note 24**).
2. Spot standards of polyprenols on a separate lane(s) beside the sample.
3. Insert the silica gel and RP HPTLC plates into a tank with S-TLC and RP-TLC solvent, respectively.
4. After approx. 15–20 min, when the solvent front arrives close to the plate edge (approx. 0.5 cm below), take the plate out, mark the solvent front, and dry inside the ventilation chamber (warm but not hot fan might be used).
5. Insert dry TLC plates into the tank (TLC tank or other container with a lid) containing iodine crystals to stain the chromatograms.

3.4.2 HPLC Analysis

1. Transfer the three obtained fractions to a pre-weighed vial or screw-cap tube for HPLC analysis.
2. Evaporate solvents under a stream of nitrogen, estimate the amount of lipids gravimetrically, and dissolve in a calculated volume of 2-propanol (*see* **Note 25**) to obtain the final lipid concentration 3 mg/mL.
3. Vortex the samples carefully and store in the refrigerator.
4. Inject 50 μL of each sample to the HPLC using a manual injector or autosampler. Quantitative (polyprenol and dolichol mixtures of known composition) and qualitative (single polyprenol solution of known concentration) standards should be injected at least once per each set of analyses.
5. The following HPLC settings are routinely used in our laboratory: RP-HPLC type column (we use Zorbax Eclipse XDB-C18; 4.6×75 mm, 3.5 μm);

   1.5 mL/min flow rate; diode array detector (DAD) or UV detector set at 210 nm; column temperature of 21 °C; a combination of linear gradients used is shown in Table 1.
6. After each run, the HPLC system is equilibrated for 5 min by elution with solvent A (*see* **Note 26**). The obtained chromatogram is shown in Fig. 3.

3.5 Result Calculation

1. Compare the analyzed chromatograms (Fig. 3) with that of qualitative standards, and identify the signals corresponding to the retention time of the polyprenols of interest.

Table 1
HPLC gradient program used for separation of polyisoprenoid lipids isolated from Arabidopsis seedlings

Time (min)	% Solvent A	% Solvent B
0	100	0
20	25	75
28	10	90
30	0	100
39	0	100
40	100	0

A combination of linear gradients is used. For details *see* Subheading 3.4.2

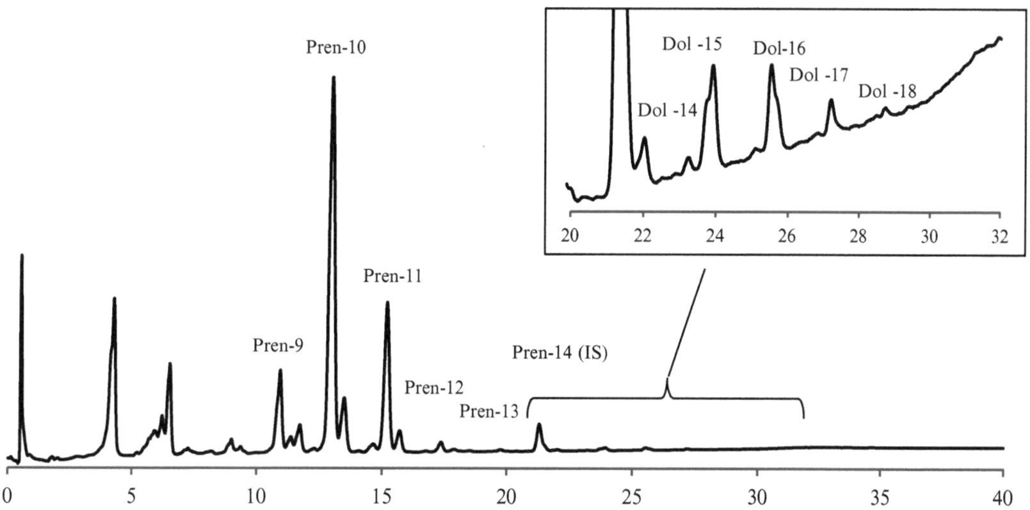

Fig. 3 HPLC/UV chromatogram of polyisoprenoid lipids isolated from 3-week-old *A. thaliana* seedlings. Analysis of a purified fraction containing polyisoprenoids (15 % E/H) revealed the presence of a family of polyprenols (Pren-9 to -13, Pren-10 dominating) and dolichols (Dol-14 to -18, Dol-15 dominating). *Insert* shows an enlarged fragment of the HPLC chromatogram with a family of dolichols. Corresponding signals as well as a signal of the internal standard—Pren-14 (IS) are indicated

2. Integrate the signals in the chromatogram using the HPLC software.
3. Calculate the amount of polyisoprenoid according to the equation:

$$m_x = \left(\frac{l_s}{l_x} \times \frac{m_s \times A_x}{A_s} \right) \div m_t \, (g / gFW)$$

m_x: amount of the analyzed compound (g)
m_s: amount of the internal standard (g)

m_t: amount of the tissue (g)
l_s: number of double bonds in the internal standard molecule
l_x: number of double bonds in the analyzed compound molecule
A_s: area of the integrated signal of the internal standard
A_x: area of the integrated signal of the analyzed compound

4 Notes

1. Add water to NaOH and vortex until a homogenous solution is obtained, then add ethanol, toluene, and pyrogallol. Upon addition of pyrogallol the solution changes color to brown. Use freshly prepared solution.
2. Prior to adding to the sample the hydrolyzing mixture has to be carefully vortexed until a well-mixed emulsion of solvents is formed. Pyrogallol has to be well dissolved.
3. Solvents A and B should be carefully mixed until a homogenous solution is obtained. If necessary (if your HPLC system is not equipped with a degasser) remove air from the solvents, e.g., bubble the solvents with helium for several minutes. If not used the HPLC solvent can be stored in tightly capped bottles for 2–3 days, preferably in the cold room. Right before used, the solvents must be pre-warmed at room temperature.
4. Make sure the TLC tank is tightly covered with a lid and introduce S-TLC or RP-TLC solvent. Control the level of liquid—the sample origin on the TLC chromatogram should be above this level. TLC solvents may be eventually used for a few days. However, in order to preserve the composition of the mixture, evaporation should be avoided.
5. Prepare a gel suspension a day before the chromatography is planned and store at room temperature. During this time the gel will swell and trapped air will be released.
6. Polyisoprenoid composition of the tissue of interest has to be estimated prior to the internal standard selection. For *Arabidopsis thaliana* seedlings and leaves we use Prenol-14.
7. Chloroform is an aggressive solvent—you should wear gloves. Make sure that the bottle for C/M mixture is tightly closed.
8. Ethyl ether is highly volatile and flammable—remember to work under the hood. Mix 2 % E/H and 15 % E/H until you get a clear solution.
9. When several samples are to be analyzed the homogenizer has to be rinsed between the samples.
10. For 3 g of fresh tissue 20 mL of C/M (1/1 V/V) should be used in order to obtain the ratio of 0.3/1/1 of water/methanol/

chloroform. Using this ratio the tissue homogenate is suspended in a homogenous mixture of solvents.

11. We recommend to continue the extraction for 24 h in darkness. Otherwise the extraction might be also performed for 2 h at 37 °C (water bath).
12. Add water and chloroform to obtain the ratio of water/methanol/chloroform solvents equal to 1/2/3. In these conditions, two solvent phases should be separated while precipitated proteins together with tissue remnants should be recovered in the solvent interphase. For better and faster phase separation, the sample might be centrifuged in a low-speed centrifuge (e.g., 950 × *g*). Alternatively, samples might be cooled down on ice and warmed again in a warm water bath. Sometimes, addition of a small aliquot of methanol or ethanol (50–100 μL) improves the separation.
13. Chosen tubes should be resistant to high temperatures. Tightness of the cap should be checked. These tubes will be used for hydrolysis.
14. During evaporation you can add an aliquot (0.2 mL) of ethanol—it will facilitate getting rid of the remaining water.
15. Remember to dissolve the sediment completely. After adding the hydrolyzing mixture vortex the tube carefully. Make sure the cap is tightly closed.
16. If the phases do not separate easily, add 2 ml of saturated NaCl solution.
17. Remember to pool all the four hexane phases—the final volume is approx. 9 mL.
18. For column purification choose regular glass tubes, e.g., with a volume of 25–30 mL. Mark them: 0, 2, 15 %.
19. Too much cotton wool can drastically decrease solvent flow and consequently stop or delay the separation on silica gel. Before adding gel inside the pipette, add some hexane and check the flow. If it is too low, discard this pipette and make a new column.
20. Calculate gel volume using the equation:

$$V = \frac{\Pi \times d^2 \times h}{4}$$

V: volume
d: column diameter
h: gel height

21. The column is well-packed when the level of gel does not move upon touching, gently tap the column to speed up gel sedimentation. To avoid drying of the column, keep it standing on the bottom of the tube and fill it regularly with an eluting solvent.

22. Load the sample very carefully, to avoid damaging the gel surface touch the inner surface of the Pasteur pipette.
23. When purifying polyisoprenoids extracted from plant photosynthetic tissues, continue the elution with 2 % E/H until the first yellow band (carotenoids) is eluted out from the column.
24. We recommend 10 cm long pre-coated glass TLC plates. If required cut the plate prior to spotting the sample. The spots of analyzed lipids ("origin") should be located on the TLC plate above (approx. 0.3–0.5 cm) the solvent level in the TLC tank. A warm but not hot air fan (hair-drier) might be used to accelerate sample loading. When the iodine-stained TLC plate is removed from the iodine tank it should be immediately "sandwiched" between two clean glass plates or in a transparent plastic envelope. The chromatogram has to be immediately scanned since the iodine desorption is quite fast. Use gloves when working with iodine, and avoid contact with cloth.
25. Some compounds might not be easily soluble in 2-propanol. To get a clear solution, warm the tube (use a water bath or a hair dryer).
26. In order to keep the HPLC column ready-to-use, flush the system with solvent A and subsequently solvent B for 10 min at the end of the analysis of a sample set. To avoid sample contamination, the HPLC injector should be rinsed (use a syringe) with 2-propanol (e.g., 100 μL) after each run.

Acknowledgements

This investigation was supported by a grant funded by the Polish National Cohesion Strategy Innovative Economy [UDA-POIG 01.03.01-14-036/09] and the School of Molecular Biology IBB PAS to K.G. We would like to thank Agata Lipko, IBB PAS for her help with TLC analyses.

References

1. Swiezewska E, Danikiewicz W (2005) Polyisoprenoids: structure, biosynthesis and function. Prog Lipid Res 44:235–258
2. Jozwiak A, Ples M, Skorupinska-Tudek K, Kania M, Dydak M, Danikiewicz W, Swiezewska E (2013) Sugar availability modulates polyisoprenoid and phytosterol profiles in *Arabidopsis thaliana* hairy root culture. Biochim Biophys Acta 1831:438–447
3. Ciepichal E, Wojcik J, Bienkowski T, Kania M, Swist M, Danikiewicz W, Marczewski A, Hertel J, Matysiak Z, Swiezewska E, Chojnacki T (2007) Alloprenols: novel α-*trans*-polyprenols of *Allophylus caudatus*. Chem Phys Lipids 147:103–112
4. Surmacz L, Swiezewska E (2011) Polyisoprenoids—secondary metabolites or physiologically important superlipids? Biochim Biophys Res Commun 407:627–632
5. Skorupinska-Tudek K, Poznanski J, Wojcik J, Bienkowski T, Szostkiewicz I, Zelman-Femiak M, Bajda A, Chojnacki T, Olszowska O, Grunler J, Meyer O, Rohmer M, Danikiewicz W, Swiezewska E (2008) Contribution of the

mevalonate and methylerythritol phosphate pathways to the biosynthesis of dolichols in plants. J Biol Chem 283:21024–21035
6. Schwarz F, Aebi M (2011) Mechanisms and principles of N-linked protein glycosylation. Curr Opin Struct Biol 21:576–582
7. Chojnacki T, Dallner G (1988) The biological role of dolichol. Biochem J 251:1–7
8. Bergamini E (2003) Dolichol: an essential part in the antioxidant machinery of cell membranes? Biogerontology 4:337–379
9. Bajda A, Konopka-Postupolska D, Krzymowska M, Hennig J, Skorupinska-Tudek K, Surmacz L, Wojcik J, Matysiak Z, Chojnacki T, Skorzynska-Polit E, Drazkiewicz M, Patrzylas P, Tomaszewska M, Kania M, Swist M, Danikiewicz W, Piotrowska W, Swiezewska E (2009) Role of polyisoprenoids in tobacco resistance against biotic stresses. Physiol Plant 135:351–364
10. Valtersson C, van Duijn G, Verkleij AJ, Chojnacki T, de Kruijff B, Dallner G (1985) The influence of dolichol, dolichyl esters, and dolichyl phosphate on phospholipid polymorphism and fluidity in model membranes. J Biol Chem 260:2742–2751
11. Wang X, Mansourian AR, Quinn PJ (2008) The effect of dolichol on the structure and phase behaviour of phospholipid model membranes. Mol Membr Biol 25:547–556
12. Ciepichal E, Jemiola-Rzeminska M, Hertel J, Swiezewska E, Strzalka K (2011) Configuration of polyisoprenoids affects the permeability and thermotropic properties of phospholipid/polyisoprenoid model membranes. Chem Phys Lipids 164:300–306
13. Rezanka T, Votruba J (2001) Chromatography of long chain alcohols (polyprenols) from animal and plant sources. J Chromatogr A 936: 95–110
14. Chojnacki T, Swiezewska E, Vogtman T (1987) Polyprenols from plants—structural analogues of mammalian dolichols. Chim Scripta 27:209–214
15. Skorupinska-Tudek K, Bienkowski T, Olszowska O, Furmanowa M, Chojnacki T, Danikiewicz W, Swiezewska E (2003) Divergent pattern of polyprenols and dolichols in different organs in *Coluria geoides*. Lipids 38:981–991
16. Kania M, Skorupinska-Tudek K, Swiezewska E, Danikiewicz W (2012) Atmospheric pressure photoionization mass spectrometry as a valuable method for the identification of polyisoprenoid alcohols. Rapid Commun Mass Spectrom 26:1705–1710

Chapter 10

Analysis of Diterpenes and Triterpenes from Plant Foliage and Roots

Qiang Wang, Reza Sohrabi, and Dorothea Tholl

Abstract

Terpene specialized metabolites exhibit multiple functions in plant–environment interactions and plant development. Molecular biologists investigating the biochemistry and molecular function of terpenes need to apply robust but yet sensitive analytical methods optimized and adapted to the structural diversity and often varying concentrations of terpene compounds in plant tissues. Here we present hands-on protocols for sample preparation and GC-MS or LC-MS/MS analysis of selected diterpene and triterpene hydrocarbons or oxygenated derivatives from roots and shoots of Arabidopsis and rice.

Key words Diterpene, Triterpene, Arabidopsis, Rice, Root, Leaves, SPME, GC-MS, LC-MS/MS

1 Introduction

The history of extracting terpene natural products from plants dates back to the first production of perfumes or pharmaceutically active oils in ancient cultures. Traditional methods for recovering terpenes and other phytochemicals from plant materials include cold pressing, steam distillation, and solvent extraction [1]. To date, developing or optimizing terpene analytical methods may be routine for the experienced phytochemist, but these protocols are often less comprehensible to molecular scientists, who study the biological function of small molecules such as terpenes. Terpenes have caught increasing attention because of their roles as plant-specific signals and phytohormones [2–4] and studies investigating these biological activities require hands-on protocols for terpene sample preparation and analysis. Methods for terpene analysis usually depend on the physicochemical properties of the various terpene metabolites, which makes it difficult to describe general analytical procedures. Volatile monoterpenes and sesquiterpenes with high vapor pressures are frequently extracted as components of essential oils by distillation or supercritical fluid extraction (e.g. [5, 6]). Additional methods to recover clean mixtures of volatile

Manuel Rodríguez-Concepción (ed.), *Plant Isoprenoids: Methods and Protocols*, Methods in Molecular Biology, vol. 1153, DOI 10.1007/978-1-4939-0606-2_10, © Springer Science+Business Media New York 2014

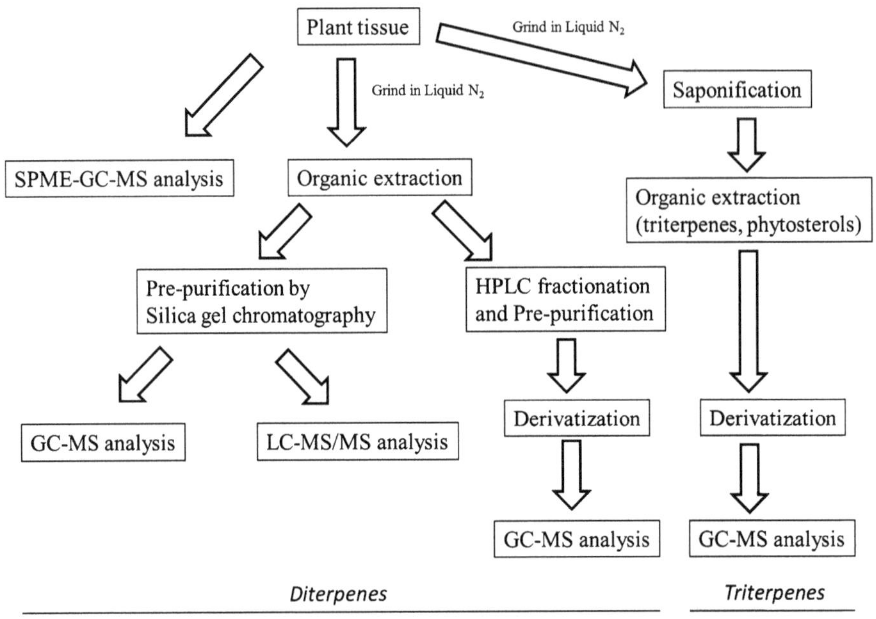

Fig. 1 Flow chart of protocols described in this chapter

terpenes are the trapping of volatile compounds on solid phase microextraction (SPME) devices or other adsorbents in combination with thermal desorption or solvent elution and gas chromatography-mass spectrometry (GC-MS) [7]. Moreover, real-time measurements of volatile terpene emissions using proton transfer reaction-mass spectrometry (PTR-MS) have been developed over the past decade [8]. While detailed protocols for the analysis of volatile terpenes are presented in Chapter 11, this chapter focuses on short protocols to extract and analyze semi-volatile and non-volatile terpenes with C20 (diterpene) and C30 (triterpene) carbon skeletons. Most diterpenes and triterpenes are extracted from plant tissues using organic solvents and subsequently analyzed by GC-MS or liquid chromatography (LC)-MS. Derivatization such as methylation or trimethylsilylation is often required for proper analysis of oxidized terpenes (e.g., [9]) and pre-purification is needed when compounds occur at low concentrations (e.g., [10]). In this chapter, we describe protocols for the qualitative analysis of selected diterpenes and triterpenes from the roots or foliage of dicots (*Arabidopsis*) and monocots (rice) (Fig. 1). Specific examples are presented for the extraction and analysis of newly discovered diterpene hydrocarbons called rhizathalenes in *Arabidopsis* roots [11] and the oxidized diterpenes, 11α-hydroxy-ent-cassadiene [12] and syn-pimaradien-19-oic acid [10] in rice. Analysis of triterpenes will be described at the example of the triterpene diol, arabidiol [13, 14], produced in *Arabidopsis* roots. We focus on the preparation and targeted analysis of individual compounds with antimicrobial, antiherbivore, or allelopathic activities or their respective precursors rather than on methods applied to high-throughput screening and metabolite profiling. The presented procedures can be generally applied for the analysis of diterpene and triterpene hydrocarbons or their derivatives with degrees of oxygenation similar to those described here and may be modified or combined with high performance, high resolution protocols. Such protocols based on ultra-high performance liquid chromatography (UPLC) or capillary electrophoresis coupled with high resolution MS/MS (e.g., triple quadrupole [TQ], Q-Trap, Q-time of flight [TOF], and Ion trap-TOF) have been reviewed in detail [15, 16]. For specific methods in triterpene and triterpene saponin extraction and purification including organic solvent choices, preparative sample preparation methods, and chromatographic purification steps the reader is also referred to protocols described in suggested references [17, 18]. We did not include quantitative protocols or preparative methods that usually require larger amounts of plant tissue or the extraction of compounds from bacterial or yeast cultures after transformation with the corresponding biosynthetic enzymes (*see* Chapters 17–19).

2 Materials

2.1 Analysis of Rhizathalenes in Arabidopsis Roots

1. *Arabidopsis thaliana* soil-free roots obtained as follows: cultivate Arabidopsis Col-0 plants for 4 weeks in potting mix (90 % Sunshine mix #1 [Sun Gro Horticulture], 10 % sand) at 22 °C under long day conditions (14 h light/10 h dark) and a photosynthetically active radiation (PAR) of 150 μmol/m^2/s. Establish hydroponic cultures by transferring 4-week-old plants to plastic containers containing Hoagland's solution under constant aeration as described in [19].
2. GC-MS: gas chromatograph coupled with a quadrupole mass spectrometer (GC-MS-QP2010S, Shimadzu), equipped with Rxi™-XLB GC column (Restek) of 30 m × 0.25 mm i.d. × 0.25 μM film thickness.
3. AOC-5000 Shimadzu autosampler for automated liquid injection and SPME to analyze volatile samples.
4. Prepurified compressed N_2 gas (Airgas).
5. Flash chromatography column with standard joint and PTFE Stopcock, 18″ with 20 mm internal diameter glass column (Kemtech America).
6. Screw top 20 mL glass headspace vials (Supelco) and magnetic screw cap with PFTE/silicon septum (1.5 mm thickness) (Supelco).
7. SPME 100 μM polydimethylsiloxane (PDMS) fiber (Supelco).
8. Mortar and pestle.

9. Silica gel (Merck grade 9385), pore size 60 Å, 230–400 mesh (Sigma).
10. Hexane (Fisher, HPLC grade).

2.2 Analysis of Syn-Pimaradien-19-oic Acid in Rice

1. *Oryza sativa* plants: Grow rice plants (Nipponbare) in a growth chamber to the sixth leaf stage under 12 h light (28 °C)/12 h dark (24 °C) conditions. To induce diterpene accumulation, treat rice plants with 0.2 % (v/v) methyl jasmonate by spraying each plant with approximately 2 mL and covering it with a plastic bag for 12 h.
2. HPLC: Agilent 1100 series instrument equipped with fraction collector, Agilent ZORBAX Eclipse XDB-C8 column (4.6 × 150 mm, 5 μm), and diode array detector.
3. GC-MS: Varian 3900 gas chromatograph with a Saturn 2100 ion trap mass spectrometer; equipped with an Agilent HP-5 column.
4. Centrifuge with fixed angle rotor.
5. Prepurified compressed N_2 gas (Airgas).
6. Mortar and pestle.
7. Ethyl acetate, acetonitrile, methanol (Fisher, HPLC grade); diazomethane saturated hexane (Sigma).

2.3 Analysis of 11α-Hydroxy-ent-Cassadiene in Rice

1. Rice plants grown and treated as described in Subheading 2.2, **item 1**.
2. LC-MS/MS: Agilent Technologies 1100 Series HPLC system coupled to an Agilent Technologies Mass Selective Trap SL detector equipped with a nanoflow electrospray ion source; equipped with Agilent ZORBAX Eclipse XDB-C8 column (4.6 × 150 mm, 5 μm).
3. Centrifuge with fixed angle rotor.
4. Prepurified compressed N_2 gas (Airgas).
5. Mortar and pestle.
6. Ethyl acetate, acetonitrile, methanol (Fisher, HPLC grade).

2.4 Triterpene Extraction and Analysis from Arabidopsis Roots

1. Arabidopsis roots from wild type (Col-0) and mutant *cyp705a1-1* (SALK_043195 from Arabidopsis Biological Resource Center) plants grown axenically.
2. Lyophilizer (Labconco).
3. Rotary evaporator (Cole-Parmer).
4. Compressed N_2 tank (Airgas).
5. Mortar and pestle.
6. Jasmonic acid (Sigma) dissolved in ethanol.
7. Hexanes (HPLC grade) (Fisher).
8. Butylated hydroxytoluene (BHT) (Fisher).

9. Ethanolic potassium hydroxide (10 % KOH (w/v), in ETOH 80 % (v/v), with 0.5 mg/mL of BHT): Make 80 % EtOH (v/v) solution, and then dissolve 5 g of KOH (final 10 % w/v) in 25 mL of 80 % EtOH. Add 0.5 mg/mL of BHT as antioxidant. Bring the volume to 50 mL.
10. Silylating reagent BSFTA (*N*,*O*-bis(trimethylsilyl)trifluoroacetamide) (Sigma).
11. Betulin (Lup-20(29)-ene-3β,28-diol) (MP Biomedicals).
12. Pyridine (Sigma).
13. GC-MS: gas chromatograph coupled with a quadrupole mass spectrometer (GC-MS-QP2010S, Shimadzu), equipped with Rxi™-XLB GC column (Restek) of 30 m × 0.25 mm i.d. × 0.25 μM film thickness.

3 Methods

3.1 Collection and SPME-GC-MS Analysis of Rhizathalenes

Rhizathalenes are semi-volatile diterpene hydrocarbons with an unusual tricyclic spiro-hydrindane structure produced by *Arabidopsis* roots [11]. Analysis of these compounds by SPME (this section) and solvent extraction (next section) coupled to GC-MS is generally applicable for the recovery and identification of plant diterpene hydrocarbons.

1. Collect 1 g of hydroponically grown root tissue and place it in a 20 mL screw cap glass vial containing 1 mL of deionized water.
2. Collect rhizathalene volatiles in the root tissue headspace by penetrating the septum of the screw cap with the SPME fiber and incubating the sample for 30 min at 30 °C in the incubator device of the autosampler (*see* **Note 1**).
3. Thermally desorb volatiles from the SPME fiber by inserting it into the GC injection port.
4. Separate compounds on the nonpolar GC column using the following conditions: Helium carrier gas flow rate of 1.4 mL/min, 2:1 split injection with an injection temperature of 240 °C and a temperature gradient of 5 °C/min from 40 °C (2 min hold) to 240 °C (2 min hold); recording of mass spectrometry data from 50 to 400 *m*/z starting 4 min after injection.

3.2 Extraction and GC-MS Analysis of Rhizathalene

1. Collect 2 g of Arabidopsis root tissue as described in Subheading 3.1, **step 1** and grind in liquid nitrogen to a fine powder (*see* **Note 2**).
2. Transfer the ground tissue into a 250 mL glass flask and add 50 mL of hexane.
3. Extract at room tissue by stirring over night.
4. Concentrate the organic extract to approximately 1 mL under a gentle N_2 stream at room temperature.

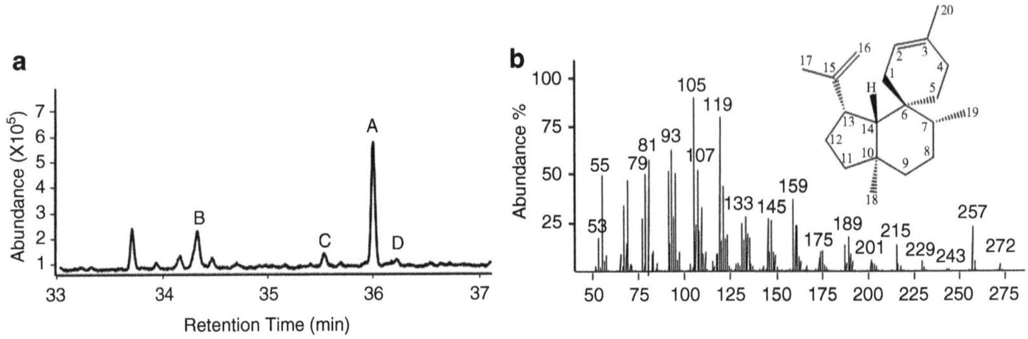

Fig. 2 GC-MS analysis of rhizathalenes from *Arabidopsis* roots. (**a**) Total ion GC/MS chromatogram of volatiles collected from 1 g of roots using SPME. Four diterpene hydrocarbons are detected designated rhizathalene A to D (peaks *A* to *D*). (**b**) EI mass spectrum of rhizathalene A [11]. *Inset*, molecular structure of rhizathalene A

5. Prepare a flash chromatography column with 5 g of silica gel (*see* **Note 3**).
6. Load the concentrated root extract on the silica gel column (*see* **Note 4**).
7. Elute the diterpene containing fraction with 25 mL of hexane and concentrate the eluate to 100 μL under a gentle N_2 stream at room temperature.
8. Perform GC-MS analysis of the pre-purified hexane extract by injecting 1 μL on the GC column.
9. Apply the same separation method as described in Subheading 3.1, **step 4** but use an injection temperature of 230 °C and a temperature gradient of 10 °C min^{-1} from 70 °C (2 min hold) to 300 °C (2 min hold). Rhizathalenes are detected at retention times between 34 and 36.5 min (Fig. 2).

3.3 Analysis of Syn-Pimaradien-19-oic Acid in Rice

Syn-pimaradien-19-oic acid is an intermediate in the biosynthesis of momilactone diterpenes in rice [10]. Derivatization (methylation) of the carboxyl group is required prior to GC-MS analysis. Analysis of this compound is representative for procedures applied for simple diterpene carboxylic acids.

1. Grind 2 g of rice leaf tissue to a powder using liquid nitrogen.
2. Extract tissue with 50 mL of ethyl acetate in a 250 mL glass flask by stirring overnight at room temperature.
3. Clarify the mixture by centrifuging at 3,000 × *g* in glass or Teflon centrifuge tubes.
4. Separate the upper organic phase and dry under a stream of N_2 gas at room temperature.
5. Dissolve the residue in 0.2 mL of 50 % methanol/water and centrifuge again to remove undissolved material.

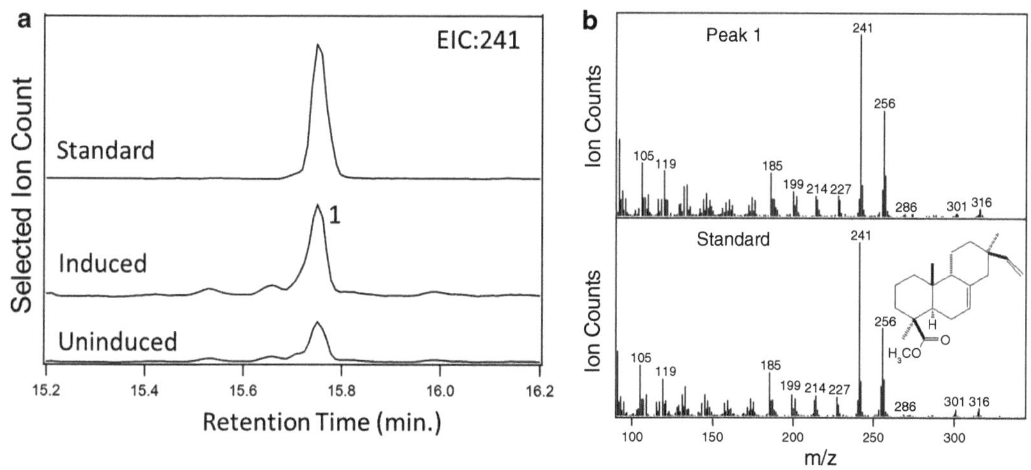

Fig. 3 Detection of *syn*-pimaradien-19-oic acid from rice (analyzed as methyl ester derivative). (**a**) Extracted Ion GC-MS chromatogram (m/z = 241) of authentic standard, methyl jasmonate-induced and uninduced fractionated rice leaf extract (from *top* to *bottom*). (**b**) MS spectrum of *syn*-pimaradien-19-oic acid methyl ester in rice extract (peak *1*) and of the corresponding authentic standard [10]. *Inset*, molecular structure of *syn-pimaradien-19*-oic acid as methyl ester

6. Load the extract on the HPLC column at a flow rate of 0.5 mL/min.
7. Wash the column with 20 % acetonitrile/dH_2O (0–2 min) and elute with 20–100 % acetonitrile (2–7 min), followed by a 100 % acetonitrile wash (7–27 min).
8. Collect fractions between 8 and 23 min of retention time over 1 min intervals using a detector wavelength of 220 nm.
9. Dry all collected fractions under N_2 and derivatize with 200 μL of diazomethane saturated hexane at room temperature in the dark for 2 h (*see* **Note 5**).
10. Dry the derivatization mixture under a N_2 stream and redissolve in 100 μL of hexane.
11. Inject 1 μL of sample in splitless mode on the GC column at 50 °C.
12. Apply the following GC-MS conditions: Helium carrier gas flow rate of 1.2 mL/min, a temperature gradient of 14 °C/min from 50 °C (3 min hold) to 300 °C (3 min hold); recording of mass spectrometry data from 90 to 600 m/z starting 12 min after injection. The *syn*-pimaradien-19-oic acid methyl ester is detected in fraction 9 corresponding to a retention time of 16–17 min (Fig. 3).

3.4 Analysis of 11α-Hydroxy-ent-Cassadiene in Rice

11α-Hydroxy-ent-cassadiene is an intermediate in phytocassane diterpene biosynthesis in rice [12]. Because of its low abundance in rice tissue, LC-MS/MS is used to analyze its presence in rice organic extracts. This protocol may be applied and modified for the analysis of diterpenes with one or more hydroxyl groups.

1. Uproot a rice plant and grind 2 g of whole plant tissue to a powder using liquid nitrogen.
2. Extract tissue with 50 mL of ethyl acetate by stirring overnight at room temperature in a 250 mL glass flask.
3. Clarify the mixture by centrifuging at 3,000×*g* (glass or Teflon tubes).
4. Separate the upper organic phase and dry under a stream of N_2 gas at room temperature.
5. Dissolve the residue in 0.5 mL of 45 % methanol/45 % acetonitrile/10 % deionized water and centrifuge again to remove undissolved material.
6. Load 200 μL of sample on the LC column at a flow rate of 0.5 mL/min.
7. Wash the column with 20 % acetonitrile/water (0–5 min) and elute with 20–100 % acetonitrile (5–15 min), followed by a 100 % acetonitrile wash (15–30 min). 11α-Hydroxy-ent-cassadiene elutes at a retention time of 21.5–22 min. The purified compound is easily detected and used to optimize selective MS/MS ion monitoring in the positive mode. Specifically, for 11a-hydroxy-entcassadiene, the base peak (m/z=271 $[MH–H_2O]^+$) from the molecular ion (m/z=289 $[MH]^+$) is selected for further MS/MS fragmentation, with the resulting mass spectra compared with the pure standard at the same retention time.

3.5 Triterpene Extraction and Analysis from Arabidopsis Roots

This protocol is used for the extraction of triterpenes and sterols from *Arabidopsis* tissues. The initial sample preparation steps include freeze-drying and pulverization of the sample. Upon saponification of the sample, the non-saponifiable lipids are extracted with hexane and finally analyzed using GC-MS. A specific example for the extraction of the triterpene diol, arabidiol [13, 14], is presented. Arabidiol is only detected in the *Arabidopsis cyp705a1-1* mutant, which is defective in arabidiol degradation leading to the accumulation of small amounts of arabidiol.

1. Treat axenically grown *Arabidopsis* roots with 100 μM jasmonic acid for 24 h (*see* **Note 2**).
2. Collect 1 g of fresh root sample and freeze in liquid nitrogen (*see* **Note 6**).
3. Grind tissue into a fine powder in the presence of liquid nitrogen.
4. Freeze-dry samples using a lyophilizer (*see* **Note 7**).
5. To saponify powdered tissue, transfer the dried powder to a screw cap flask and add 50 mL of ethanolic potassium hydroxide.

If derivatization is planned add betulin as internal standard before saponification. Incubate at 70 °C in a water bath for 2 h.

6. Mix by swirling every 30 min to ensure complete saponification.
7. After saponification, let the reaction mixture cool down to 30–40 °C.
8. Dilute saponified mixture with 50 mL of double-distilled water (*see* **Note 8**).
9. Add 50 mL of hexanes and mix vigorously. Upon phase separation transfer the organic phase to a new glass container. Repeat the extraction twice and collect the organic phase.
10. Concentrate the organic extract using a rotary evaporator at 60 °C. When the volume is down to 4–5 mL, transfer the extract to a clean glass vial and evaporate the solvent to dryness under a gentle stream of nitrogen. This extract contains non-saponifiable lipids including triterpenes and phytosterols. At this stage a small aliquot of sample can be analyzed using GC-MS with method parameters specified below (**step 14**). To ensure higher MS detector signals for hydroxylated compounds precede to the derivatization step (**steps 11–14**).
11. Use a small aliquot of extract (1–5 mg). Dry organic solvent (*see* **Note 9**).
12. Dissolve sample in 100 μL of pyridine (a catalyst for silylation reaction as well).
13. Incubate at 70 °C for 2 h or until derivatization is complete. Then, evaporate pyridine under a N_2 stream. Dissolve the sample in hexane.
14. Inject 1 μL of sample in splitless mode on the GC column at 50 °C. Use the following GC-MS conditions: Helium carrier gas flow rate of 1.22 mL/min, with a temperature gradient of 20 °C/min from 200 to 300 °C (15 min hold); recording of mass spectrometry data from 50 to 600 *m/z* starting 4 min after injection. Arabidiol is detected at retention time 17.6 only in the mutant *cyp705a1-1* background. Using this method, phytosterols including stigmasterol, campesterol, and sitosterol and other triterpenes such as thalianol [23] can be detected. Quantitative preparation of pure triterpenes for NMR analysis often requires the production of the compounds from yeast cultures expressing the respective oxidosqualene cyclase (triterpene synthase) gene (*see* **Note 10**).

4 Notes

1. Penetration of the vial septum and incubation of the fiber are performed automatically by the AOC-5000 Shimadzu autosampler. Alternatively, a handheld SPME device (Supelco) and

manual injection may be used. The incubation time may be increased if emission rates are low. The method can also be applied to other tissues such as leaves and flowers.

2. Alternatively, axenically grown root tissue or washed roots of plants grown in potting substrate can be used for extraction. A reference protocol for plant growth under axenic conditions is available [20].
3. Avoid any water contamination, which will negatively affect the chromatography on the silica gel. Standard protocols are available elsewhere [21, 22].
4. No preconditioning of the column is required.
5. Diazomethane is toxic and explosive and should therefore be handled with great care under the fume hood.
6. To avoid any possible postharvest biochemical changes, freeze tissue in liquid nitrogen immediately after collection.
7. Freeze drying can enhance the extractability of compounds using organic solvent. The procedure can be done overnight depending on the type of tissue being analyzed.
8. Alternatively, a brine solution can be used. To make brine solution, dissolve 36 g of NaCl in 100 mL of double-distilled water.
9. Use a N_2 stream to evaporate organic solvents. Residual water should be removed by lyophilization to improve the silylation reaction efficiency.
10. Detailed protocols exist for the production of an arabidiol standard [13, 14] and purification of triterpenes from yeast [14, 17].

Acknowledgement

This work was supported by a National Science Foundation grant MCB-0950865.

References

1. Bart H-J (2011) Extraction of natural products from plants—an introduction. In: Bart H-J, Pilz S (eds) Industrial scale natural products extraction. Wiley-VCH, Weinheim, Germany
2. Tholl D (2006) Terpene synthases and the regulation, diversity and biological roles of terpene metabolism. Curr Opin Plant Biol 9: 297–304
3. Brewer PB, Koltai H, Beveridge CA (2013) Diverse roles of strigolactones in plant development. Mol Plant 6:18–28
4. Chaturvedi R, Venables B, Petros RA, Nalam V, Li MY, Wang XM, Takemoto LJ, Shah J (2012) An abietane diterpenoid is a potent activator of systemic acquired resistance. Plant J 71:161–172
5. Erkan N, Tao Z, Rupasinghe HP, Uysal B, Oksal BS (2012) Antibacterial activities of essential oils extracted from leaves of *Murraya koenigii* by solvent-free microwave extraction and hydro-distillation. Nat Prod Commun 7:121–124

6. Fornari T, Vicente G, Vazquez E, Garcia-Risco MR, Reglero G (2012) Isolation of essential oil from different plants and herbs by supercritical fluid extraction. J Chromatogr A 1250:34–48
7. Tholl D, Boland W, Hansel A, Loreto F, Rose USR, Schnitzler JP (2006) Practical approaches to plant volatile analysis. Plant J 45:540–560
8. Schaub A, Blande JD, Graus M, Oksanen E, Holopainen JK, Hansel A (2010) Real-time monitoring of herbivore induced volatile emissions in the field. Physiol Plant 138:123–133
9. Magome H, Nomura T, Hanada A, Takeda-Kamiya N, Ohnishi T, Shinma Y, Katsumata T, Kawaide H, Kamiya Y, Yamaguchi S (2013) *CYP714B1* and *CYP714B2* encode gibberellin 13-oxidases that reduce gibberellin activity in rice. Proc Natl Acad Sci U S A 110:1947–52
10. Wang Q, Hillwig ML, Peters RJ (2011) CYP99A3: functional identification of a diterpene oxidase from the momilactone biosynthetic gene cluster in rice. Plant J 65:87–95
11. Vaughan MM, Wang Q, Webster FX, Kiemle D, Hong YJ, Tantillo DJ, Coates RM, Wray AT, Askew W, O'Donnell C, Tokuhisa JG, Tholl D (2013) Formation of the unusual volatile diterpene rhizathalene by the *Arabidopsis* class I terpene synthase TPS08 in the root stele is involved in defense against belowground herbivory. Plant Cell 25:1108–1125
12. Swaminathan S, Morrone D, Wang Q, Fulton DB, Peters RJ (2009) CYP76M7 is an ent-cassadiene C11α-hydroxylase defining a second multifunctional diterpenoid biosynthetic gene cluster in rice. Plant Cell 21:3315–3325
13. Xiang T, Shibuya M, Katsube Y, Tsutsumi T, Otsuka M, Zhang H, Masuda K, Ebizuka Y (2006) A new triterpene synthase from *Arabidopsis thaliana* produces a tricyclic triterpene with two hydroxyl groups. Org Lett 8:2835–2838
14. Kolesnikova MD, Obermeyer AC, Wilson WK, Lynch DA, Xiong Q, Matsuda SP (2007) Stereochemistry of water addition in triterpene synthesis: the structure of arabidiol. Org Lett 9:2183–2186
15. Wu H, Guo J, Chen S, Liu X, Zhou Y, Zhang X, Xu X (2013) Recent developments in qualitative and quantitative analysis of phytochemical constituents and their metabolites using liquid chromatography-mass spectrometry. J Pharm Biomed Anal 72:267–291
16. Gotti R (2011) Capillary electrophoresis of phytochemical substances in herbal drugs and medicinal plants. J Pharm Biomed Anal 55: 775–801
17. Morlacchi P, Wilson WK, Xiong Q, Bhaduri A, Sttivend D, Kolesnikova MD, Matsuda SP (2009) Product profile of PEN3: the last unexamined oxidosqualene cyclase in *Arabidopsis thaliana*. Org Lett 11:2627–2630
18. Harinantenaina L, Kasai R, Yamasaki K (2002) Cussosaponins A-E, triterpene saponins from the leaves of *Cussonia racemosa*, a Malagasy endemic plant. Chem Pharm Bull 50:1290–1293
19. Gibeaut DM, Hulett J, Cramer GR, Seemann JR (1997) Maximal biomass of *Arabidopsis thaliana* using a simple, low-maintenance hydroponic method and favorable environmental conditions. Plant Physiol 115:317–319
20. Hetu MF, Tremblay LJ, Lefebvre DD (2005) High root biomass production in anchored *Arabidopsis* plants grown in axenic sucrose supplemented liquid culture. Biotechniques 39:345–349
21. Leonard J, Lygo B, Procter G (2013) Advanced practical organic chemistry, 3rd edn. CRC, Boca Raton, FL
22. Still WC, Kahn M, Mitra A (1978) Rapid chromatographic technique for preparative separations with moderate resolution. J Org Chem 43:2923–2925
23. Field B, Osbourn AE (2008) Metabolic diversification-independent assembly of operon-like gene clusters in different plants. Science 320:543–547

Chapter 11

Gas Chromatography–Mass Spectrometry Method for Determination of Biogenic Volatile Organic Compounds Emitted by Plants

Astrid Kännaste, Lucian Copolovici, and Ülo Niinemets

Abstract

Gas chromatography–mass spectrometry (GC-MS) is one of the most widely used methods for analyzing the emissions of biogenic volatile organic compounds (VOCs) from plants. Preconcentration of VOCs on the cartridges filled with different adsorbents is a well-accepted method for sampling of headspace. Here, we describe a gas-chromatographic method for determination of different isoprenoids (isoprene, monoterpenes, homoterpenes, and sesquiterpenes). The technique is based on adsorption of compounds of interest on multibed adsorbent cartridges followed by thermodesorption, and detection and analysis by GC-MS.

Key words Cartridge sampling, Desorption, Gas chromatography, Green leaf volatiles, Isoprene, Mass spectrometry, Monoterpenes, Sesquiterpenes

1 Introduction

Plants odors are complex mixtures of chemically heterogeneous volatile organic compounds (VOCs). Plants can synthesize, store, and emit in physiological conditions more than 30,000 different compounds [1–4]. Common plant volatiles include various volatile lipoxygenase pathway products (LOX, also called green leaf volatiles, GLV), terpenes, phenylpropanoids, and/or benzenoids [5]. For example, a single plant species often emits more than 20 different monoterpenes [6, 7]. In addition, various sesqui- and oxygenated sesquiterpene alcohols, aldehydes, and ketone derivatives have been identified in the emissions from natural vegetation [8–10] and from crops [11, 12]. All of these compounds have widely differing chemical reactivity and ozone-forming potentials in the atmosphere [13, 14].

Simultaneous measurement of plant volatiles with different physicochemical characteristics, including differences in low temperature vapor pressure, partitioning in gas, liquid, and lipid phases

Manuel Rodríguez-Concepción (ed.), *Plant Isoprenoids: Methods and Protocols*, Methods in Molecular Biology, vol. 1153, DOI 10.1007/978-1-4939-0606-2_11, © Springer Science+Business Media New York 2014

(volatility), is an analytical challenge. Various methods such as solid phase microextraction [15], washing of plant leaves with solvents [16], sorption of VOCs into liquid coatings [17], or collecting the VOCs into coated capillary columns, multichannel tubes or solvent traps [17] have been developed for collecting volatile compounds prior to their analysis. Pre-concentration of the VOCs on solid adsorbents followed by thermal desorption [16, 18–20] has become one of the standard methods in field and laboratory studies [8, 21–23]. Nevertheless, the applicability of thermal desorption method has received much less attention [8, 23, 24], although true volatile profile of a plant may not always be obtained due to artifacts resulting from problems during sampling, cartridge storage, and desorption [16, 23, 25–27]. The most common adsorbents include diverse carbon-based adsorbents (graphitized carbon, and carbon molecular sieves such as Carbotrap®, Carbograph® Carboxen®) and polymeric adsorbents (e.g., Tenax®) that need to be carefully selected to correspond to the physicochemical characteristics of the compounds sampled [28], but also considering other aspects such as possible presence of water vapor and reactive gases such as ozone in the sampled gas stream (*see* ref. 23 for a review of adsorbent selection and sampling caveats). In the past, activated carbon was often used as an adsorbent, but it is not recommended any longer due to its high surface activity and because high temperatures needed for desorption of VOCs can cause their degradation [29].

The quantitative and qualitative detection of VOCs depends on various aspects of the compound adsorption and thermal desorption (TD) methods. Special attention must be paid on packing of tubes, avoiding contamination of tubes and standardization of desorption procedures. Additionally, calibration of GC-MS systems using sample tubes and thermal desorption procedures exactly as with the tubes containing gas samples is essential for quantitative assessment of plant VOC emissions. Hence, in the present chapter we characterize our adsorption and thermal desorption method for simultaneous determination of volatile products of lipoxygenase pathway (LOXs), isoprene, and various monoterpenes, sesquiterpenes, and homoterpenes in the headspace of plants. The method is based on preconcentration of VOCs on solid absorbents coupled with the gas chromatography mass-spectrometry with thermal desorption. Application of the outlined method has neem described in several primary research articles investigating volatiles released constitutively and in response to various biotic and abiotic stresses (e.g., refs. 30–34).

2 Materials

1. Electropolished stainless steel tubes (10.5 cm of length, 3 cm of inner diameter, Supelco) with Teflon® tops and bottoms were filled with Carbotrap C 20–40 mesh (0.2 g), Carbopack

B 40–60 mesh (0.1 g), and Carbotrap X 20–40 mesh (0.1 g) adsorbents (Supelco).

2. Standard compounds and solvents (Sigma-Aldrich) at highest purity (>98 %) were used for preparing calibration solutions.
3. ZB-624 capillary column 0.32 mm i.d. × 60 m, 1.8 μm film thickness Zebron (Phenomenex) was used for sample separation.

3 Methods

1. There is no adsorbent that can both retain and release all the plant key VOCs either due to too strong adsorption requiring excessive temperatures for desorption or due to breakthrough (*see* **Notes 1** and **2**). For this reason, the adsorption tubes were filled with three different solid adsorbents (multibed tubes).
2. Carbotrap C 20–40 mesh have Brunauer, Emmett, Teller (BET) surface area of 10 m^2/g, BET surface for Carbopack B 40–60 mesh is 100 m^2/g while Carbotrap X 20–40 mesh has a BET surface area 240 m^2/g (*see* **Note 1**). Carbotrap C is used for adsorption of compounds which have number of atoms (without H) between 12 and 20 (sesquiterpenes, homoterpenes), Carbopack B is used for adsorption of compounds which have number of atoms (without H) between 6 and 12 (LOX compounds, monoterpenes) and Carbotrap X is used for adsorption of compounds which have number of atoms (without H) between 3 and 9 (isoprene, LOX; *see* **Note 3**). The packing of stainless steel tubes (*see* **Note 4**) is finished by layers of glass wool at the top and bottom.
3. In the multi-bed adsorbent tubes, the adsorbents are arranged in order of increasing adsorbent strength, from sample inlet to sample outlet. The sampled VOCs with the highest molecular masses are trapped by the first adsorbent bed. Smaller molecules are trapped by the succeedingly stronger beds (*see* **Note 5**). Thus, the flow direction during sampling should be from the weakest towards the strongest adsorbent. During desorption, the carrier gas flow must be reversed, i.e., from the strongest to the weakest adsorbent.
4. Prior to sampling, the multi-bed tubes filled with adsorbents are conditioned by flowing He of purity 99.9999 % (He 6.0) at 40 ml/min through the tube for 1 h at 250 °C (*see* **Note 6**).
5. For calibrating the thermal desorption (TD) method, the calibration solutions are prepared by dissolving 50 μl of each commercially available reference compound at the highest purity available in 5 ml of GC grade hexane.

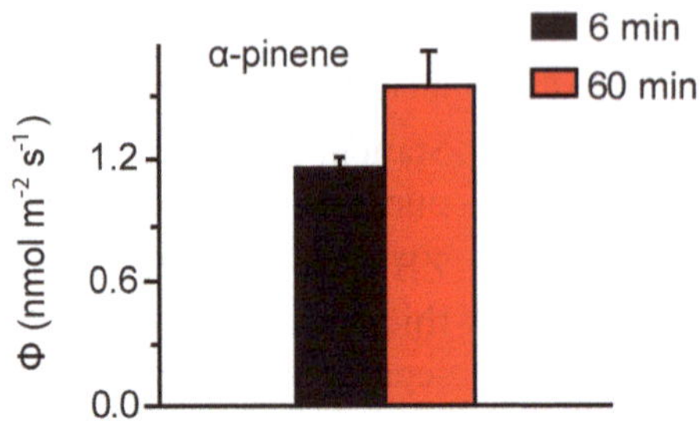

Fig. 1 Artifacts resulting from the discrepancy of desorption times during sample analysis and calibration in quantification of α-pinene emission rates (mean ± SE) from the needle emissions of the conifer *Pinus mugo*. The primary desorption times in the thermal desorber units were either 6 or 60 min for the sample, while the desorption time during calibration was 6 min (*see* **steps 5–7** of Subheading 3 for details of GC-MS system calibration). As this experiment demonstrates, not all α-pinene is released during the desorption time of 6 min, but consistent use of the same desorption time for sample and calibration tube results in correct estimation of the emission flux rate. Use of 60 min desorption time for the sample and 6 min for the calibration tube would have overestimated the emission rate (*see* **steps 7** and **10** of Subheading 3 and **Note 10**; unpublished observations of Kännaste, Copolovici, and Niinemets). The mean amounts of α-pinene estimated using different desorption times are statistically different ($P<0.01$, one-way ANOVA)

6. 1 μl of final sample of calibration solution (*see* **Note 7**) is injected into the tube filled with adsorbents. From each solution, at least three replicate tubes are prepared. The tube is connected to the pressurized nitrogen gas (99.999 % purity) bottle and nitrogen gas is passed at 300 ml/min through the tube for 5 min in order to evaporate the solvent and trap the volatiles on the adsorbent. Finally, the tube is ready to be analyzed in GC-MS.
7. During the analysis, the same desorption time must be used for the sample and calibration tubes (*see* **step 10** of Subheading 3 and Fig. 1).
8. While collecting the VOCs of plants, attention must be paid on the tubing and connection materials used for connecting different parts of the plant gas exchange systems, both outside and inside the measurement console. It is recommended to replace standard tubing of commercially available gas-exchange systems used for CO_2 and H_2O exchange measurements (e.g., PVC tubes) with more inert fluorinated hydrocarbon tubing such as Teflon (*see* **Note 8**).
9. After the analysis of plant volatiles, the tubes must be reconditioned (Fig. 2, *see* **step 4** of Subheading 3 and **Note 9**).
10. Two-stage desorption is used. He purge flow is set to 40 ml/min, primary desorption temperature to 250 °C, and primary

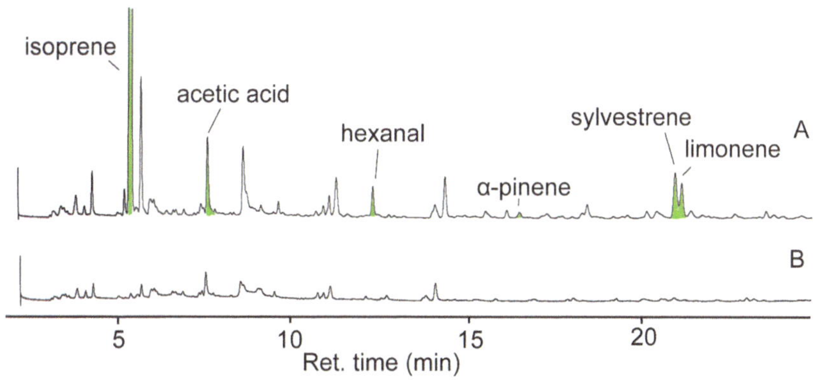

Fig. 2 Chromatograms demonstrating the backgrounds for a non-conditioned tube (**a**) and a conditioned tube (**b**) (*see* **steps 4** and **9** of Subheading 3 and **Note 9**). In (**a**), plant volatiles that will potentially disturb the quantification of plant compounds in the subsequent sample are colored in *green*. In (**a**), the tube was analyzed immediately after the first cycle of desorption (6 min desorption time), while in (**b**), the tube was reconditioned as stated in **step 9** of Subheading 3 (1 h at 250 °C with 40 ml/min He flow through the tube; unpublished observations of Kännaste, Copolovici, and Niinemets)

desorption time to 6 min (*see* **Note 10**). Second stage trap temperature during primary desorption is set to −20 °C, and second stage trap desorption temperature to 280 °C with hold time of 6 min (*see* **Notes 11** and **12**).

11. GC oven program used for separation of the compounds is as follows: 40 °C for 1 min, 9 °C/min to 120 °C, 2 °C/min to 190 °C, 20 °C/min to 250 °C, and 250 °C for 5 min.
12. The mass spectrometer of GC-MS instrument is operated in electron-impact mode (EI) at 70 eV, in the scan range of *m/z* 30–400. The transfer line temperature is set at 240 °C and ion-source temperature at 150 °C.

Compounds are identified by the NIST library of mass spectra and based on retention time identity with the authentic standards (GC purity). The absolute concentrations of individual compounds are determined using the calibration coefficients determined as described in **steps 5–7** of Subheading 3.

4 Notes

1. Traps packed with solid adsorbents should retain the largest number of VOCs ranging from C3 to C20 present in 5 L air samples collected at ambient temperatures between 10 and 35 °C in field conditions (typically 20–25 °C for lab sampling). At the same time, the adsorbents must quantitatively release all VOCs at temperatures that prevent possible decomposition of VOCs.

2. If the choice of the adsorbent is not correct, the measurements are not quantitative due to partial losses of some compounds which cannot be adsorbed and/or due to too strong adsorption of other compounds that cannot be released from the adsorbents at temperatures below the compound thermal destruction.
3. The mesh size of adsorbents for tubes with small inner diameter is generally less or equal to 40–60 mesh to reduce backpressure and avoid flow reduction through the tube, while for tubes with greater diameter, the mesh size of adsorbents can be larger. Excessive pressure during sampling can result in more wear on the sampling pump, and can also lead to bulk flow leaks.
4. Glass cartridges perform similarly to the stainless steel cartridges, and have the advantage that the packing quality can be visually better assessed, but they are more fragile and therefore more difficult to handle in the field. In rare cases, we have had glass cartridges broken in the desorber unit due to the failure of the autosampler.
5. If the first adsorbent adsorbs the smallest molecules, the absorption of higher molecules is disturbed due to irreversible adsorption of the smaller molecules, which is particularly problematic for repeated use of adsorbent tubes.
6. The purity of the gas must be at least 99.999 %. Otherwise, gas impurities will contaminate the adsorbents in the tube. Depending on the concentration of impurities and sampled VOCs, gas impurities adsorbed in the tube will interfere with the identification and/or the measurement of the peak areas of compounds. It is not recommended to use compressed air prepared from ambient air by an air compressor. In addition to impurities, ambient air can contain significant amounts of water, which can generate problems with compound sampling by filling adsorption sites, in particular for stronger adsorbents, but also can lead to difficulties with thermal desorption (clogging cryotraps).
7. For calibrating the GC-MS method, the volume of 1 μl of calibration sample is enough for injectin into the multi-bed tube. If too much of the calibration sample is injected into the tube, it will take longer to fully evaporate the solvent in the tube. The time period for gas flow through the tube must also not be too long. Otherwise lighter VOCs are partially evaporated together with the solvent. However, if the flow time is not enough to fully evaporate the solvent, the solvent residues will disturb identification of VOCs.
8. The connections used in plant gas-exchange systems should have low adsorption capacity for plant VOCs. For instance, PVC (polyvinylchloride), or especially silicon or rubber tubes can absorb and release volatiles from the ambient air, leading

to memory effects. Hence during the collection of VOCs, the volatiles released from the surface of tubing will contaminate the tubes of adsorbents (*see* **step 4** of Subheading 3). Fluorinated hydrocarbons or stainless steel or glass are the recommended materials, but the seals should be carefully made to avoid bulk flow leaks [23].

9. After the tube is analyzed, it must be reconditioned by letting the gas to flow at least for 1 h through the tube. Otherwise the VOCs of previous sample can partly disturb (depending on VOC concentrations in the sample) identification and analysis of the VOCs in the following samples (Fig. 2).
10. During the analysis, use of different primary desorption times can cause considerable differences among the amounts of volatiles detected (Fig. 1). Therefore, it is particularly important to use the same desorption times for calibration and sample tubes.
11. The cooling temperature of the TD is suggested to be as low as possible to best cryofocus the VOCs into the GC column and to avoid broad peaks. Modern TD systems usually can cool down the cryotrap temperatures to lower than -20 °C [35].
12. Excessively high temperature of desorption used can cause formation of artifacts. For example, heat-catalyzed isomerization of (*Z*)-*β*-ocimene [36] can lead to formation of allo-ocimene, a thermal desorption artifact [16].

References

1. Borg-Karlson A-K, Eidmann HH, Lindström M, Norin T, Wiersma N (1985) Odoriferous compounds from the flowers of the conifers *Picea abies*, *Pinus sylvestris* and *Larix sibirica*. Phytochemistry 24(3):455–456
2. Banthorpe DV (1991) Classification of terpenoids and general procedures for their characterisation. In: Charlwood BV, Banthorpe DV, Harborne JB (eds) Terpenoids. Methods in plant biochemistry, vol 7. Academic, London, pp 1–41
3. Peñuelas J, Llusià J, Estiarte M (1995) Terpenoids: a plant language. Trends Ecol Evol 10(7):289
4. Ormeño E, Goldstein A, Niinemets Ü (2011) Extracting and trapping biogenic volatile organic compounds stored in plant species. Trends Anal Chem 30:978–989
5. Dudareva N, Negre F, Nagegowda DA, Orlova I (2006) Plant volatiles: recent advances and future perspectives. Crit Rev Plant Sci 25(5):417–440
6. Niinemets Ü, Reichstein M (2002) A model analysis of the effects of nonspecific monoterpenoid storage in leaf tissues on emission kinetics and composition in Mediterranean sclerophyllous *Quercus* species. Glob Biogeochem Cycle 16(1):1110. doi:1110.1029/2002GB001927
7. Theis N, Lerdau M (2003) The evolution of function in plant secondary metabolites. Int J Plant Sci 164(3):S93–S102
8. Ortega J, Helmig D (2008) Approaches for quantifying reactive and low-volatility biogenic organic compound emissions by vegetation enclosure techniques – part A. Chemosphere 72(3):343–364. doi:10.1016/j.chemosphere.2007.11.020
9. Bouvier-Brown NC, Holzinger R, Palitzsch K, Goldstein AH (2009) Large emissions of sesquiterpenes and methyl chavicol quantified from branch enclosure measurements. Atmos Environ 43:389–401
10. Sakulyanontvittaya T, Duhl T, Wiedinmyer C, Helmig D, Matsunaga S, Potosnak M, Milford J, Guenther A (2008) Monoterpene and sesquiterpene emission estimates for the United States. Environmental Science & Technology 42(5):1623–1629. doi:10.1021/es702274e

11. Bartelt RJ, Wicklow DT (1999) Volatiles from *Fusarium verticillioides* (Sacc.) Nirenb. and their attractiveness to nitidulid beetles. J Agric Food Chem 47(6):2447–2454
12. Ciccioli P, Brancaleoni E, Frattoni M, Di Palo V, Valentini R, Tirone G, Seufert G, Bertin N, Hansen U, Csiky O, Lenz R, Sharma M (1999) Emission of reactive terpene compounds from orange orchards and their removal by within-canopy processes. J Geophys Res 104(D7):8077–8094
13. Arneth A, Niinemets Ü (2010) Induced BVOCs: how to bug our models? Trends Plant Sci 15:118–125
14. Fuentes JD, Lerdau M, Atkinson R, Baldocchi D, Bottenheim JW, Ciccioli P, Lamb B, Geron C, Gu L, Guenther A, Sharkey TD, Stockwell W (2000) Biogenic hydrocarbons in the atmospheric boundary layer: a review. Bull Am Meteorol Soc 81(7):1537–1575
15. Cavalli JF, Fernandez X, Lizzani-Cuvelier L, Loiseau AM (2003) Comparison of static headspace, headspace solid phase microextraction, headspace sorptive extraction, and direct thermal desorption techniques on chemical composition of French olive oils. J Agric Food Chem 51(26):7709–7716
16. Griffiths DW, Robertson GW, Birch ANE, Brennan RM (1999) Evaluation of thermal desorption and solvent elution combined with polymer entrainment for the analysis of volatiles released by leaves from midge (*Dasineura tetensi*) resistant and susceptible blackcurrant (*Ribes nigrum* L.) cultivars. Phytochem Anal 10(6):328–334
17. Pillonel L, Bossett JO, Tabacchi R (2002) Rapid preconcentration and enrichment techniques for the analysis of food volatiles. A review. Food Science and Technology 35(1):1–14
18. Agelopoulos NG, Pickett JA (1998) Headspace analysis in chemical ecology: effects of different sampling methods on ratios of volatile compounds present in headspace samples. J Chem Ecol 24(7):1161–1172
19. Agelopoulos NG, Hooper AM, Maniar SP, Pickett JA, Wadhams LJ (1999) A novel approach for isolation of volatile chemicals released by individual leaves of a plant in situ. J Chem Ecol 25(6):1411–1425
20. Brancaleoni E, Scovaventi M, Frattoni M, Mabilia R, Ciccioli P (1999) Novel family of multi-layer cartridges filled with a new carbon adsorbent for the quantitative determination of volatile organic compounds in the atmosphere. J Chromatogr A 845(1–2):317–328
21. Helmig D, Bocquet F, Pollmann J, Revermann T (2004) Analytical techniques for sesquiterpene emission rate studies in vegetation enclosure experiments. Atmos Environ 38: 557–572
22. Tholl D, Boland W, Hansel A, Loreto F, Rose USR, Schnitzler J-P (2006) Practical approaches to plant volatile analysis. Plant J 45(4):540–560
23. Niinemets Ü, Kuhn U, Harley PC, Staudt M, Arneth A, Cescatti A, Ciccioli P, Copolovici L, Geron C, Guenther AB, Kesselmeier J, Lerdau MT, Monson RK, Peñuelas J (2011) Estimations of isoprenoid emission capacity from enclosure studies: measurements, data processing, quality and standardized measurement protocols. Biogeosciences 8: 2209–2246
24. Ortega J, Helmig D, Daly RW, Tanner DM, Guenther AB, Herrick JD (2008) Approaches for quantifying reactive and low-volatility biogenic organic compound emissions by vegetation enclosure techniques – part B: applications. Chemosphere 72:365–380
25. Helmig D, Greenberg J (1995) Artifact formation from the use of potassium-iodide-based ozone traps during atmospheric sampling of trace organic gases. J High Resolut Chromatogr 18:15–18
26. Calogirou A, Larsen BR, Brussol C, Duane M, Kotzias D (1996) Decomposition of terpenes by ozone during sampling on Tenax. Anal Chem 68:1499–1506
27. Coeur C, Jacob V, Denis I, Foster P (1997) Decomposition of α-pinene and sabinene on solid sorbents, Tenax TA and Carboxen. J Chromatogr A 786:185–187
28. Harper M (2000) Sorbent trapping of volatile organic compounds from air. J Chromatogr A 885:129–151
29. Hori H, Tanaka I, Akiyama T (1989) Thermal desorption efficiencies of two-component organic solvents from activated carbon. Am Ind Hygiene Assoc J 50:24–29
30. Copolovici L, Kännaste A, Niinemets Ü (2009) Gas chromatography-mass spectrometry method for determination of monoterpene and sesquiterpene emissions from stressed plants. Studia Universitatis Babes-Bolyai, Chemia 54:329–339
31. Toome M, Randjärv P, Copolovici L, Niinemets Ü, Heinsoo K, Luik A, Noe SM (2010) Leaf rust induced volatile organic compounds signalling in willow during the infection. Planta 232:235–243
32. Copolovici L, Kännaste A, Pazouki L, Niinemets Ü (2012) Emissions of green leaf volatiles and terpenoids from *Solanum lycopersicum* are quantitatively related to the severity

of cold and heat shock treatments. J Plant Physiol 169:664–672

33. Opriş O, Copaciu F, Soran ML, Ristoiu D, Niinemets Ü, Copolovici L (2013) Influence of nine antibiotics on key secondary metabolites and physiological characteristics in *Triticum aestivum*: leaf volatiles as a promising new tool to assess toxicity. Ecotoxicol Environ Safety 87:70–79
34. Niinemets Ü, Copolovici L, Hüve K (2010) High within-canopy variation in isoprene emission potentials in temperate trees: implications for predicting canopy-scale isoprene fluxes. J Geophys Res Biogeosci 115, G04029. doi:10.1029/2010JG001436
35. Ciccioli P, Brancaleoni E, Frattoni M (2002) Sampling of atmospheric volatile organic compounds (VOCs) with sorbent tubes and their analysis by GC-MS. In: Burden FR, Kelvie IM, Forstner U, Guenther A (eds) Environmental monitoring handbook. McGraw-Hill, New York, pp 21.21–21.85
36. He J, Xie M, Tang X, Qi X (2012) Kinetic and mechanistic study on the thermal isomerization of ocimene in the liquid phase. J Phys Org Chem 25:373–378

Chapter 12

Analysis of Steroidal Alkaloids and Saponins in *Solanaceae* Plant Extracts Using UPLC-qTOF Mass Spectrometry

Uwe Heinig and Asaph Aharoni

Abstract

Plants of the *Solanaceae* family are renowned for the production of cholesterol-derived steroidal glycosides, including the nitrogen containing glycoalkaloids and steroidal saponins. In this chapter we describe the use of UPLC (Ultra Performance Liquid Chromatography) coupled with qTOF (Quadrupole Time-of-Flight) mass spectrometry for profiling of these two large classes of semipolar metabolites. The presented method includes an optimized sample preparation protocol, a procedure for high resolution chromatographic separation and metabolite detection using the TOF mass spectrometer which provides high resolution and mass accuracy. A detailed description for non-targeted data analysis and a strategy for putative identification of steroidal glycosides from complex extracts based on interpretation of mass fragmentation patterns is also provided. The described methodology allows profiling and putative identification of multiple steroidal glycoside compounds from the assortment of *Solanaceae* species producing these molecules.

Key words UPLC/qTOF mass spectrometry, *Solanaceae*, Steroidal alkaloids, Steroidal saponins, Metabolomics

1 Introduction

Plants of the *Solanaceae* family are well known for the production of numerous natural products, of which sapogenins are one major compound class. Sapogenins are plant terpenoid secondary metabolites, derived from the cytosolic mevalonic acid pathway. Two molecules of farnesyldiphosphate (FPP), formed from the universal terpene precursors isopentenyldiphosphate (IPP) and dimethylallyldiphosphate (DMAPP), are condensed to squalene followed by oxidation toward the precursor for all phytosteroids, 2,3-oxidosqualene. Oxidosqualene is then either converted to triterpenes (C30), such as amyrins or to lanosterol the precursor of cholesterol (C27) [1].

Manuel Rodríguez-Concepción (ed.), *Plant Isoprenoids: Methods and Protocols*, Methods in Molecular Biology, vol. 1153, DOI 10.1007/978-1-4939-0606-2_12,

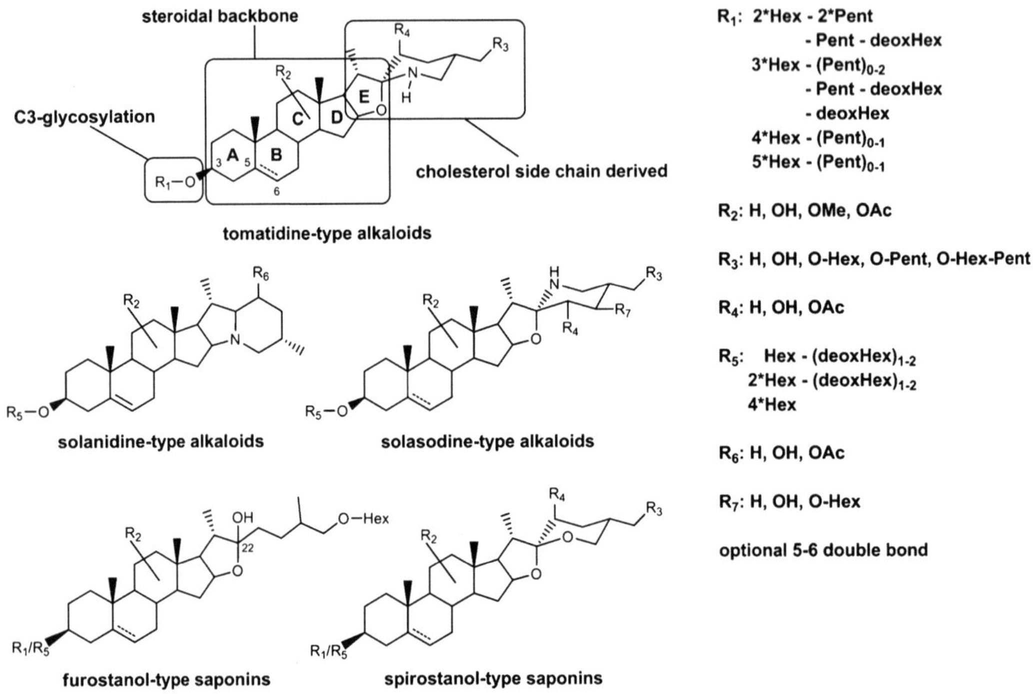

Fig. 1 Diversity of penta-cyclic steroidal saponins and alkaloids. All molecules of these two classes generally consist of a penta-cyclic steroidal backbone, a modified side chain (cyclic, open-chain, nitrogen-containing in case of alkaloids) originated from the side chain of cholesterol and are highly glycosylated at the hydroxyl-group at C3 ($R_{1,5}$). Further substitutions like hydroxylations, acetylations, and O-glycosylations can occur on the backbone or more likely on the side-chain ($R_{2,3,4,6,7}$). *Hex* hexose, *Pent* pentose, *deoxHex* deoxyhexose

In *Solanaceae* species, including tomato (*Solanum lycopersicum*), potato (*Solanum tuberosum*), eggplant (*Solanum melongena*), or black nightshade (*Solanum nigrum*), the predominant semipolar compounds detected are cholesterol-derived steroidal glycoalkaloids and steroidal saponins [2, 3]. Their biosynthesis starting from cholesterol involves hydroxylations, oxidations and in case of the alkaloids a transamination step, followed by extensive glycosylation, resulting in the observed diversity of detected end products. In Fig. 1, the five major aglycone structures observed in the *Solanaceae* are presented: tomatidine, solasodine and solanine (alkaloids) and furostanol and spirostanol (saponins). These carbon backbones can be substituted in different positions (R_{1-7} in Fig. 1) [2, 4–6].

By and large *Solanaceae* steroidal glycosides consist of three major structural parts: (a) a penta-cyclic steroidal backbone (rings marked with A–E in Fig. 1) that can contain a double bond at C5, (b) a modified side-chain derived from C20–C27 cholesterol side-chain, and (c) a glycosyl-moiety at the C3 hydroxyl-group (upper structure in Fig. 1). The occurrence of specific compounds thereby depends strongly on the plant species, the analyzed tissue and the developmental stage (e.g., fruit maturation in tomato [4, 6]).

In this chapter we describe a UPLC/qTOF-MS method for simultaneous chromatographic separation and mass spectrometric detection of the major steroidal alkaloids and saponins found in the *Solanaceae*. Besides sample preparation and chromatography, we focused here on the putative identification of compounds using a non-targeted approach based on accurate mass measurements and interpretation of mass fragmentation patterns.

2 Materials

2.1 Material and Reagents for Sample Preparation

1. Water, double deionized, from a Milli-Q purification system (Millipore), resistivity 18.2 MΩ cm, filtered through a 0.22-μm membrane filter.
2. Methanol, gradient grade for liquid chromatography (e.g., Merck KGaA).
3. Acetonitrile, gradient grade for liquid chromatography (e.g., Merck KGaA).
4. Liquid nitrogen for grinding and freezing plant samples.
5. Standards for QC (quality control) samples: L–Tryptophan (Sigma), L–Phenylalanine (Sigma), Chlorogenic acid (Fluka), Caffeic acid (Sigma), *p*-Coumaric acid (Sigma), Ferulic acid (Aldrich), Sinapic acid (Sigma), Rutin hydrate (Sigma), Quercetin dihydrate (Sigma), Tomatine (Apin), Naringenin (Fluka), Kaempferol (Fluka) (*see* **Note 1**).
6. Mortar and pestle or a ball mill (e.g., Retsch MM301).
7. Balance.
8. Screw-cap Polypropylene (PP) tubes (15 mL, e.g., Greiner) or 2 mL PP safe-lock eppendorf tubes for storage of tissue and extraction, suitable for centrifugation at 3,000 × *g* and 15,000 × *g*, respectively.
9. Ultrasonic bath.
10. Vortex.
11. Single-use sterile latex-free 1 mL syringes (e.g., 0.5 mm × 16 mm syringe, BD Plastipak™).
12. Single-use, 0.22 μm membrane syringe filters (e.g., 4 mm diameter PVDF (Polyvinylidenefluoride, Millex-GV) or 12 mm diameter PTFE (Polytetrafluoroethylene, PALL).
13. Amber-glass 2 mL autosampler vials and caps with a PTFE/Silicone septum. Use suitable 250 μl glass inserts for small extract volumes injection (*see* **Note 2**).

2.2 UPLC-qTOF and Data Analysis

1. UPLC/qTOF system: e.g., a UPLC Waters Acquity instrument connected in-line to a Synapt HDMS detector (tandem quadrupole/time-of-flight mass spectrometer).

The MS detector is equipped with an electrospray ion source (ESI). The Synapt HDMS system is operated in the standard qTOF mode, without using the ion mobility capabilities (*see* **Note 3**).

2. UPLC BEH C18 column (Waters Acquity), 100 × 2.1 mm i.d., 1.7 μm, with a column pre-filter.
3. Solvents for chromatography: mobile phase A, 5 % acetonitrile (ACN) in ddH_2O + 0.1 % formic acid (FA); mobile phase B, ACN + 0.1 % FA.
4. Washing solutions: strong needle wash, 80 % methanol (to remove organic components); weak needle wash, 5 % acetonitrile (re-equilibration to chromatographic starting conditions); seal wash, 10 % methanol.
5. MassLynx 4.1 instrument software (Waters).
6. DataBridge program (Waters).
7. XCMS package for R for peak picking and retention time correction between replicate samples [7].
8. Software for mass peak clustering and further statistical analysis, e.g., Principal component analysis (PCA).

3 Methods

3.1 Sample Preparation

The most widely used method for extraction of semipolar compounds for LC-MS analysis, such as glycoalkaloids and saponins, is sample preparation using acidic methanol [8, 9]. A final methanol content of 75 and 0.1 % formic acid (FA) was found to be a most efficient extraction solution when applied in various plant species and different tissues (*see* **Note 4**). A detailed description of the sample preparation procedure for tomato can also be found in [10] and is carried out as follows:

1. Homogenize frozen tissue using mortar and pestle pre-cooled in liquid nitrogen, or when starting with limited amount of material add a metal ball into the 2 mL eppendorf tube which contains the tissue, place the tubes into the pre-cooled ball mill tray and disrupt the material by shaking with 20 Hz for 2 min. After shaking, immediately place the tray into liquid nitrogen. Make sure that samples do not begin to thaw during the whole procedure.
2. Prepare appropriate number of clean tubes (15 or 2 mL), pre-cool in liquid nitrogen and weigh them on a balance.
3. Transfer the powder from **step 1** to 15 mL PP screw-cap tube (>350 mg of material) or in case of using less than 350 mg into a 2 mL eppendorf tube (*see* **Note 5**).
4. Weigh the biomaterial containing tubes and calculate the weight of the samples. In case of using the ball mill, determine the weight before adding the metal ball and grinding.

5. Add the required amount of acidified MeOH or MeOH/water (ratio *see* **Note 4**) to the frozen samples, to a final MeOH concentration of 75 % (*see* **Note 6**).
6. Vortex until all frozen powder is completely re-suspended in the solvent.
7. Sonicate for 20 min.
8. Vortex for several seconds.
9. Remove debris by centrifugation for 10 min with 3,000 × *g* when using 15 mL tubes or 15,000 × *g* in case of using 2 mL eppendorf tubes.
10. Carefully transfer the supernatant into a fresh tube.
11. Filter extract through a 0.22 μm PTFE (or PVDF) disposable filter into either another 2 mL eppendorf tube or directly into an autosampler vial using a 1 mL syringe. When volume is limited (less than 500 μL), filter an aliquot of the supernatant into an autosampler vial insert (approximately 150 μL). Use not less than 50 μL and make sure that no air bubbles remain at the bottom of the insert. Close vials tightly with PTFE/Silicone septum-caps to avoid evaporation of solvent. In case of storage of the extracts for later injection at −20 °C, check if precipitates are formed and eventually repeat filtering directly before injection.
12. As a blank sample use 75 % MeOH/ddH_2O supplemented with 0.1 % FA.
13. Place vials for injection into the UPLC autosampler, cooled to 12 °C.
14. The remaining extract (e.g., for MS/MS analysis) should be stored at −20 °C immediately.

3.2 UPLC/qTOF-MS Analysis

Chromatographic conditions are chosen according to the amphiphilic nature of the analytes (glycosylated cholesterol-derived molecules). The polarity during chromatographic separation is thereby mainly determined by the sugar moieties, hence, gradient elution starting with 5 % ACN in ddH_2O supplemented with 0.1 % FA to 31.6 % ACN (= 28 % B) over 22 min was found to give good separation and peak shape [10]. Steroidal alkaloids and saponins elute typically between 10 min and the end of the chromatogram. Thereby they are well separated from more polar compounds, like for example organic acids which elute earlier.

Column preparation, system stability testing and chromatographic conditions for analysis are listed below.

1. Equilibrate new columns in 50 % A/50 % B for at least 60 min, followed by equilibration to chromatographic starting conditions (5–10 column volumes A). Columns already used and stored in 50 % A/50 % B are equilibrated with at least 5 column volumes of A.

2. Performance of the system (stability of retention times, signal intensities, mass accuracy) can be tested by several injections of standard QC-mix (*see* **Notes 1** and 7).
3. UPLC conditions: column oven, 35 °C; flow rate, 0.3 mL/min; autosampler temperature, 12 °C; gradient, initial conditions 100 % A, 22 min 72 % A, 22.5 min 60 % A, 23 min 0 % A, 26.5 min 0 % A, 27 min 100 % A, 28 min 100 % A.
4. Inject 4–10 μL of biological sample (*see* **Note 8**).
5. Wash needle between injections with 200 μL strong needle wash solution and 600 μL weak needle wash (*see* **item 4** of Subheading 2.2) solution.
6. For longer data-sets, the QC-mix should be injected at least every ten samples to control the stability of the system (*see* **step 2** of Subheading 3.2).

For mass spectrometric detection, we use the Synapt HDMS detector equipped with an ESI source. "Soft" ESI ionization allows detection of intact metabolites (molecular ions), necessary for calculation of elemental composition and putative identification in a non-targeted manner. The mass detector part (Time of Flight, TOF) is operated in V mode with a mass resolution of 9,000. Spectra are recorded from *m/z* 50 to 1,500 with scan duration of 0.25 s and an interscan delay of 0.02 s in centroid mode. Mass spectrometer parameters are set to: capillary voltage = 3.4 kV (ESI-: 3 kV), cone voltage = 24 eV (ESI-: 28 eV), source temperature = 125 °C, desolvation temperature = 275 °C, cone gas flow = 25 L/h, desolvation gas flow = 650 L/h and collision energy = 4 eV. Argon is used as a collision gas and the mass spectrometer is calibrated with leucine enkaphalin ($[M+H]^+$: *m/z* 556.2771, $[M-H]^-$: *m/z* 554.2620). In a second channel the collision energy is ramped from 10 to 30 eV in the positive mode and from 15 to 35 eV in the negative mode. For MS/MS experiments, product ion spectra of selected masses are recorded with various collision energies (15–50 eV) with scan duration of 0.4 s and an interscan delay of 0.02 s in the same mass range.

3.3 Data Analysis Workflow

Data analysis is performed in two different ways. Compounds like for example α-tomatine that are commercially available can be identified directly by comparison with an authentic standard. However most of the metabolites formed are not known and have to be putatively assigned in a non-targeted manner. Therefore, information obtained from the known compounds, like polarity hence retention on the column, ionization behavior, and especially compound class specific mass fragmentation are necessary for putative identification. Non-targeted data analysis involves the following steps:

1. Control system stability manually by comparison of QC-control injections at the beginning, during, and at the end of the acquisition sequence (*see* **Note 9**).
2. Convert raw data files to NetCDF format using the MassLynx Databridge program and organize files into sample groups (replicate groups, e.g., Species 1, Species 2). Treat data acquired in positive and negative ionization mode separately. For further analysis use only NetCDF files created from the first channel of raw data files (first channel: mass data collision energy = 4 eV; second channel: mass data collision energy ramp; third channel: lock mass calibration; fourth channel: PDA detector, absorption spectra recorded from 210 to 500 nm).
3. Run mass peak detection and retention time correction with XCMS package for R [7]. Parameters thereby depend on the performance of chromatography as well as mass spectrometer specifications (*see* **Note 10**).
4. Perform quality control for XCMS using the MetaboQC program [11] or manually. XCMS creates different types of outputs, in our case a peak table including all detected mass traces, integrated intensity values and a number of extracted ion chromatograms (comparison between two replicate groups) that also show the borders for integration (*see* **Note 11**).
5. Perform mass peak clustering using software such as "CAMERA" for XCMS [12], MZ-mine [13], or Metalign [14].
6. Filter and sort the output data. Set for example intensity thresholds or sort according to retention time, mass or intensity.
7. Identify masses of peak groups/clusters (= chromatographic peaks) in raw data file.
8. Identify molecular ions of compounds; compare therefore positive and negative ionization datasets in order to avoid selection of the wrong ion, due to for example neutral losses during ionization or formation of adducts.
9. Calculate elemental composition of compounds using the elemental composition calculator of MassLynx 4.1. Molecular formulas are determined using accurate mass and isotopic pattern.
10. Search databases, e.g., the Natural product database or Scifinder for possible structures.
11. Verify putative structures by analysis of mass fragmentation patterns obtained by "collision energy ramp" or MS/MS experiments.

3.4 Identification of Steroidal Alkaloids and Saponins

Whereas the general procedure described above is especially useful to identify differences between samples according to all masses detected in the analysis, putative identification of steroidal alkaloids and saponins has to be done according to their mass spectral characteristics (Fig. 2). Thereby, the procedure can be divided into several steps, including (a) identification of the molecular ions of present compounds, (b) analysis of the glycosylation pattern, (c) analysis of substitution of the aglycones by determination of specific neutral losses and the fragmentation of the aglycones in order to distinguish between different types of saponins and alkaloids as presented in Fig. 3. We suggest doing the analysis in the following manner:

1. Identify molecular ions of compounds by comparison between the mass peak clustering analysis and raw data. It is highly recommended to compare data-sets obtained by positive and negative ionization. Steroidal glycoalkaloids tend to form formic acid adducts in the negative ionization mode ($[M-H+46]^-$), due the presence of formic acid in the solvent. In positive mode $[M+H]^+$ is the most commonly detected ion, although in case of substitutions on the aglycone [M+H-substituent]$^+$, e.g., minus H_2O or acetate, are observed, too. Steroidal saponins of the furostanol-type loose easily their 22-OH group resulting in exclusive detection of $[M+H-H_2O]^+$ in the positive mode, whereas $[M-H]^-$, and in minor amounts the FA adducts are observed, in the negative mode (Fig. 2). In low energy, doubly charged ions can occur ($M/2+H^+$).
2. Divide compounds into putative alkaloids (even mass, compounds contain one nitrogen atom) and saponins (odd mass, no nitrogen atom in the molecule) (*see* **Note 12**).
3. Filter detected compound ions for masses higher than the mass of possible aglycones, e.g., *m/z* 413 and higher for steroidal saponins (*see* **Note 13**).
4. Perform MS/MS analysis for all detected putative steroidal glycosides (alkaloids and saponins) using different collision energies in order to fragment only the glycosyl moieties (15–30 eV) or the entire molecule including the aglycone (40–50 eV). Especially for the analysis of neutral losses from the aglycone and glycosylation pattern, data obtained by applying the collision energy ramp (*see* Subheading 3.2) can be used (*see* **Note 14**).
5. Analyze glycosylation pattern of compounds. Typically steroidal glycosides contain up to seven glycosyl residues, which are mainly hexoses (galactosyl-, glucosyl-, loss of 162 Da), pentoses (xylosyl-, loss of 132 Da) or deoxyhexoses (rhamnosyl-, loss of 146 Da; Figs. 1 and 2d). In rare cases, other substituents occur (*see* **Note 15**). In the example shown in Fig. 2d, the glycosylation pattern of α-tomatine from *Solanum lycopersicum*

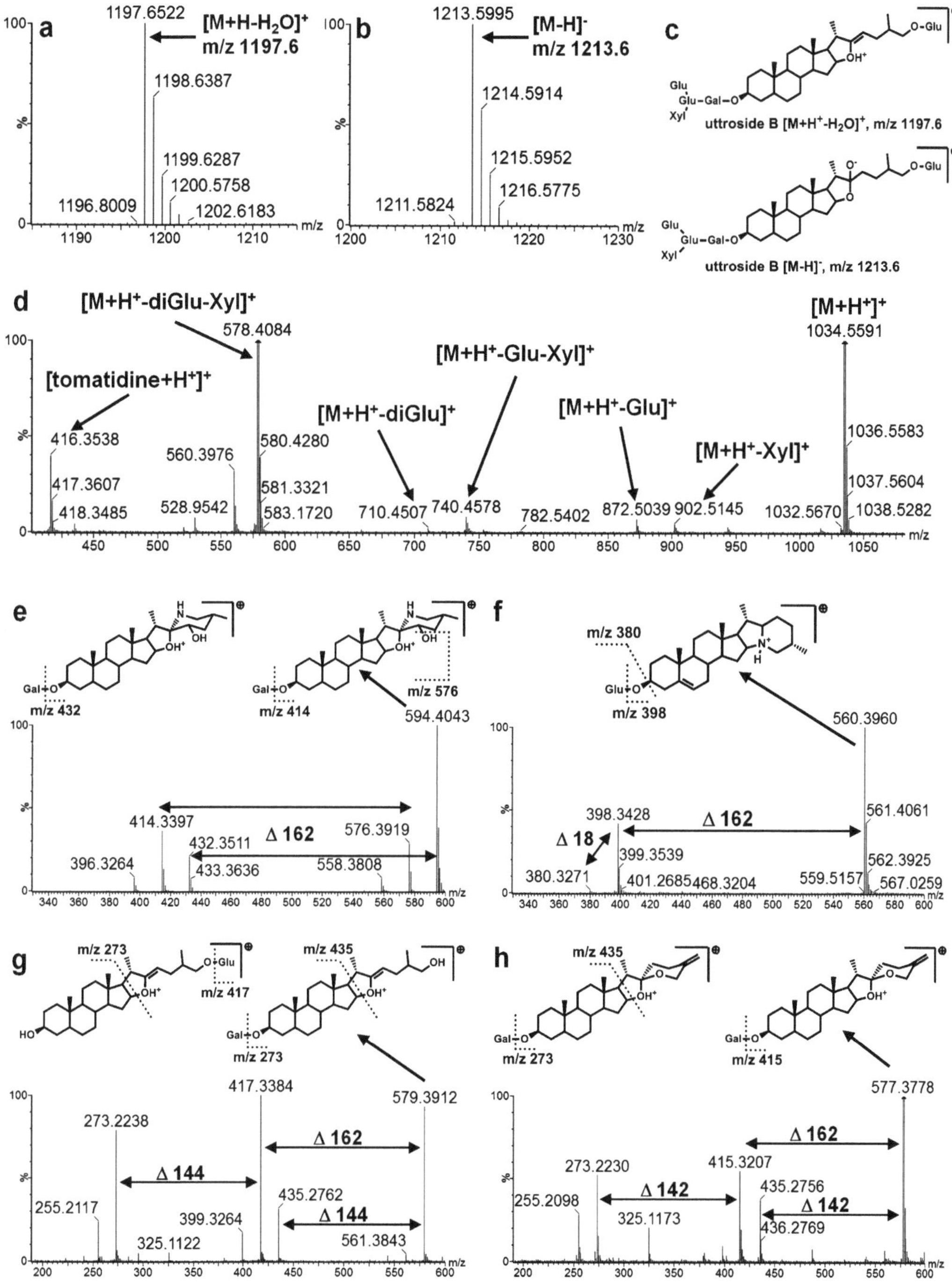

Fig. 2 Mass spectrometric characteristics of steroidal saponins and steroidal glycoalkaloids. (**a**) Molecular ion of uttroside B from *Solanum nigrum* detected in the positive ionization mode. (**b**) Molecular ion of uttroside B detected in the negative ionization mode. (**c**) structures of detected molecular ions. (**d**) fragmentation pattern of the C3-glycosyl-chain of α-tomatine from *Solanum lycopersicum*. Characteristic fragmentation patterns and structures of: (**e**) hydroxyl-tomatine as an example for tomatidine-type aglycones, (**f**) α-solanine from *Solanum tuberosum* as an example for solanidine-type aglycones, (**g**) uttroside B from *Solanum nigrum* as an example for furostanol-type saponins, and (**h**) a putative hydroxylated spirostanol-type saponin from *Capsicum* sp.; *m/z*: mass to charge, *M* molecular ion, *Glu* glucosyl, *Xyl* xylosyl, *Gal* galactosyl

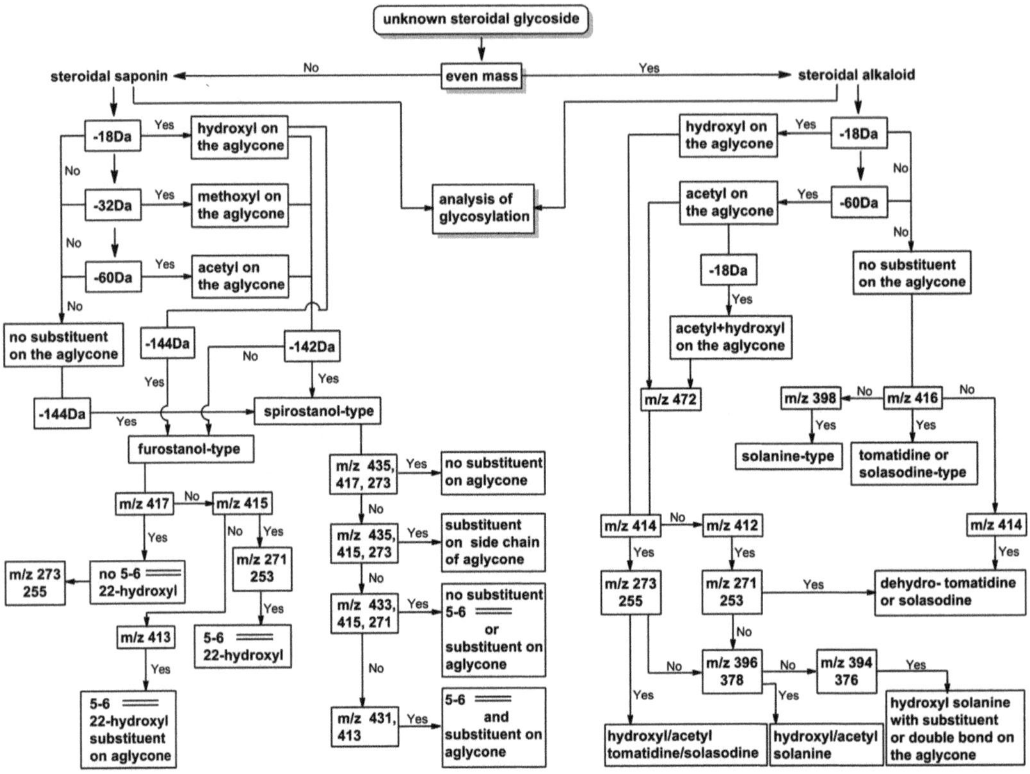

Fig. 3 Scheme for putative identification of steroidal saponins and alkaloids via interpretation of characteristic mass fragmentation patterns. Alkaloids and saponins can be first differentiated according to the occurrence of odd or even (one nitrogen containing compounds) molecular ion masses. Besides analysis of glycosylation pattern (mainly from C3) hydroxyl-, acetyl- and methoxyl-substitutions lead to specific neutral losses of 18, 32, and 60 Da (most common aglycone substituents, in rare cases also other groups can occur, e.g., carboxyl). Discrimination between different aglycone structures can be done either according to specific neutral losses (e.g., −144 Da for furostanol-type saponins, −142 Da for spirostanol-type saponins) or by analysis of the aglycone mass and masses of fragmentation products of these aglycones. *m/z* mass to charge, *Da* Dalton, = double bond

is determined to 3*hexose + 1*pentose. Loss of xylose as well as glucose directly from the molecular ion indicates that both sugar residues are located at the end of the glycosyl chain.

6. Identify characteristic neutral losses, besides sugars, that provide a hint to the substitution of the aglycone. As shown in Figs. 1 and 3, hydroxyl, acetyl and methoxyl substituents are common for steroidal glycosides. Neutral losses of 18, 32, or 60 Da from the molecular ion or a difference of 16 Da between positive and negative ionization (*see* **step 1** of Subheading 3.4) (Figs. 2 and 3) are characteristic for these substitutions.
7. To distinguish between different types of aglycones (Fig. 1), analyze the mass and the fragmentation pattern of the carbon backbone. In case of steroidal saponins, two major types are

formed in *Solanaceae* species, furostanol- and spirostanol-type saponins. Dependent on the substitution pattern determined before, specific neutral loss of the side-chain can be used for differentiation. If the putative compound does not have substitutions on the aglycone, a loss of 144 Da is characteristic for the spirostanol-type, resulting in fragment masses of *m/z* 435 and 433 which are the ions for rings A–D (Fig. 1) plus a hexose moiety at C3, without or with the 5–6 double bond. If the molecules are substituted on the aglycone side-chain a characteristic loss of 142 Da is observed, resulting in fragment masses of *m/z* 435, 433, or 431 depending on the number and position of the substitution (Figs. 2h and 3) [15]. For confirmation of the putative assignment, marker ions *m/z* 417–413 (aglycone without sugars) and *m/z* 273/271 (ions of rings A–D) can be used to determine position of the substitution and the possibility of a double bond in the molecule. Furostanol-type compounds typically loose water by loss of the free 22-OH group (Fig. 2a–c). Hence water loss and loss of the resulting side chain (−144 Da, Fig. 2g) is a clear indication for the furostanol-type. As described for spirostanol-type saponins the number of substitutions and saturation of the aglycone can be determined by analysis of the fragments of the aglycone (Fig. 3).

8. Steroidal alkaloids show a similar fragmentation behavior, although loss of the side chain is not observed as in the case of the saponins. When compounds are not substituted, aglycone masses of *m/z* 416 and *m/z* 398 (carbon backbone without sugars) are detected for tomatidine/solasodine-type and solanine-type alkaloids, respectively (Fig. 2d, f). If *m/z* 414 is observed as the dominant aglycone-fragment ion for non-substituted molecules together with fragment ions of *m/z* 271 and 253, the compounds are putative dehydro-tomatidine/solasodine-type alkaloids.

9. In the case of substituted compounds (e.g., acetyl or hydroxyl), elemental compositions and hence the detected ions, can be identical for different types of structures (for example hydroxyl-solanine and dehydro-tomatidine). To assign the correct structure, combinations of mass fragments have to be taken into consideration as well as their relative abundance.

10. These ions include typical aglycone masses like *m/z* 414/412 (hydroxyl-tomatidine after water loss/hydroxyl-dehydro-tomatidine after water loss, but also possibly hydroxyl-solanine) in combination with further fragments of these, for example *m/z* 273, 271 (A–D ring fragment), *m/z* 255, 253 (A–D ring fragment minus water). Characteristic ions of solanidene-type molecules are for example *m/z* 396/394 (desaturated solanidene) and *m/z* 378/376 (desaturated solanidene minus water).

Although these last mentioned ions can occur in tomatidine/solasodine-alkaloids' spectra, too, higher relative intensity compared to *m/z* 414, 412 supports a solanidene-type aglycone (*see* **Note 16**).

11. The scheme shown in Fig. 3 summarizes the differences in fragmentation behavior that can be used for and efficient putative assignment of unknown steroidal glycosides structures (saponins and alkaloids) according to the proposed stepwise analysis strategy of mass fragmentation spectra.

4 Notes

1. Prepare stock solutions with a concentration of 1 mg/mL in MeOH for all standards except tomatine and tryptophan. Tomatine can be dissolved in MeOH in a concentration of 0.5 mg/mL by sonication. Tryptophan is dissolved in 80 % MeOH supplemented with 2 % FA by sonication. It is recommended to start with a higher percentage of water, because of the poor solubility of the compound in MeOH. A QC-mix is prepared by combination of equal amounts of all stock solutions (final concentration 83 μg/mL, tomatine 42 μg/mL) and is stored at −20°C. Prior to injection this mix is further diluted 1:10 with MeOH.
2. All plastic materials have to be resistant to 75 % MeOH/0.1 % FA.
3. Ion mobility separation technology can be used to separate isobaric compounds (co-eluting compounds, e.g., isomers) and hence to reduce noise and complexity of MS and MS/MS spectra that is due to overlapping isotope and fragmentation patterns.
4. The ratio of MeOH/ddH_2O should be adjusted according to the biological material used. For tissues such as tomato fruit that have a high water content, addition of three volumes of MeOH was found to be optimal to obtain a final MeOH/water ratio of about 75 %. Furthermore some tissues are rich in organic acids, hence no additional acidification is necessary [10].
5. Pre-cool spatula in liquid nitrogen to avoid sticking and thawing of material on it.
6. A solvent/biomaterial ratio of 3:1 is used normally. Depending on the tissue and the abundance of the compounds of interest this ratio can be changed (e.g., 2:1 to increase concentration of trace metabolites).
7. To control system stability we use a 12-compound standard mix (8 μg/mL each, tomatine 4 μg/mL). The compounds are selected substances known from plants and cover the polarity

range of the chromatographic run from 2 min (L-phenylalanine) to 19.5 min (kaempferol) as well as a broad mass range (e.g., *m/z* 179.03 for caffeic acid, *m/z* 1,078.5 for tomatine $[M-H+FA]^-$ in negative ionization mode). This allows its use for stabilization of retention times and control of mass accuracy. Furthermore, the column performance over time can be monitored by comparison of peak resolution and shape of QC-mix injections on new and old columns.

8. Injection of higher volumes is not recommended because it leads to bad peak shape and loss in resolution. For analysis of low abundant compounds, concentrate the extract, e.g., via lyophilization.
9. Compare QC-mix samples to control stability of retention times, intensity, and mass accuracy.
10. An example of a script for peak detection and processing is shown below. Different parameters thereby depend strongly on the quality of the raw data or on the general specifications of the analytical setup. Most of these parameters are defined in line 1, including mass accuracy (20 ppm), peakwidth (5–20 s), prefilter (three consecutive scans have to display an intensity higher than 15 IPS), the signal to noise threshold and the integration method. For example, mass accuracy could be lowered when using an instrument with a higher resolution. When using HPLC instead of UPLC peaks will be much wider. Also further steps depend on the quality of the entire dataset, because grouping (group) and retention time correction (retcor) is done for masses that appear in all or many samples (chromatograms). In cases of large retention time shifts from sample to sample; this will result in a low number of peaks in the final output. The last line in the given example script defines the given output after peak picking and processing of the dataset.

Example-Script for XCMS Peak Picking

- Name = xcmsSet(method="centWave", ppm=20, peakwidth=c(5,20), prefilter=c(3,15), snthresh=10, integrate=1, mzdiff=-0.001)
- Name = group(Name, bw = 5, mzwid = 0.06, minsamp = 1)
- Name = retcor(Name_cor, method = "loess", plottype = "none", span = 2)
- Name_cor = group(Name_cor, bw = 4, mzwid = 0.06, minsamp = 1)
- Name_fill = fillPeaks(Name_cor)
- diffreport(Name_fill, 'SG1', 'SG2', 'Name', sortpval = TRUE, 200, metlin = -0.05)

11. Typically an XCMS output table contains between several hundred and several thousand mass traces. If this number is significantly lower, the parameters need to be adjusted in cases when in the chromatogram more compounds/masses are clearly observed. In the table generated each line represents a mass trace and the intensity in each sample for this mass is given. The replicate to replicate reproducibility (e.g., WT1, WT2, WT3) can be checked by comparison of these values. Differences thereby can be due to different extraction efficiency between samples, but also to chromatographic variations between injections. This variance can lead to missing of a mass trace at a retention time in the chromatogram of a sample in which the peak is shifted and hence does not fulfill the criteria defined for XCMS. Analysis of the extracted ion chromatograms allows control of integration borders as well as alignment of mass trace peaks between the samples.
12. Steroidal alkaloids typically contain one nitrogen atom; hence they show an even mass. This observation is of course also true for compounds containing three or five nitrogen (or odd mass for two nitrogen atoms). Therefore this differentiation of glycoalkaloids and saponins is preliminary and putative assignment always has to be validated by calculation of elemental composition and analysis of the fragmentation pattern.
13. Filtering of data for masses higher than the possible aglycone removes all mass peaks that originate from fragmentation that already occurred at low collision energy and can reduce the complexity of the XCMS output table significantly.
14. For high abundant compounds, like for example uttroside B from *Solanum nigrum* [16] or α-tomatine from tomato [4], analysis of the collision energy ramp data can be sufficient for putative identification, since here the signal to noise ratio is sufficient to assign all fragments to the compounds. For many low intensity signals, mass fragments of co-eluting substances interfere with the compound specific ions, e.g., for two differently glycosylated but co-eluting alkaloids it is not possible to determine the origin of the sugar residue loss. In these cases only MS/MS data can lead to a putative identification.
15. Rare substituents on the glycosyl-chain include for example organic acids (e.g., ferulic acid, coumaric acid). Differentiation between losses of sugar residues and acids can be done either according to their accurate mass or by looking for glycosylated aromatic acids in the product ion spectrum, that are formed via alternative fragmentation with the steroidal molecule part as a neutral loss. Another indication for the presence of aromatic substituents is a higher double bond equivalent than usual when calculating the elemental composition using MassLynx 4.1 elemental composition calculator.

16. Although shown to be a powerful tool for putative assignment of steroidal glycosides the presented method of mass spectra interpretation cannot completely solve the structure of newly detected molecules. The correct order of sugar residues can only be partially determined and the exact position of substituents/double bonds on the aglycones becomes not completely clear. However, such information is most helpful in discriminating between the (often) large numbers of possible structures obtained for a given calculated elemental composition.

Acknowledgements

We thank A. Tishbee for operating the UPLC/qTOF instrument, I. Rogachev and S. Meir for assistance in XCMS analysis, mass spectra interpretation and putative assignment of alkaloids from potato and tomato and the European Research Council (SAMIT-FP7 program) for supporting the work in the A. A. laboratory. A. A. is an incumbent of the Peter J. Cohn Professorial chair.

References

1. Vincken JP, Heng L, de Groot A et al (2007) Saponins, classification and occurrence in the plant kingdom. Phytochemistry 68:275–297
2. Nohara T, Ikeda T, Fujiwara Y et al (2007) Physiological functions of solanaceous and tomato steroidal glycosides. J Nat Med 61:1–13
3. Friedman M (2002) Tomato glycoalkaloids: role in the plant and in the diet. J Agric Food Chem 50:5751–5780
4. Itkin M, Rogachev I, Alkan N et al (2011) Glycoalkaloid metabolism1 is required for steroidal alkaloid glycosylation and prevention of phytotoxicity in tomato. Plant Cell 23:4507–4525
5. Itkin M, Heinig U, Tzfadia O et al (2013) Biosynthesis of antinutritional alkaloids in solanaceous crops is mediated by clustered genes. Science 341:175–179
6. Mintz-Oron S, Mandel T, Rogachev I et al (2008) Gene expression and metabolism in tomato fruit surface tissues. Plant Physiol 147: 823–851
7. Smith CA, Want EJ, O'Maille G et al (2006) XCMS: processing mass spectrometry data for metabolite profiling using nonlinear peak alignment, matching, and identification. Anal Chem 78:779–787
8. De Vos RC, Moco S, Lommen A et al (2007) Untargeted large-scale plant metabolomics using liquid chromatography coupled to mass spectrometry. Nat Protoc 2:778–791
9. Moco S, Bino R, De Vos R et al (2007) Metabolomics technologies and metabolite identification. Trends Analyt Chem 26: 855–866
10. Rogachev I, Aharoni A (2012) UPLC-MS-based metabolite analysis in tomato. Methods Mol Biol 860:129–144
11. Brodsky L, Moussaieff A, Shahaf N et al (2010) Evaluation of peak picking quality in LC-MS metabolomics data. Anal Chem 82: 9177–9187
12. Kuhl C, Tautenhahn R, Neumann S (2010) CAMERA: collection of annotation related methods for mass spectrometry data. R package (version 1.4.2.). http://msbi.ipb-halle.de/msbi/CAMERA/
13. Katajamaa M, Oresic M (2005) Processing methods for differential analysis of LC/MS profile data. BMC Bioinforma 6:179
14. Lommen A (2009) MetAlign: interface-driven, versatile metabolomics tool for hyphenated full-scan mass spectrometry data preprocessing. Anal Chem 81:3079–3086
15. Li R, Zhou Y, Wu Z et al (2006) ESI-QqTOF-MS/MS and APCI-IT-MS/MS analysis of steroid saponins from the rhizomes of *Dioscorea panthaica*. J Mass Spectrom 41:1–22
16. Sharma SC, Chand R, Sati OP et al (1983) Oligofurostanosides from *Solanum nigrum*. Phytochemistry 22:1241–1244

Part III

Isoprenoid Profiling in Specialized Organs

Chapter 13

Isoprenoid and Metabolite Profiling of Plant Trichomes

Gerd U. Balcke, Stefan Bennewitz, Sebastian Zabel, and Alain Tissier

Abstract

Plant glandular trichomes are specialized secretory structures located on the surface of the aerial parts of plants with large biosynthetic capacity, often with terpenoids as output molecules. The collection of plant trichomes requires a method to separate trichomes from leaf epidermal tissues. For metabolite profiling, trichome tissue needs to be rapidly quenched in order to maintain the indigenous state of intracellular intermediates. Appropriate extraction and chromatographic separation methods must be available, which address the wide-ranging polarity of metabolites. In this chapter, a protocol for trichome harvest using a frozen paint brush is presented. A work flow for broad-range metabolite profiling using LC-MS^2 analysis is described, which is applicable to assess very hydrophilic isoprenoid precursors as well as more hydrophobic metabolites from trichomes and other plant tissues.

Key words Glandular, Trichome, Metabolomics, Metabolite, Metabolism, Isoprenoids, Quenching

1 Introduction

1.1 Trichome Harvest

Multiple methods to isolate glandular trichomes have been developed. A common method is the mechanical abrading of trichomes from leaves pre-equilibrated in an aqueous buffer in a cell disrupter filled with buffer and 0.5 mm glass beads [1, 2]. Another method uses vortexing of flash-frozen trichome bearing tissue with finely ground dry ice [3]. This method works well for robust tissue like geranium pedicels or rosemary leaves but more fragile samples tend to break down and contaminate the separated trichomes. Epidermal peels have been used to enrich for glandular trichomes [4]. Trichomes can be also enriched by a Percoll density gradient centrifugation of mechanical fragmented tissue [5]. Further methods to isolate glandular heads and exudate involve gently touching the leaf surface with microscope slides [6], wiping cotton swabs over the leaf [7], collecting trichomes using adhesive tape [8, 9], gently brushing with a toothbrush across the surface of leaves submerged in buffer [10] or simply scraping the surface of frozen leaves with a scalpel, spatula, or brush [11]. Finally, laser micro-

Manuel Rodríguez-Concepción (ed.), *Plant Isoprenoids: Methods and Protocols*, Methods in Molecular Biology, vol. 1153, DOI 10.1007/978-1-4939-0606-2_13, © Springer Science+Business Media New York 2014

dissection and pressure catapulting has also been successfully used to isolate trichomes [12, 13]. Since most methods are not suitable for metabolomics analysis due to limited yield and insufficient quenching, a new method for trichome harvest in conjunction with metabolite profiling was developed and is presented here.

1.2 Metabolome Analysis

Metabolites are small molecules with molar masses between 50 and about 1,250 Da. This mass range is typically covered by modern quadrupole, time of flight (TOF), orbitrap, or Fourier transform ion cyclotron resonance (FT-ICR) mass detectors. Few milligrams of extracted plant tissue may contain thousands of metabolites. However, due to disparate mass resolution and scanning speed of the spectrometer used, the number of corresponding mass signals can be very different. Furthermore, as shown for *Escherichia coli*, quantitative inspection of central carbon metabolites demonstrated that their intracellular abundance can be quite diverse, ranging over six molar orders of magnitude [14]. Similarly, in plant leaves, the intracellular level of some plant metabolites can accumulate to millimolar concentrations, e.g., chlorogenic acid or malate [15] or trichome-derived terpenoids for example, which can interfere with accurate quantification.

Apart from their diverse abundance, small molecule metabolites can have very different chemical properties with respect to volatility, polarity, or stability upon extraction. Primarily, under normal conditions, phosphorylated, carboxylated, or aminated metabolites appear as charged forms. While carrying charged functional groups, isoprenoid precursors of both pathways (mevalonate and non-mevalonate) and most central carbon metabolites are very hydrophilic and do not contain sufficiently large hydrophobic moieties which could be exploited analytically. On the other hand, in isoprenoid biosynthesis, the isoprenyl diphosphate precursors are of increasing length by units of five carbon atoms, from isopentenyl diphosphate (IPP, C5) to geranylgeranyl diphosphate (GGPP, C20) and feature a relatively wide range of hydrophobicity. Isoprenyl diphosphates are then converted to extremely hydrophobic compounds, namely, olefinic terpenes, through the loss of diphosphate. These terpenes can then be further oxidized and/or cleaved, in turn leading to more hydrophilic compounds, such as for example the gibberellic acids. An overview of the varying degrees of hydrophobicity/hydrophilicity along the course of isoprenoid biosynthesis is presented in Fig. 1.

Moreover, next to their presence as aglycones, many secondary metabolites can become conjugated with sugars, malonylsugars, short acyl chains, and many other functional groups in order to regulate their availability and transport in the plant or trichome [16]. From an analytical point of view, conjugation will also greatly alter the polarity of a metabolite, requiring different methods to assess aglycone and conjugate. Taking the octanol–water coeffi-

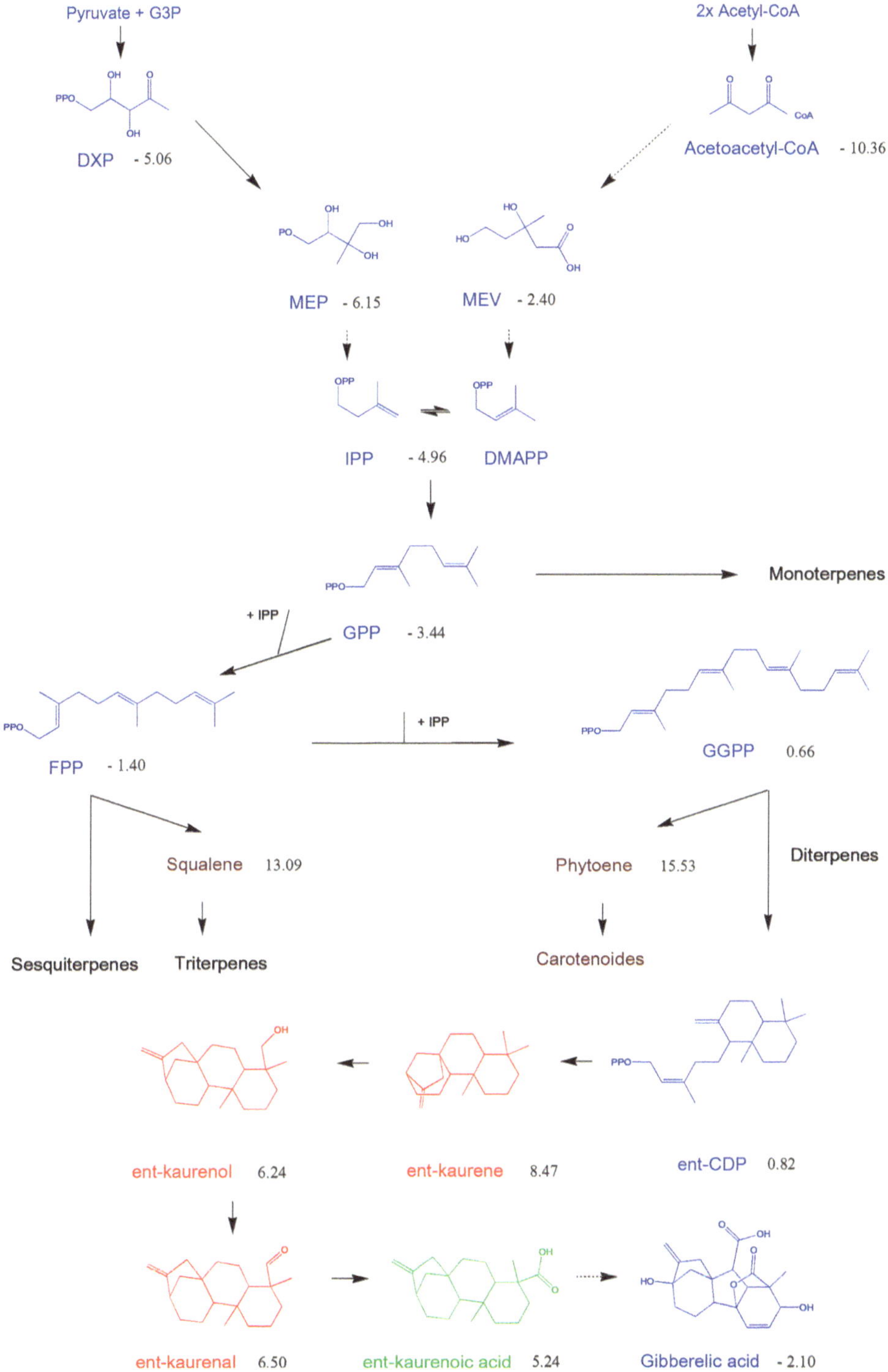

Fig. 1 Schematic overview of isoprenoid biosynthesis, with the example of gibberellic acids, to illustrate the varying degrees of polarity at different steps. The integration into different chromatographic separation approaches to assess individual metabolites relates to their polarity, which is expressed as the logarithmic octanol–water distribution coefficient (log K_{OW}). Negatively charged metabolites being very hydrophilic (*blue*) to polar (*green*) have log K_{OW} values below 1 and between 1 and 6, respectively, and can be separated by the method described in Section 3.3.2. Depending on their volatility and stability in the gas phase, polar to medium hydrophobic metabolites (*red*/*black*; log K_{OW} 6–9) can be assayed by methods described in Section 4.3.1. or by GC-MS, with polyhydroxylated and polycarboxylated intermediates being preferentially analyzed by LC-MS. Hydrophobic metabolites (*brown*) with log $K_{OW} > 9$ are analyzed by method described in Section 3.3.3

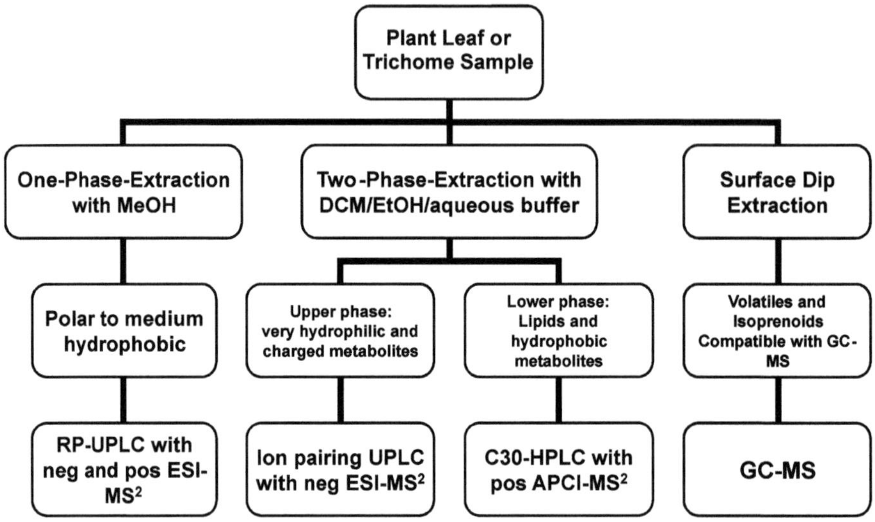

Fig. 2 Schematic overview of extraction and analysis methods for global metabolite profiling of isoprenoids and precursors of the isoprenoid biosynthesis in glandular trichomes. Details are provided in the text

cient as a polarity index, the difference in the polarity between most hydrophilic, e.g., dihydroxyacetone phosphate, and most hydrophobic metabolites, e.g., lycopene, of the same sample can vary over more than 21 orders of magnitude (based on data from Chemspider, http://www.chemspider.com). Thus, comprehensive assaying of intracellular metabolites requires suitable extraction procedures as well as specific chromatographic methods. With these issues in mind, metabolomic methods for plant leaf and trichome analysis were developed and are presented below. An overview of the whole process is presented in Fig. 2. Regarding the metabolite analysis, we do not present the analysis of leaf surface metabolites by GC-MS since this is already described in detail in a number of publications [17–20].

2 Materials

2.1 Isolation of Glandular Trichomes

1. Healthy plants grown in a greenhouse or in growth chambers.
2. Liquid nitrogen.
3. Paintbrush (*see* **Note 1**).
4. Personal protection equipment.
5. Spatula.
6. Microcentrifuge (2 mL) tubes.
7. Collection vessel (e.g., mortar).
8. Steel sieve or nylon mesh (pore size of 100 μm).

2.2 Metabolite Extraction

1. Teflon® cassette for 2 mL centrifuge tubes to be cooled down to liquid nitrogen temperatures.
2. Steel beads (3 mm, AISI 304 Grade 100, Spherotec AG).
3. Reinforced 2 mL cryo-tubes (e.g., Biozym).
4. Liquid nitrogen and dry ice.
5. FastPrep 24 extraction device (MP Biomedicals LLC) with dry ice cooling option.
6. Polar to medium hydrophobic metabolite extraction buffer: 99.875 % methanol, 0.125 % formic acid (both HPLC grade).
7. Dichloromethane:ethanol (2:1, v/v) (both HPLC grade).
8. Tetrahydrofuran (THF) (non-stabilized, HPLC grade).
9. 50 mM aqueous ammonium acetate in deionized water.
10. Polypropylene tubes (15 mL) for diluted extracts prior to the SPE step.
11. PVDF 0.2 μm hydrophilic membrane filtration plates (96 well, Corning).
12. PTFE 0.2 μm hydrophobic membrane filtration plates (96 well, Acroprep Advance PTFE, VWR).
13. 96 deep-well filter plate filled with 50 mg per well HR-XC polymeric SPE material (Macherey & Nagel) (*see* **Note 2**).
14. 96 polypropylene deep-well receiving plate (1–2 mL well volume).
15. Centrifuge for 96-well plates (e.g., Beckman-Coulter, Avanti J-E).

2.3 Metabolite Analysis

1. UPLC (e.g., Acquity, Waters) consisting of sample manager, column oven and a binary solvent pump system capable of supplying solvents at 0.4 mL/min at pressures up to 1,000 atm.
2. For polar to medium hydrophobic metabolites: C18 RP UPLC column (Waters BEH type) or equivalent.
3. For very hydrophilic and charged metabolites: C18 RP column with a carbon load of ≥10 %, e.g., C18 RP UPLC column Vision HT HL 100 × 2.1 × 1.5 μm (Grace Alltech Inc.) or equivalent.
4. For lipids and hydrophobic metabolites: Prontosil C30 100 × 2.1 × 3.5 μm (Bischoff GmbH) or equivalent.
5. Time of flight (TOF) mass spectrometer with data dependent MS^2 or data-independent SWATH-MS^2 features [21, 22] (e.g., TripleToF 5600, AB Sciex).
6. Alternatively, a triple quadrupole or other mass selective detector for targeted metabolite analysis.
7. Water, methanol (MeOH), acetonitrile (ACN), tetrahydrofurane (THF), glacial acetic acid, formic acid, ammonium formate, and tetrabutylamine (TBA) (all LC-MS grade).

3 Methods

3.1 Isolation of Glandular Trichomes

The method has been tested on tomato, tobacco and rosemary leaves. It is however applicable to many different samples. For plants with softer leaves, like tomato and tobacco, the younger leaves are best suited for trichome collection. An illustration of the method is provided in Fig. 3.

1. Cool down the collection vessel and paintbrush in liquid nitrogen (*see* **Note 3**).
2. Cut a trichome bearing leaf at the base, hold it on one end and brush several times across the trichome bearing surface while keeping right above the collection vessel (*see* **Notes 4** and **5**).
3. Trichomes will flash freeze and fall into the vessel while the leaf itself remains flexible and thus will not break down.
4. Work quickly to minimize the formation of ice on the collection vessel (*see* **Note 6**).

Fig. 3 (**a**) Trichome harvest of a tomato leaflet utilizing the Frozen Paintbrush Method. (**b**) Isolated glandular trichome heads (*yellow*) and stalks under a Nikon AZ100 Stereomicroscope. No contamination by *green* leaf fragments is observed. Scale bar, 100 μm. (**c**, **d**) Tomato leaflet before (**c**) and after (**d**) harvest. Note that the majority of non-glandular type II trichomes remain attached to the leaflet while all glandular trichomes are removed. Scale bar, 1 mm

5. Repeat with a sufficient number of leaves to collect enough material for analysis. Typically, 10–25 tobacco or tomato plants are required to collect 100 mg of trichome material.
6. Once the collection is finished the harvested material can easily be filtered with liquid nitrogen through nylon mesh of a steel sieve or even a series of different pore sizes to further purify the sample, in particular to eliminate long hairs which are not glandular or pieces of the leaf.
7. Let the liquid nitrogen evaporate, scrape out the collection vessel with a cooled down spatula, collect the sample in a 2 mL microcentrifuge tube, and store at −80 °C until downstream processing.

3.2 Metabolite Extraction and Enrichment

3.2.1 One-Phase-Extraction for Medium Hydrophobic Secondary Metabolites

1. Grind and homogenize trichome material in liquid nitrogen.
2. Cool down 2 mL cryo-tubes containing five steel beads on dry ice.
3. Weigh 100 mg of ground trichome or other plant material, transfer into cryo-tubes pre-cooled to liquid nitrogen temperatures, and immediately add 500 μL of cold extraction buffer (−80 °C) (*see* **Note 7**).
4. Bead-mill at the highest possible frequency while maintaining low ambient temperatures (i.e., <−40 °C). The milling time depends on the machine being used. Using a FastPrep 24 (MP Biomedicals) 1 min milling time at an elongation speed of 6.5 m/s is sufficient to rupture plant cell walls.
5. Centrifuge for 4 min at 4 °C and 20,000 × *g* and dilute 450 μL of the supernatant with water to give a final volume of 5 mL.
6. Condition a polymeric mixed mode HR-XC SPE column with 1 mL of MeOH followed by 1 mL of water.
7. Pass the dilute extract in 1 mL portions over the HR-XC SPE column by spinning a 96-well plate centrifuge slowly (e.g., 150 rpm).
8. Rinse the SPE column with 1 mL of water.
9. Desorb with 900 μL of ACN and then either inject a small volume, e.g., ≤3 μL directly onto a C18 RP UPLC column or evaporate the sample to dryness in a nitrogen stream and resuspend the dried sample first in 40 μL of MeOH followed by 60 μL of water (*see* **Notes 8** and **9**).

3.2.2 Two-Phase Extraction for Very Polar and Apolar Metabolites

1. Grind and homogenize trichome or other plant material in liquid nitrogen (*see* **Note 10**).
2. Cool down 2 mL cryo-tubes with five steel beads on dry ice.
3. Weigh 100 mg of ground plant material into cooled cryo-tubes, immediately add 900 μL of cold DCM/EtOH extraction solution, and chill on dry ice.

4. Add 150 μL of 50 mM aqueous ammonium acetate buffer to the organic phase (*see* **Note 11**).
5. Bead-mill at the highest possible frequency while maintaining decreased ambient temperatures (i.e., <–40 °C). The milling time depends on the machine being used. Using a FastPrep 24 (MP Biomedicals) 1 min milling time at an elongation speed of 6.5 m/s is sufficient to rupture plant cell walls (*see* **Note 12**).
6. Centrifuge at 4 °C and 20,000×*g* for 4 min and transfer 200 μL of the upper phase to a 0.2 μm PVDF spin filter.
7. Add another 150 μL of 50 mM aqueous ammonium acetate buffer to the extraction remainder and repeat **steps 5** and **6** but combining the aqueous extracts from the first and second extraction.
8. Filter the aqueous extract over PVDF 0.2 μm.
9. Store the aqueous extract at –80 °C until analysis or, if further enrichment is required, lyophilize the extract overnight and resuspend in 100 μL of water (*see* **Note 13**).
10. For collection of hydrophobic metabolites, sample 700 μL of the organic phase and transfer to a 2 mL centrifuge tube. Carefully avoid picking up any cell debris from the interface layer.
11. Add 500 μL of THF and repeat **step 5**.
12. Centrifuge at 0 °C and 20,000×*g* for 4 min and combine 450 μL of the organic phase with the first 700 μL of organic extract.
13. Repeat **steps 11** and **12**.
14. Wash the PTFE 0.2 μm filters twice with THF and apply the organic extract for spin-filtration (*see* **Note 13**).
15. Rinse the filter with another 200 μL of THF and combine all filtrates (*see* **Notes 14** and **15**).
16. Evaporate the organic extract to dryness in a nitrogen stream.
17. Resuspend in 100 μL of THF and store at –80 °C until analysis.

3.3 Metabolite Analysis

3.3.1 Polar to Medium Hydrophobic Metabolites

1. Solvent A is 500 μL of 300 mM aqueous ammonium formate buffer and 12.5 μL of formic acid in 500 mL of water (pH 3.5) and solvent B is ACN.
2. Inject 5 μL of the sample on a C18 RP UPLC column (Waters BEH type). Chromatography gradient details are given in Table 1.
3. Analyze by positive and negative mode electrospray ionization and MS^1-TOF plus MS^2-QToF detection. Further mass spectrometry details are given in Table 2. A typical profile of hydrophilic precursors of the isoprenoid backbone biosynthesis is given in Fig. 4.

Table 1
Chromatographic conditions

Method	Column [$L \times ID \times$ particle size] in [mm × mm × μm]	Solvent A	Solvent B	Chromatographic conditions
Polar to medium hydrophobic metabolites	BEH (Waters Inc.) [100 × 2.1 × 1.7]	0.3 mM aqueous ammonium formate pH = 3.5	ACN	0–2 min: 5, 19 min: 95, 22 min: 95, 22.01 min: 5, 24 min: 5. Flow rate: 0.4 mL/min. Column temperature: 40 °C. Autosampler temperature: 4 °C
Very hydrophilic and charged metabolites	Vision HT HL (Grace Alltech Inc.) [100 × 2.1 × 1.5]	10 mM aqueous tributylamine acidified to pH = 6.2	ACN	0–2 min: 2, 18 min: 36, 21 min: 95, 22.5 min: 95, 22.52 min: 2, 24 min: 2. Flow rate: 0.4 mL/min. Column temperature: 40 °C. Autosampler temperature: 4 °C
Lipids and hydrophobic metabolites	Prontosil C30 (Bischoff GmbH) [100 × 2.1 × 3.5]	5 mM methanolic ammonium acetate	THF	0–2 min: 5, 17 min: 95, 19 min: 95, 19.01 min: 5, 24 min: 5. Flow rate: 0.2 mL/min. Column temperature: 25 °C. Autosampler temperature: 4 °C

Table 2
Mass spectrometric conditions

Method		3.3.1.	3.3.2.	3.3.3.
Mode		(+,–) ESI	(–) ESI	(+) APCI
Source temperature	[°C]	600	600	300
Source gases 1/2	[arbitrary units]	60/70	60/70	60/70
Curtain gas	[arbitrary units]	35	35	35
Ion spray voltage floating	[V]	5,500/–4,500	–4,500	5,500
Declustering potential	[V]	±35	–35	35
Collision energy in MS^1/MS^2	[V]	10/rce[a]	10/rce[a]	10/rce[a]
Accumulation time in MS^1/MS^2	[ms]	10/20[b]	10/20[b]	10/20[b]
Nebulizer current	[mA]	–	–	3

[a]rce: MS^2 scans are acquired in SWATH mode with a rolling collision energy (rce) and a collision energy spread of 15 V
[b]MS^2 data are acquires in SWATH mode scanning a mass range from 65 to 1,250 Da in Q1 pockets of 30 Da for the precursor ion mass

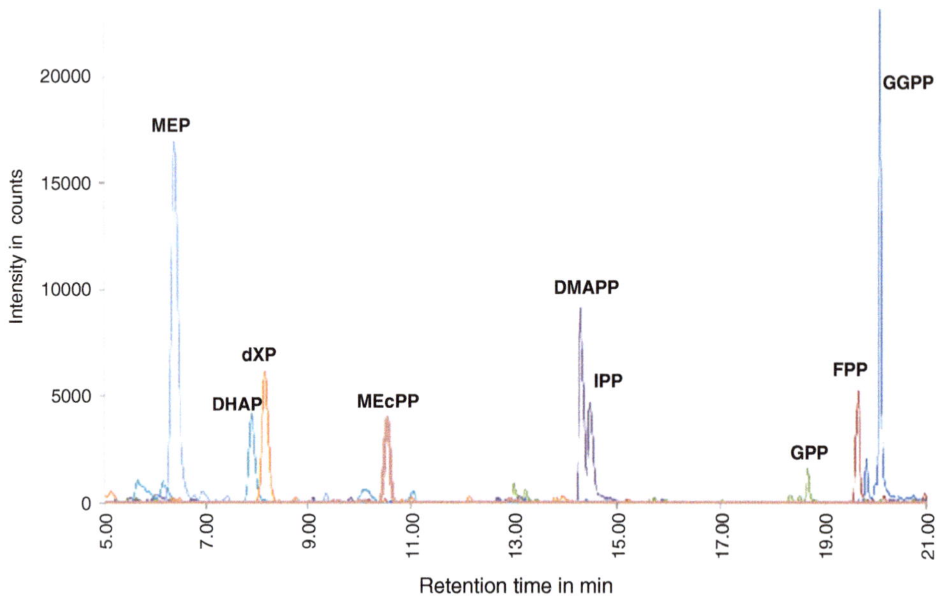

Fig. 4 Separation of hydrophilic isoprenoid precursors from 100 mg of trichome of *N. sylvestris* as detected by the method described in Section 3.3.2. DXP, deoxyxylulose phosphate; DHAP, dehydroxyacetone phosphate; MEcPP, methylerythritol cyclodiphosphate; MEP; methylerythritol phosphate, IPP; isopentenyl diphosphate, DMAPP; dimethallyl diphosphate; GPP, geranyl diphosphate; FPP, farnesyl diphosphate; GGPP, geranylgeranyl diphosphate

3.3.2 Very Hydrophilic and Charged Metabolites

1. Solvent A consists of 2,385 μL of tributylamine and 575 μL of glacial acetic acid added to 1,000 mL of water. Before use, the mixture is vigorously shaken for 5 min. Prepare new solution daily. Solvent B is ACN (*see* **Note 16**).
2. Inject 5 μL of the sample on a C18 RP column with a carbon load of ≥10 %. Chromatography gradient details are given in Table 1 (*see* **Note 17**).
3. Analyze by negative mode electrospray ionization and MS^1-TOF plus MS^2-QToF detection. Mass spectrometric details are given in Table 2 (*see* **Note 18**).

3.3.3 Lipids and Hydrophobic Metabolites

1. Solvent A is a 5 mM ammonium acetate methanolic solution. Solvent B is THF.
2. Inject 5 μL onto a C30 RP column. Further chromatographic details are given in Table 1 (*see* **Notes 19** and **20**).
3. Analyze by positive atmospheric pressure ionization and MS^1-TOF plus MS^2-QToF detection. Further mass spectrometric details are given in Table 2 (*see* **Note 21**).

4 Notes

1. Different plants require paint brushes of different sizes. While a 5 cm wide painters brush is optimal to harvest large tobacco leaves, small rosemary leaves require a small diameter artists brush. Test the brush by undergoing multiple freeze–thaw cycles, some brush hair material breaks when cooled down by liquid Nitrogen.
2. SPE enrichment is preferentially done in 96-well formats using low spin centrifugation of the entire plate. Some vendors sell loose SPE material which can be packed in 50 mg per well portions to reduce costs.
3. The brush needs to be dry; a wet brush will harden and freeze, leading to increased tissue breakage.
4. Hold the leaf at the tip, usually having the lowest trichome density and brush towards the stalk.
5. Previous wounding significantly increases the trichome number and thus the yield per plant. As between wounding and harvest several days of time have passed by, the metabolomes of wounded and not wounded plants will be comparable.
6. Trichome harvest of hundreds of mg fresh weight nitrogen takes several hours. As the plant matter is brushed and intermittently stored in a mortar filled with liquid nitrogen ambient humidity condenses at the cold surface of the porcelain. The latter will affect the percent share of trichome dry weight and needs to be determined if fresh weight is used to normalize metabolomic data of different samples.
7. An alternative extraction method is provided in ref. 23. This method assumes a 85 % water content of the plant matter and uses a fresh weight to extractant ratio of 1:3 between plant matter and acidic methanol. This ratio was optimized to get maximum extraction efficiency for medium hydrophobic metabolites while avoiding chlorophyll contamination. We found that the water content of trichomes can vary more widely and can go down to 60 %. In our approach using nearly pure methanol, chlorophyll and other more hydrophobic contaminants are first co-extracted but do not elute from the SPE column in the following step.
8. Omitting intermediate drying of the SPE eluate saves time and results in improved recovery for metabolites. However, for very low abundant metabolites drying and further concentration of the extract before injection is mandatory.
9. Samples should not be refrozen because precipitations are observed. If the measurement needs to be repeated store the samples at 4 °C until analysis.

10. It is of extreme importance that the samples do not warm up before extraction since the levels of many central carbon metabolites very rapidly respond, i.e., within seconds, to environmental perturbation.
11. For extraction of reduced forms of NAD cofactors, acidify the extraction buffer to pH = 3. For extraction of isoprenyl diphosphates, the extraction buffer should be brought to pH = 10 and weakly alkaline conditions should be maintained throughout.
12. Before bead-milling, the upper (aqueous) phase will be frozen. After arrest of metabolic activity in the plant matter by low temperature and subsequent denaturing by organic solvents, it is intended to perform a cryo-extraction. This is achieved as the covering ice layer is smashed by the spinning of the steel beads. During the extraction with a bead beater the temperature inside a well increases to about 13 °C within 20 s if the tubes are not cooled in a dry ice cryo-adapter. Low quantities of dry ice added to a cooling sink around the cryo-tubes will maintain lower temperatures. Importantly, ensure that the water phase will melt during extraction.
13. Before analysis, the aqueous phase should be kept frozen as long as possible. With older samples we observe hydrolysis of diphosphate compounds and losses of NADH, NADPH, and GSH due to oxidative processes.
14. While preparing filtrates of the organic phase containing very hydrophobic metabolites, hydrophobic PTFE filters tend to release fine particles, which can be removed by previous rinsing with solvent.
15. Very hydrophobic metabolites tend to stick to PTFE filters and can only be fully removed by extra solvent washes. The extra solvent can be removed by evaporation later on.
16. Ensure measurement within 24 h after thawing the samples. High loads of inorganic salts, especially inorganic phosphate bias the chromatography. Typically, 100 mM salt background is acceptable, except for Pi, which should not exceed 5 mM.
17. After having finished a batch of samples rinse the C18 column with organic solvent. The LC column must not be left submerged in high loads of aqueous TBA, since this drastically reduces performance and column lifetime.
18. Working with TBA will irreversibly contaminate the LC system and leave a trace of $m/z = 186$ as a background in positive mode. Ideally, reserve a LC device for work with TBA and conduct measurements in positive mode exclusively on different machines never in contact with TBA.
19. Separation of carotenoids works best under decreased temperatures. Under these conditions the C30 chains of the RP column

become more rigid, which improves peak separation. Thus, if possible use a column oven with cooling option.

20. For respiratory quinone and carotenoid analysis, several solvent systems have been proposed, among them MTBE and other non-water-miscible solvents. We found THF to be superior as it is highly water-miscible. If the analytical LC method often needs to be switched from aqueous systems to the analysis of very hydrophobic compounds, working with THF may reduce the risk to clogging narrow bore tubes which are used in UPLC.
21. The hydrophobic nature of carotenoids makes them “sticky” and sorb to any surface material in the LC. To reduce carry over between consecutive samples it is recommended to conduct an intermediate wash injection.

References

1. Gershenzon J, McCaskill D, Rajaonarivony JI et al (1992) Isolation of secretory cells from plant glandular trichomes and their use in biosynthetic studies of monoterpenes and other gland products. Anal Biochem 200: 130–138
2. Gershenzon J, Duffy MA, Karp F et al (1987) Mechanized techniques for the selective extraction of enzymes from plant epidermal glands. Anal Biochem 163:159–164
3. Yerger EH, Grazzini RA, Hesk D et al (1992) A rapid method for isolating glandular trichomes. Plant Physiol 99:1–7
4. Croteau R (1977) Site of monoterpene biosynthesis in Majorana hortensis leaves. Plant Physiol 59:519–520
5. Slone JH, Kelsey RG (1985) Isolation and purification of glandular secretory cells from Artemisia tridentata (ssp vaseyana) by Percoll density gradient centrifugation. Am J Bot 72: 1445–1451
6. Keene CK, Wagner GJ (1985) Direct demonstration of duvatrienediol biosynthesis in glandular heads of tobacco trichomes. Plant Physiol 79:1026–1032
7. Sterling TM, Houtz RL, Putnam AR (1987) Phytotoxic exudates from velvetleaf (Abutilon theophrasti) glandular trichomes. Am J Bot 74:543–550
8. Yamaura T, Tanaka S, Tabata M (1992) Localization of the biosynthesis and accumulation of monoterpenoids in glandular trichomes of thyme. Planta Med 58:153–158
9. Gopfert J, Conrad J, Spring O (2006) 5-deoxynevadensin, a novel flavone in sunflower and aspects of biosynthesis during trichome development. Nat Prod Commun 1:935–940
10. Croteau R, Winters JN (1982) Demonstration of the intercellular compartmentation of l-menthone metabolism in peppermint (Mentha piperita) leaves. Plant Physiol 69:975–977
11. Wang E, Wang R, DeParasis J et al (2001) Suppression of a P450 hydroxylase gene in plant trichome glands enhances natural-product-based aphid resistance. Nat Biotechnol 19:371–374
12. Olsson ME, Olofsson LM, Lindahl AL et al (2009) Localization of enzymes of artemisinin biosynthesis to the apical cells of glandular secretory trichomes of Artemisia annua L. Phytochemistry 70:1123–1128
13. Olofsson L, Lundgren A, Brodelius PE (2012) Trichome isolation with and without fixation using laser microdissection and pressure catapulting followed by RNA amplification: expression of genes of terpene metabolism in apical and sub-apical trichome cells of Artemisia annua L. Plant Sci 183:9–13
14. Bennett BD, Yuan J, Kimball EH et al (2008) Absolute quantitation of intracellular metabolite concentrations by an isotope ratio-based approach. Nat Protoc 3:1299–1311
15. Schauer N, Zamir D, Fernie AR (2005) Metabolic profiling of leaves and fruit of wild species tomato: a survey of the Solanum lycopersicum complex. J Exp Bot 56:297–307
16. Schilmiller A, Shi F, Kim J et al (2010) Mass spectrometry screening reveals widespread diversity in trichome specialized metabolites of tomato chromosomal substitution lines. Plant J 62:391–403
17. Rontein D, Onillon S, Herbette G et al (2008) CYP725A4 from yew catalyzes complex structural rearrangement of taxa-4(5),11(12)-

diene into the cyclic ether 5(12)-oxa-3(11)-cyclotaxane. J Biol Chem 283:6067–6075

18. Ennajdaoui H, Vachon G, Giacalone C et al (2010) Trichome specific expression of the tobacco (Nicotiana sylvestris) cembratrien-ol synthase genes is controlled by both activating and repressing cis-regions. Plant Mol Biol 73:673–685
19. Sallaud C, Giacalone C, Töpfer R et al (2012) Characterization of two genes for the biosynthesis of the labdane diterpene Z-abienol in tobacco (Nicotiana tabacum) glandular trichomes. Plant J 72:1–17
20. Tissier A, Sallaud C, Rontein D (2013) Tobacco trichomes as a platform for terpenoid biosynthesis engineering. In: Bach TJ, Rohmer M (eds) Isoprenoid synthesis in plants and microorganisms. Springer, New York, NY, pp 271–283
21. Hopfgartner G, Tonoli D, Varesio E (2012) High-resolution mass spectrometry for integrated qualitative and quantitative analysis of pharmaceuticals in biological matrices. Anal Bioanal Chem 402:2587–2596
22. Gillet LC, Navarro P, Tate S et al (2012) Targeted data extraction of the MS/MS spectra generated by data-independent acquisition: a new concept for consistent and accurate proteome analysis. Mol Cell Proteomics 11 (O111):016717
23. De Vos RC, Moco S, Lommen A et al (2007) Untargeted large-scale plant metabolomics using liquid chromatography coupled to mass spectrometry. Nat Protoc 2:778–791

Chapter 14

Sample Preparation for Single Cell Transcriptomics: Essential Oil Glands in *Citrus* Fruit Peel as an Example

Siau Sie Voo and Bernd Markus Lange

Abstract

Many plant natural products are synthesized in specialized cells and tissues. To learn more about metabolism in these cells, they have to be studied in isolation. Here, we describe a protocol for the isolation of epithelial cells that surround secretory cavities in *Citrus* fruit peel. Cells isolated using laser microdissection are suitable for RNA isolation and downstream transcriptome analyses.

Key words Laser microdissection, Microarray, Oil gland, Single cell, Transcriptome

1 Introduction

Plants have evolved specialized organs, tissues, and cell types for the synthesis and deposition of biomolecules important for storage or defense. For example, seeds may contain oils and/or proteins, tubers accumulate carbohydrate polymers, and resin ducts can be filled with insect-deterring terpenoid oleoresins [1]. Storage organs are generally large enough to be directly amenable to experimental studies. However, specialized tissues and cell types may be embedded into an organ where they contribute only a small (or even very small) fraction of the total biomass. To learn more about the biosynthesis of terpenoid natural products, it is often crucial to investigate the unique capabilities of these cells in isolation. This knowledge can then be used to develop molecular breeding and metabolic engineering approaches for a higher level production of desirable metabolites. An example would be the enhancement of monoterpene biosynthesis in glandular trichome cells of peppermint [2]. Alternatively, the relevant pathways can be transferred to microbial platform strains generated using approaches commonly referred to as synthetic biology. Amorphadiene, a precursor to the antimalarial sesquiterpene artemisinin, is an example of a plant metabolite that has been accumulated to high levels in yeast [3].

Manuel Rodríguez-Concepción (ed.), *Plant Isoprenoids: Methods and Protocols*, Methods in Molecular Biology, vol. 1153, DOI 10.1007/978-1-4939-0606-2_14, © Springer Science+Business Media New York 2014

As one of the first steps toward understanding the biosynthesis of plant cell specialization, it is desirable to acquire information about global transcript levels at cellular resolution. There are two main strategies for single cell transcriptomics: (1) in situ methods that enable the monitoring of gene expression patterns within intact cells and tissues, and (2) methods that require the removal of cells of interest from tissue samples [4]. In situ techniques, such as *f*luorescence *i*n *s*itu *h*ybridization (FISH) and mRNA *i*n *s*itu *h*ybridization (ISH), are highly sensitive techniques for visualizing the localization of a limited number of genes at the subcellular, cellular, and tissue level [5], but they are not suitable for genome-wide investigations [6]. Approaches that require the removal of cells from tissues include *f*luorescence-*a*ctivated *c*ell *s*orting (FACS), *i*solation of *n*ucleic *ta*gged in specific *c*ell *t*ypes (INTACT), and *l*aser *c*apture *m*icrodissection (LCM) [7–13]. FACS and INTACT are rapid and robust techniques for sorting heterogeneous mixtures of cells [8, 14], but both of these methods require the prior introduction of a fluorescent tag into the cells of interest [11], which complicates genome-scale analyses. LCM allows the selection of cells from tissue samples solely based on morphological differences, and therefore has potential applicability to all cell types [11, 15]. The only disadvantage is that, for most applications with specialized plant cell types, the repeated collection of hundreds or thousands of cells can require a substantial time commitment. Once methods for the collection of cells and preservation of their contents have been developed for an experimental system, however, LCM can be followed up with powerful post-genomic technologies such as cDNA library construction and gene cloning, microarray hybridization, reverse transcription PCR, next-generation sequencing, metabolite analysis, or proteomics [16–27]. In this chapter, we provide a detailed protocol for obtaining epithelial cells of secretory glands in grapefruit peel by LCM and isolating RNA for subsequent transcriptome analysis [28]. This method can be modified for transcriptome investigations of other specialized cell types [29] and we are commenting on opportunities and limitations.

2 Materials

Prepare all buffers and solutions with nuclease-free water (treated with diethyl pyrocarbonate or purchased as nuclease-free) (*see* **Note 1**). Follow appropriate regulations when handling chemicals (fume hood, protective clothing) and disposing waste.

2.1 Sample Fixation

1. Use scalpel with new blade or a used blade that was cleaned, wrapped in aluminum foil and autoclaved.

2. Fixative: ethanol (EtOH)–acetic acid (5:1; v/v) [30], pre-cooled on ice for 30 min before use. Both reagents should be of high quality (ACS grade or higher). Alternatively, the prepared fixative can be stored in refrigerator at 4 °C.
3. EtOH/xylene gradient series: EtOH–xylene (3:1; v/v); EtOH–xylene (1:1; v/v); EtOH–xylene (1:3; v/v). Both reagents should be of high quality (ACS grade or higher).
4. Xylene–paraffin gradient series: xylene (100 %); xylene–liquid paraffin (3:1; v/v); xylene–liquid paraffin (1:1; v/v); xylene–liquid paraffin (1:3; v/v); liquid paraffin (100 %). Use low melting point (45 °C) paraffin (Sigma-Aldrich) (*see* **Note 2**).
5. Clean and autoclave forceps for tissue embedding. Pre-warm forceps and aluminum boat to 45 °C on a heating plate before embedding. Melt the low-melting point paraffin in an oven set to 45 °C.

2.2 Sectioning and De-paraffinizing

1. Irradiate polyethylene naphthalene (PEN) membrane-coated slides (MembraneSlide NF 1.0 PEN, Zeiss) with UV (254 nm) light for 30 min before use.
2. Fill water bath with DEPC-treated (nuclease-free) water and heat to 40 °C.

2.3 Cell Isolation Using LCM

1. Apply the following setting to the PALM MicroBeam system (P.A.L.M. Microlaser Technologies): (1) tune down brightness of the light source so that section can still be seen but as little light as possible reaches the specimen; (2) set the laser energy to 74 (scale from 0 to 100) and the laser focus to 57 (scale from 0 to 100) (*see* **Note 3**).
2. Transfer 20 μL of RLT lysis buffer (RNeasy Micro kit, Qiagen) to the cap of a 0.5 mL microfuge vial and mount on PALM MicroBeam system (*see* **Note 3**).

2.4 RNA Extraction from LCM-Isolated Cells

1. Materials and reagents supplied with Qiagen RNeasy Micro kit: RNeasy spin column (pink), 2 mL nuclease-free collection tube, 1.5 mL nuclease-free collection tube; extraction buffer RLT; buffer RW1; RNase-free water.
2. NanoDrop spectrophotometer (Thermo Scientific).
3. Additional materials and supplies needed: 2 mL microcentrifuge tubes (Axigen Scientific); RNase-free plastic pipette tips with filters (Rainin); ß-mercaptoethanol (Sigma-Aldrich); 96–100 % EtOH (200 proof, Decon Laboratories).
4. Add 10 μL of β-mercaptoethanol to buffer RLT before first use. Dispense the chemical in a fume hood and wear appropriate protective clothing. The modified buffer can be stored at room temperature for up to 1 month.

5. Prepare DNase I stock solution: inject 550 μL of RNase-free water (with an RNase-free needle and syringe) directly into the vial (through the cap) containing the lyophilized DNase I (1,500 Kunitz units). Mix gently by inverting the vial but do not vortex. For long-term storage of DNase I, remove the stock solution from the glass vial, divide it into single-use aliquots, and store at –20 °C for up to 9 months. Thawed aliquots can be stored at 2–8 °C for up to 6 weeks. Do not refreeze the aliquots after thawing.
6. Spray working area with 70 % EtOH and wipe surface with paper towel.

3 Methods

3.1 Sample Preparation

The two most commonly used methods for tissue pretreatment are chemical fixation and cryofixation [17, 18, 25]. The main goal is to strike a balance between preserving cell morphology by tissue fixation and retaining RNA integrity. This process needs to be optimized for the smallest possible number of washing steps and the shortest period of time exposing samples to elevated temperatures and light. In our hands, chemical fixation using a mixture of ethanol and acetic acid gave excellent results. Ethanol penetrates most tissues rapidly, which is desirable for fixation, but can cause tissue shrinking, which is counteracted by the addition of acetic acid.

3.1.1 Fixation and Embedding

1. Using a sterile and RNase-free scalpel, slice the peel of *Citrus* fruit (flavedo) into roughly 3 × 3 mm squares (*see* **Note 4**).
2. Infiltrate cut tissue squares with pre-cooled, RNase-free fixative for 1 h at 4 °C in a vacuum oven at 0.78 bar (*see* **Note 5**). Repeat this step twice with new pre-cooled fixative.
3. Incubate sections with fixative for 12 h at 4 °C (*see* **Note 6**).
4. Replace fixative with EtOH and incubate sections for 30 min at room temperature (*see* **Note 7**). Repeat this step twice with new EtOH.
5. Infiltrate sections with a gradient series of solvents (*see* **Note 7**):
   (a) 3:1 EtOH:xylene for 30 min at room temperature.
   (b) 1:1 EtOH:xylene for 30 min at room temperature.
   (c) 1:3 EtOH:xylene for 30 min at room temperature.
   (d) 100 % xylene for 1 h at room temperature.
   (e) 3:1 xylene–paraffin for 1 h at 48 °C in a vacuum oven at 0.78 bar.
   (f) 1:1 xylene–paraffin for 1 h at 48 °C in a vacuum oven at 0.78 bar.

(g) 1:3 xylene–paraffin for 1 h at 48 °C in a vacuum oven at 0.78 bar.

(h) 100 % liquid paraffin for 1 h at 48 °C in a vacuum oven at 0.78 bar.

(i) repeat the last step twice.

6. Using pre-warmed, sterile forceps transfer the tissue section into a pre-warmed mold and turn the tissue section so that the cut side faces down.
7. Fill the mold entirely with liquid paraffin (*see* **Note 8**).
8. Remove mold from heating plate and allow to cool down to room temperature for roughly 30 min (until the paraffin block is completed solidified). Put paraffin blocks into a self-seal plastic bag, add a pouch of desiccant, close bag and store at 48 °C until further use (*see* **Note 9**).

3.1.2 Sectioning and De-paraffinizing

1. Mount paraffin block onto a rotary microtome.
2. Cut paraffin block into ribbons of 15 μm thickness (*see* **Note 10**).
3. Expand paraffin ribbons by floating on the surface of DEPC-treated water.
4. Transfer the expanded paraffin ribbon onto UV-treated PEN-membrane slide (*see* **Note 11**).
5. Air-dry the slide with paraffin ribbon for 20 min at room temperature in a fume hood, and then place the slide on a slide warmer set to 45 °C for 30 min.
6. Store the dried sample slides in a slide box filled with desiccant. Store the slide box at −80 °C.
7. Before de-paraffinization, warm the slide box up to room temperature (*see* **Note 12**).
8. In a fume hood, de-paraffinize samples by incubation in 100 % xylene for 3 min at room temperature. Repeat this step with fresh xylene three times.
9. Air-dry sample slides in a fume hood for 15 min at room temperature.

3.2 Cell Isolation Using LCM

1. Excise cells of interest using the laser of the PALM MicroBeam system according to the manufacturer's instructions and catapult excised cells into the collection cap moistened with RLT lysis buffer (Fig. 1). Exchange collection caps every 30 min to minimize exposure of sample to ribonucleases. Continue to collect cells until roughly 2,500 cells of the desired cell type have been obtained.
2. Pool cells from all collection caps in a microfuge vial place on ice during the LCM session.
3. At the end of the session, store the capped vial containing isolated cells at −80 °C until further use.

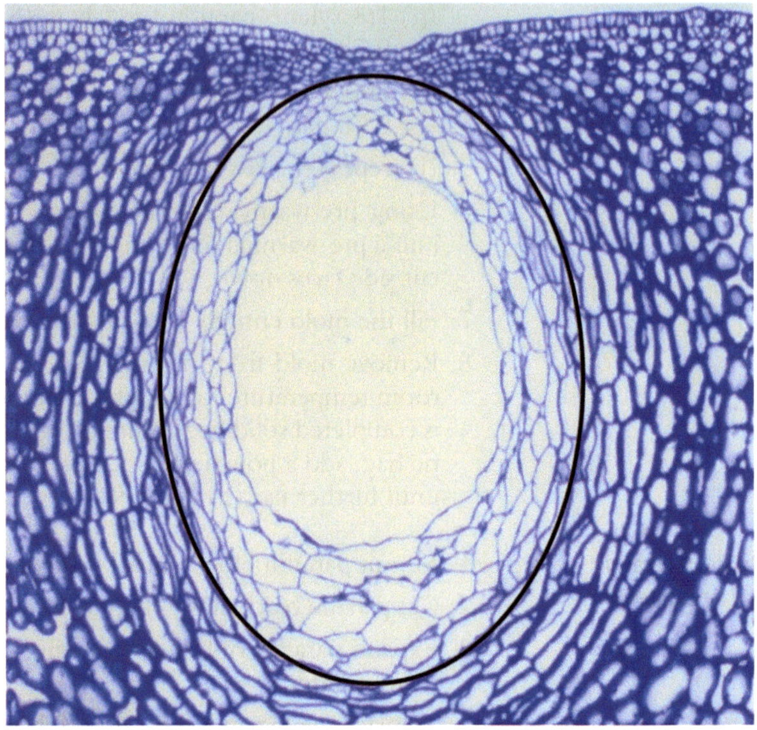

Fig. 1 Cross section of grapefruit peel, with an oil gland in the center. The oil cavity is lined by epithelial cells with thin cell walls. These cells are responsible for the biosynthesis and secretion of essential oil into the cavity [28]. The *inserted oval* indicates the path of the laser for cutting out the cells of interest. Note that, for transcriptome analyses, it is essential to also cut reference cells. We selected parenchyma cells that are located several cell layers away from epithelial cells

3.3 RNA Extraction from LCM-Isolated Cells

RNA is extracted using the RNeasy Micro kit according to the manufacturer's instructions (for buffer designations and other reagents, please refer to the manual) with modifications:

1. Thaw cell suspension on ice. Adjust the total volume to 350 μL with buffer RLT.
2. Incubate tube with collected *Citrus* cells at 37 °C for 5 min, followed by vigorous mixing for 1 min at room temperature (*see* **Note 13**).
3. Centrifuge the lysis mixture at 15,000 × *g* for 1 min. Transfer the supernatant into a new vial. Be careful not to touch the bottom of the vial with the pipette tip. The transferred volume should be slightly less than 350 μL.
4. Add 1 vol of 96–100 % EtOH to the homogenized lysate, and mix well by pipetting up and down (*see* **Note 14**). Do not centrifuge and proceed immediately to the next step.

5. Transfer 700 μL of the sample suspension, including any precipitate that may have formed, to an RNeasy MinElute spin column (pink). Place spin column in 2 mL collection vial (supplied with kit), close the lid gently, and centrifuge for 15 s at 8,000 × *g*. Discard the flow-through and reuse the collection vial (*see* **Note 15**). Repeat this step with the remainder of the sample.
6. Add 350 μL of buffer RW1 to the RNeasy MinElute spin column. Place spin column in 2 mL collection vial (reused from **step 5**), close the lid gently, and centrifuge for 15 s at 8,000 × *g* (*see* **Note 15**). Discard the flow-through.
7. Prepare DNase I incubation mix: add 70 μL of buffer RDD to 10 μL of DNase I stock solution (pre-thawed on ice). Mix by gently tapping the tube.
8. Place RNeasy MinElute spin column in 2 mL collection vial (reused from **step 5**), add the DNase I mix (80 μL) directly to the RNeasy MinElute spin column membrane, and leave on bench-top for 15 min at room temperature.
9. Add 350 μL of buffer RW1, close the lid gently, and centrifuge for 15 s at 8,000 × *g*. Discard the flow-through.
10. Place the RNeasy MinElute spin column in 2 mL collection vial (reused from **step 5**), add 500 μL of buffer RPE to the spin column, close the lid gently, and centrifuge for 15 s at 8,000 × *g*. Discard the flow-through.
11. Place the RNeasy MinElute spin column in a new 2 mL collection tube. Add 500 μL of 80 % EtOH, close the lid gently, and centrifuge for 2 min at 8,000 × *g* to wash the spin column membrane. Discard the flow-through and the collection tube.
12. Place the RNeasy MinElute spin column in the same 2 mL collection tube. Centrifuge at 15,000 × *g* for 5 min. Discard the flow-through and the collection tube.
13. Place the RNeasy MinElute spin column in a new 1.5 mL collection tube. Add 14 μL of RNase-free water directly onto the center of the spin column membrane, and leave spin column on bench-top for 5 min at room temperature. Close the lid gently, and centrifuge for 1 min at 15,000 × *g* to elute RNA.
14. Add the eluate back to the center of the spin column membrane and leave on bench-top for 1 min. Close the lid gently, and centrifuge for 1 min at 15,000 × *g* to elute RNA.
15. Determine the RNA yield and quality using a NanoDrop spectrophotometer.
16. Aliquot and store RNA sample at −80 °C until further use. cDNA synthesis and amplification can be performed using commercially available kits.

4 Notes

1. To avoid contamination with nucleases, follow general guidelines for handling RNA. For example, wear gloves at all times, pretreat all glassware with 0.1 % DEPC overnight and autoclave sterilize all plasticware, use RNase-free buffers and reagents, and avoid breathing directly onto your samples.
2. The lower viscosity of low melting-temperature paraffin results in a faster tissue penetration compared to paraffin of a higher melting temperature. Hence, the infiltration time is shortened and the sample exposure to elevated temperatures, which can lead to RNA degradation, is reduced.
3. Follow the protocols for instrument use and sample handling as outlined in the manual for the PALM MicroBeam system.
4. In our hands, the recommended size for tissue sections (3 mm × 3 mm) has given excellent results during fixation. This reflects a compromise between keeping the number of sections manageable and achieving complete diffusion of fixative.
5. The use of a pre-cooled fixative reduces the impact of omnipresent nucleases.
6. The period of each infiltration step can be adjusted depending on the tissue types. For example, the fixation time can be reduced by 50 % (to 30 min) for soybean leaves [29]. Use an excess of fixative (>20:1 ratio of fixative to tissue volume).
7. Place sample vials on single-tube mixer (VWR Vortexer 2G560, Scientific Industries) during the dehydration (100 % EtOH) and xylene infiltration steps. If needed, repeat last step (100 % liquid paraffin infiltration) until odor of xylene is undetectable.
8. Ensure an even distribution of paraffin to avoid cracking of paraffin block and thus breaking of the specimen during sectioning. There should also be no air trapped in the paraffin blocks.
9. If the section should "roll" during the solidification of the paraffin, reorient the tissue so that the cut side faces down. Paraffin-embedded tissues can be stored for up to 1 year. However, we recommend proceeding with sectioning within 2 months of specimen embedding.
10. The thickness of sections for LCM depends on the properties of the samples. In our hands, 15 μm sections worked well for both Citrus peel and soybean leaves [28, 29].
11. Paraffin ribbons need to be mounted on the PEN-membrane glass slide within 30 min after completing the treatment with UV light (254 nm). A longer wait may result in an insufficient

adherence of the specimen to the PEN-membrane, which could lead to sample losses during the de-paraffinization step. Re-irradiate the PEN-membrane glass slide if not being used within 30 min of UV exposure.

12. Open the slide box only after it has reached room temperature. This is to avoid the formation of moisture on the surface of the slides during the warming process.
13. In the Qiagen manual the use of carrier RNA is mentioned. In our hands, a treatment of LCM-isolated cells with RLT lysis buffer while mixing followed by vigorous mixing, without the addition of carrier RNA, gives excellent RNA yields.
14. A precipitate may be visible after addition of EtOH. This does not affect the procedure.
15. After centrifugation, carefully remove the RNeasy spin column from the collection tube so that the column does not get in contact the flow-through. Be sure to empty the collection tube completely before reusing.

Acknowledgements

The methods development for the protocol outlined here was funded by the Division of Chemical Sciences, Geosciences, and Biosciences, Office of Basic Energy Sciences of the US Department of Energy through Grant DE-FG02-09ER16054 (to B. M. L.).

References

1. Buchanan BB, Gruissem W, Jones RL (2002) Biochemistry and molecular biology of plants. Wiley, Hoboken, NJ, p 1408
2. Lange BM, Mahmoud SS, Wildung MR, Turner GW, Davis EM, Lange I, Baker RC, Boydston RA, Croteau RB (2011) Improving peppermint essential oil yield and composition by metabolic engineering. Proc Natl Acad Sci U S A 108:16944–16949
3. Westfall PJ, Pitera DJ, Lenihan JR, Eng D, Woolard FX, Regentin R, Horning T, Tsuruta H, Melis DJ, Owens A, Fickes S, Diola D, Benjamin KR, Keasling JD, Leavell MD, McPhee DJ, Renninger NS, Newman JD, Paddon CJ (2012) Production of amorphadiene in yeast, and its conversion to dihydroartemisinic acid, precursor to the antimalarial agent artemisinin. Proc Natl Acad Sci U S A 109:E111–E118
4. Nelson T, Gandotra N, Tausta SL (2008) Plant cell types: reporting and sampling with new technologies. Curr Opin Plant Biol 11:567–573
5. Ohmido N, Fukui K, Kinoshita T (2010) Recent advances in rice genome and chromosome structure research by fluorescence in situ hybridization (FISH). Proc Jpn Acad Ser B Phys Biol Sci 86:103–116
6. Karlgren A, Carlsson J, Gyllenstrand N, Lagercrantz U, Sundström JF (2009) Nonradioactive in situ hybridization protocol applicable for Norway spruce and a range of plant species. J Vis Exp 26: pii: 1205.
7. Baginsky S, Hennig L, Zimmermann P, Gruissem W (2010) Gene expression analysis, proteomics, and network discovery. Plant Physiol 152:402–410
8. Pu L, Brady S (2010) Systems biology update: cell type-specific transcriptional regulatory networks. Plant Physiol 152:411–419
9. Ckurshumova W, Caragea AE, Goldstein RS, Berleth T (2011) Glow in the dark: fluorescent proteins as cell and tissue-specific markers in plants. Mol Plant 4:794–804

10. Deal RB, Henikoff S (2011) The INTACT method for cell type–specific gene expression and chromatin profiling in *Arabidopsis thaliana*. Nat Protoc 6:56–68
11. Long TA (2011) Many needles in a haystack: cell-type specific abiotic stress responses. Curr Opin Plant Biol 14:325–331
12. Taylor-Teeples M, Ron M, Brady SM (2011) Novel biological insights revealed from cell type-specific expression profiling. Curr Opin Plant Biol 14:601–607
13. Ehrhardt DW, Frommer WB (2012) New technologies for 21st century plant science. Plant Cell 24:374–394
14. Takahara H, Endl E, O'Connell R (2011) Isolation of fungal infection structures from plant tissue by flow cytometry for cell-specific transcriptome analysis. Methods Mol Biol 729:3–13
15. Day RC (2010) Laser microdissection of paraffin-embedded plant tissues for transcript profiling. Methods Mol Biol 655:321–346
16. Schneider B, Hölscher D (2007) Laser microdissection and cryogenic nuclear magnetic resonance spectroscopy: an alliance for cell type specific metabolite profiling. Planta 225:763–770
17. Portillo M, Lindsey K, Casson S, García-Casado G, Solano R, Fenoll C, Escobar C (2009) Isolation of RNA from laser-capture-microdissected giant cells at early differentiation stages suitable for differential transcriptome analysis. Mol Plant Pathol 10:523–535
18. Thiel J, Müller M, Weschke W, Weber H (2009) Amino acid metabolism at the maternal-filial boundary of young barley seeds: a microdissection-based study. Planta 230:205–213
19. Abbott E, Hall E, Hamberger B, Bohlmann J (2010) Laser microdissection of conifer stem tissues: isolation and analysis of high quality RNA, terpene synthase enzyme activity and terpenoids metabolites from resin ducts and cambial zone tissue of white spruce (*Picea glauca*). BMC Plant Biol 10:106
20. Barcala M, García A, Cabrera J, Casson S, Lindsey K, Favery B, García-Casado G, Solano R, Fenoll C, Escobar C (2010) Early transcriptomic events in microdissected *Arabidopsis* nematode-induced giant cells. Plant J 61:698–712
21. Kaspar S, Matros A, Mock HP (2010) Proteome and flavonoid analysis reveals distinct responses of epidermal tissue and whole leaves upon UV-B radiation of barley (*Hordeum vulgare* L.) seedlings. J Proteome Res 9:2402–2411
22. Matsumura H, Yoshida K, Luo S, Kimura E, Fujibe T, Albertyn Z, Barrero RA, Krüger DH, Kahl G, Schroth GP, Terauchi R (2010) High-throughput SuperSAGE for digital gene expression analysis of multiple samples using next generation sequencing. PLoS ONE 5:e12010
23. Tang X, Zhang ZY, Zhang WJ, Zhao XM, Zhang D, Liu QQ, Tang WH (2010) Global gene profiling of laser-captured pollen mother cells indicates molecular pathways and gene subfamilies involved in rice meiosis. Plant Physiol 154:1855–1870
24. Hogekamp C, Arndt D, Pereira PA, Becker JD, Hohnjec N, Küster H (2011) Laser microdissection unravels cell-type-specific transcription in arbuscular mycorrhizal roots, including CAAT-box transcription factor gene expression correlating with fungal contact and spread. Plant Physiol 157:2023–2043
25. Schiebold S, Tschiersch H, Borisjuk L, Heinzel N, Radchuk R, Rolletschek H (2011) A novel procedure for the quantitative analysis of metabolites, storage products and transcripts of laser microdissected seed tissues of *Brassica napus*. Plant Methods 7:19
26. Schmid MW, Schmidt A, Klostermeier UC, Barann M, Rosenstiel P, Grossniklaus U (2012) A powerful method for transcriptional profiling of specific cell types in eukaryotes: laser-assisted microdissection and RNA sequencing. PLoS ONE 7:e29685
27. Torti S, Fornara F, Vincent C, Andrès F, Nordström K, Göbel U, Knoll D, Schoof H, Coupland G (2012) Analysis of the Arabidopsis shoot meristem transcriptome during floral transition identifies distinct regulatory patterns and a leucine-rich repeat protein that promotes flowering. Plant Cell 24:444–462
28. Voo SS, Grimes HD, Lange BM (2012) Assessing the biosynthetic capabilities of secretory glands in *Citrus* peel. Plant Physiol 159:81–94
29. Voo SS, Grimes HD, Lange BM (2013) Cell type-specific transcriptome analysis of the soybean leaf paraveinal mesophyll layer. Plant Mol Biol Rep 31:210–221
30. Deeken R, Ache P, Kajahn I, Klinkenberg J, Bringmann G, Hedrich R (2008) Identification of *Arabidopsis thaliana* phloem RNAs provides a search criterion for phloem-based transcripts hidden in complex datasets of microarray experiments. Plant J 55:746–759

Chapter 15

Prenylquinone Profiling in Whole Leaves and Chloroplast Subfractions

Felix Kessler and Gaetan Glauser

Abstract

Prenylquinones are indispensable molecules in plants and animals. In plants, phylloquinone (vitamin K) and plastoquinone are electron carriers during photosynthesis in chloroplasts, whereas tocopherol (vitamin E) functions as a lipid antioxidant. The biosynthetic pathways of the prenylquinones have been largely characterized but the mechanisms regulating their production and distribution in various subcompartments of the chloroplast are only starting to emerge. Research on chloroplast lipid droplets (plastoglobules) has unraveled a complex network of intersecting prenylquinone metabolic pathways that are providing unprecedented insight into the regulatory processes. In this chapter, we describe how to isolate chloroplast membrane fractions, in particular the plastoglobule lipid droplets, and how to profile the prenylquinones that are contained in these fractions.

Key words Prenylquinones, Chloroplast preparation, Plastoglobules, Sucrose gradient, Ultracentrifugation, Ultrahigh-pressure liquid chromatography, Atmospheric pressure chemical ionization, Quadrupole time-of-flight mass spectrometry

1 Introduction

Chloroplast biogenesis is characterized by the rapid emergence of the extensive thylakoid membrane system. The key lipid components of this membrane system are the galactolipids, the carotenoids, and the prenylquinones. The latter act as Redox components in the electron transport chain or as antioxidants: Plastoquinone acts as a free electron transporter between Photosystem II and cytochrome b_6f-complex. Phylloquinone (vitamin K) acts as a Redox compound within Photosystem I. Tocopherols function as antioxidants at the thylakoid membrane. Members of the prenylquinone family are indispensable for plastid function and their absence independently results in albino or even embryo lethal phenotypes.

While a large proportion of the prenylquinones is present and active in the thylakoid membrane, recent research has shown that

Manuel Rodríguez-Concepción (ed.), *Plant Isoprenoids: Methods and Protocols*, Methods in Molecular Biology, vol. 1153, DOI 10.1007/978-1-4939-0606-2_15,

a separate pool of prenylquinones exists in chloroplast lipid droplets that are also known as plastoglobules (PG). These lipid droplets are easily visible by electron microscopy due to their osmiophile properties that have been attributed to their contents of non-saturated lipid species (such as plastoquinone and members of the triacylglycerol family). Moreover, PG can easily be isolated by flotation centrifugation on linear or discontinuous sucrose gradients, due to their low density lipid droplet properties. Together with the emerging "-omics" technologies, this method has greatly enhanced our understanding of the biological function of PG.

PG were long believed to simply function as lipid storage particles (i.e., for prenylquinones, triacylgylcerols, fatty acid phytyl esters, carotenoid derivatives, and likely others). Our research has confirmed the storage aspect of PG function but has also linked it to the direct involvement of PG in a variety of lipid metabolic pathways [1–3]. It was known for a long time that PG also contain proteins but only recent proteomics studies led to the identification of a relatively limited number of PG proteins [1, 4, 5]. Quite surprisingly, a large part of these proteins are known or predicted enzymes and their roles in chloroplast lipid metabolism have been characterized in recent years. For example, VTE1 (tocopherol cyclase) participates in the production of plastochromanol-8 (the chromanol-derivative of plastoquinone) and tocopherol redox recycling [2]. PES1 and PES2, two members of the esterase/lipase/thioesterase family, are responsible for fatty acid phytyl ester and triacylglycerol accumulation in PG during senescence and nitrogen starvation [6, 7], two conditions that coincide with thylakoid membrane degradation. NDC1 regulates the redox state of plastoquinone in the chloroplast and through an unknown mechanism is also required for phylloquinone accumulation [3]. These results, which clearly show a direct involvement of PG in prenylquinone metabolism, would not have been possible without the development of new techniques to determine and quantify prenylquinones in different types of plant extracts and chloroplast fractions.

The quantitative analysis of prenylquinones in plants is a challenging task. Traditionally, prenylquinones are separated by high-performance liquid chromatography (HPLC) and detected by ultraviolet (UV) or fluorescence (FD) detection [8, 9]. Both normal and reverse-phase modes may be employed using hydrophobic mobile phase solvents such as chloroform, *tert*-butylmethyl ether, or hexane (*see* Chapter 7). Recently, we have developed an ultrafast method based on ultrahigh-pressure liquid chromatography—atmospheric pressure chemical ionization—quadrupole time-of-flight mass spectrometry (UHPLC-APCI-QTOFMS) for the determination of prenylquinones in entire leaves [10]. This method enabled the profiling of eleven prenylquinones, five of which could be absolutely quantified, in only 4.5 min and provided unprecedented limits of quantification. In this chapter, we propose an adapted and detailed protocol for the analysis of both entire leaves and subcellular distributions of prenylquinones, based on ultracentrifugation of subcellular organelles combined with UHPLC-APCI-QTOFMS measurements.

2 Materials

2.1 Whole Leaf Extraction

1. Extraction solvent: tetrahydrofuran (THF) (*see* **Note 1**) containing decylplastoquinone at 0.5 μg/mL as internal standard (IS) (*see* **Note 2**).
2. Liquid nitrogen.
3. Mortars, pestles, and spatula.
4. Microcentrifuge (1.5 mL) tubes.
5. Analytical balance.
6. Pipettes and pipette tips.
7. Glass beads (2 mm).
8. Vortex mixer.
9. Mixer mill (e.g., Retsch MM300, Haan, with holders for 1.5/2 mL microcentrifuge tubes).
10. Refrigerated centrifuge.
11. HPLC vials and caps.

2.2 Plastoglobule Isolation

1. Homogenization buffer (HB buffer): 450 mM sorbitol, 20 mM Tricine-KOH pH 8.4, 10 mM ethylenediaminetetraacetic acid (EDTA), 10 mM $NaHCO_3$, 1 mM $MnCl_2$, 5 mM Na-ascorbate, 0.05 % (w/v) bovine serum albumin (BSA) fraction V, and 1 mM phenylmethylsulfonyl fluoride (PMSF) (*see* **Note 3**). Add ascorbate and BSA immediately before use.
2. Resuspension buffer (RB buffer) (8×) stock: 2.4 M sorbitol, 160 mM Tricine-KOH pH 7.6, 20 mM EDTA, and 40 mM $MgCl_2$. Store at −20 °C. For use, dilute 100 mL of RB buffer (8×) in 700 mL of water.
3. 40 % (v/v) and 85 % (v/v) Percoll solutions: mix 40 or 85 mL of Percoll pH 8.5–9.5 (Sigma) with 12.5 mL of RB buffer (8×) and adjust volume to 100 mL with water. These solutions are stable for several months at −20 °C.
4. 80 % (v/v) acetone in water (*see* **Note 4**).
5. TrE buffer (10×) stock: 500 mM Tricine-KOH, pH 7.5, 20 mM EDTA, and 20 mM dithiothreitol (DTT). Store at −20 °C. For use, dilute 100 mL of TrE buffer (10×) in 900 mL of water.
6. Sucrose solution (0.6 M): dissolve sucrose in TrE buffer (10×) and water to reach desired volume in TrE (1×). Store at −20 °C.

7. Sucrose solutions of 45, 38, 20, 15, and 5 % (w/v): dissolve sucrose in TrE buffer (10×) and autoclaved deionized water to obtain desired sucrose concentration in TrE (1×). Store at −20 °C.
8. Centrifuges: preparative refrigerated centrifuge (e.g., Sorvall RC-5B, Thermo Scientific), with fixed-angle rotor (e.g., Sorvall SLA1500) and corresponding plastic 250 mL bottles (Nalgene), or with swinging-bucket rotor (e.g., Sorvall HB-6) and corresponding open polycarbonate 50 mL tubes (Nalgene); refrigerated benchtop centrifuge (e.g., Eppendorf 5810R), with swinging-bucket rotor (e.g., Eppendorf A4-62) and capped polypropylene 50 mL tubes (Falcon, BD biosciences); ultracentrifuge (e.g., Beckman L7), with swinging-bucket rotor (e.g., Beckman SW 28) and UltraClear™ SW28 tubes (25 × 89 mm, Beckman).
9. Waring blender.
10. Miracloth and cheesecloth.
11. Potter-Elvehjem tissue grinders (15 and 50 mL) with corresponding Teflon pestles.
12. UV–Vis Spectrophotometer.

2.3 UHPLC-APCI-QTOFMS Analysis

1. Infusion solution containing all available prenylquinone standards at 5 μg/mL in methanol.
2. Calibration solution 1: α-tocopherol, α-tocopherol quinone, γ-tocopherol, δ-tocopherol, plastochromanol-8, and plastoquinone-9 at 0.05 μg/mL in THF–water (85:15, v/v) containing decylplastoquinone at 0.5 μg/mL (*see* **Note 5**).
3. Calibration solution 2: α-tocopherol, α-tocopherol quinone, γ-tocopherol, δ-tocopherol, plastochromanol-8, and plastoquinone-9 at 0.2 μg/mL and phylloquinone at 0.1 μg/mL in THF–water (85:15, v/v) containing decylplastoquinone at 0.5 μg/mL.
4. Calibration solution 3: α-tocopherol, α-tocopherol quinone, γ-tocopherol, δ-tocopherol, plastochromanol-8, and plastoquinone-9 at 1.0 μg/mL and phylloquinone at 0.25 μg/mL in THF–water (85:15, v/v) containing decylplastoquinone at 0.5 μg/mL.
5. Calibration solution 4: α-tocopherol, α-tocopherol quinone, γ-tocopherol, δ-tocopherol, plastochromanol-8, and plastoquinone-9 at 2.0 μg/mL and phylloquinone at 1.0 μg/mL in THF–water (85:15, v/v) containing decylplastoquinone at 0.5 μg/mL.
6. Calibration solution 5: α-tocopherol at 10.0 μg/mL, α-tocopherol quinone, γ-tocopherol, δ-tocopherol, plastochromanol-8, and plastoquinone-9 at 5.0 μg/mL and phylloquinone at 2.5 μg/mL in THF–water (85:15, v/v) containing decylplastoquinone at 0.5 μg/mL.
7. Mobile phases: A = water, B = methanol (*see* **Note 6**).

8. Ultrahigh-pressure liquid chromatography (UHPLC) system interfaced to a quadrupole-time-of-flight mass spectrometer (QTOF) using atmospheric pressure chemical ionization (APCI). The UHPLC pump may be either a binary high-pressure mixing system or a quaternary low-pressure mixing system but must resist a pressure of at least 800 bars. The mass spectrometer should be a last generation QTOF providing high dynamic range (>10^4) (*see* **Note 7**).
9. Acquity BEH C18 UHPLC column (50 × 2.1 mm i.d., 1.7 μm particle size, Waters).

3 Methods

The methods described herein have been so far applied to the prenylquinone analysis of whole leaves of both dicot (i.e., *Arabidopsis thaliana* and *Pisum sativum*) and monocot plants (i.e., *Zea mays*), and to the analysis of Arabidopsis chloroplast subfractions. For analysis of other plant species, preliminary assays should be performed to verify the applicability of the method and slight adjustments might be needed.

3.1 Whole Leaf Extraction

1. Harvest leaf tissues and immediately dip them in liquid nitrogen (*see* **Note 8**). Store samples at −80 °C until extraction (*see* **Note 9**).
2. Grind leaf tissues into a fine powder and weigh approximately 100 mg of powder in a microcentrifuge tube. Take care that no thawing occurs during this step (*see* **Note 10**).
3. Add 500 μL of extraction solvent and 5–10 glass beads.
4. Vortex for approximately 10 s and extract the sample for 3 min in the mixer mill at a frequency of 30 Hz.
5. Centrifuge the tube for 3 min at 14,000 × *g* at 4 °C.
6. Transfer the supernatant in an HPLC vial (*see* **Notes 11** and **12**).

3.2 Preparation of Intact Chloroplasts and Purification of Plastoglobules

This procedure describes the routine procedure to isolate Arabidopsis chloroplasts and plastoglobules and is adapted from [11]. Each step of the protocol below should be performed at 4 °C to preserve the chloroplast integrity, and to minimize protein degradation. If the starting material is limited, smaller sucrose gradients can be prepared in appropriate tubes by proportionally reducing the volume of each of the different sucrose steps, as exemplified in [12].

1. Prior to harvest, place plants in the dark for 24–48 h to deplete starch (*see* **Note 13**).
2. Harvest Arabidopsis leaves with scissors or a scalpel, weigh them in a beaker, and place them on chilled water for 30 min (*see* **Note 14**).

3. Prepare six Percoll gradients as follows. First pour 15 mL of 40 % Percoll solution into a 50 mL open tube. Then, using a glass Pasteur pipette, carefully underlay 5 mL of 85 % Percoll solution below the 40 % Percoll layer. Keep the gradients at 4 °C.
4. Using a Waring blender, homogenize the leaves three times in 500 mL of cold HB buffer in bursts of 5 s at high strength and then two times 3 s at low strength (*see* **Note 15**).
5. Filter the homogenate immediately through two layers of cheesecloth and one layer of Miracloth placed in a funnel on top of an Erlenmeyer flask. Gently squeeze (*see* **Note 16**) the homogenate inside the cheesecloth to drain most of the liquid suspension.
6. Divide the filtrate between three or four 250 mL bottles and centrifuge for 10 min at 1,075 × *g* in an SLA1500 fixed-angle rotor. This step results in a crude chloroplast pellet.
7. By swirling the tube gently resuspend each pellet in 1–2 mL of 1× RB buffer and pool the crude chloloroplast extracts in a 50 mL Falcon tube. If required, add additional RB buffer to resuspend any residual crude chloroplast pellet, and finally, rinse the tube with a small volume of RB. Gently mix the suspension by inverting the tube and then load 2–3 mL of resuspended crude chloroplasts onto each Percoll gradient (*see* **Note 17**).
8. Centrifuge the gradients for 10 min at 13,600 × *g* in an HB-6 swinging-bucket rotor. At the end of the centrifugation (*see* **Note 18**), intact chloroplasts are located at the interface between the 85 % and 40 % Percoll phases. The green layer at the top of the 40 % sucrose consists of damaged chloroplasts. It should be discarded if intact chloroplasts are required for further experimentation.
9. Remove most of the 40 % Percoll layer with a vacuum aspirator and carefully collect the band of intact chloroplasts with a disposable plastic Pasteur pipette (with large opening). Transfer to a 50 mL Falcon tube.
10. Distribute pooled intact chloroplasts collected from the gradients in two 50 mL Falcon tubes and dilute with 10 volumes of RB buffer (1×). Gently invert the tube to mix.
11. Centrifuge the suspension for 2 min at 2,600 × *g* in a swinging-bucket rotor (A4-62).
12. Carefully decant the supernatant as the pellet is very loose.
13. Resuspend the pellet in 5 mL of 1× TrE bufferand quantify the chlorophyll (*see* **Note 19**).
14. Adjust the sample volume to 50 mL with 1× TrE buffer to wash the chloroplasts, and then centrifuge for 10 min at 2,600 × *g* in a swinging-bucket rotor (A4-62).

15. Carefully decant the supernatant and then resuspend the chloroplast pellet (*see* **Note 20**) with 0.6 M sucrose/TrE to a calculated concentration of 2–3 mg/mL of chlorophyll.
16. Incubate on ice for 10 min and then freeze at −80 °C for at least 1 h.
17. Thaw the chloroplast suspension and dilute with 2 volumes of 1× TrE buffer (*see* **Note 21**).
18. Homogenize chloroplast suspension by at least 20 strokes in a 50 mL Potter homogenizer and then transfer the homogenate into UltraClear SW28 tubes.
19. Carefully balance the tubes to the specified values and then perform ultracentrifugation at 100,000×*g* in the SW28 swinging-out rotor for 1 h.
20. Remove the supernatant, which consists of the soluble stromal fraction (*see* **Note 22**), and resuspend the total membrane (that also contains plastoglobules) in 45 % sucrose/TrE to reach a concentration of 2–6 mg of chlorophyll per mL (typically, ~10–20 mL 45 % sucrose/TrE solution will be required) (*see* **Note 23**).
21. Homogenize the resuspended membranes by 20 strokes in a 15 mL Potter homogenizer (*see* **Note 24**).
22. Pour 5 mL aliquots of the membrane homogenate (*see* **Note 25**) into the required number of UltraClear SW28 tubes, and carefully overlay each aliquot with the sucrose/TrE solutions in the following sequence: 6 mL of 38 % sucrose, 6 mL of 20 %, 4 mL of 15 % and finally 8 mL of 5 % sucrose to the top of the tube (*see* **Note 26**).
23. Precisely balance the tubes by adding or removing 5 % sucrose solution. Centrifuge overnight at 100,000×*g* (*see* **Note 27**). A typical gradient after the overnight centrifugation is shown in Fig. 1.
24. From each gradient, collect 1 mL fractions with a micropipette, from the top (fraction 1) to the bottom of the gradient (approximately 32 fractions), and store them at −80 °C. Typically, plastoglobules are present in fractions 1–6, envelopes in fractions 14–18, and thylakoid membranes in fractions 25–32 (*see* **Note 28**).
25. Dilute 200–400 μL of purified fractions with water to a final volume of 1 mL and extract three times with an equal volume of ethylacetate (*see* **Note 29**). Pool organic phases, evaporate and redissolve the residue in 50–100 μL of THF–water (85:15, v/v). Transfer the solution in an appropriate HPLC vial.

3.3 UHPLC-APCI-QTOFMS Analysis and Processing

1. The UHPLC gradient conditions presented below are appropriate for the Acquity UPLC™ system (Waters) using an Acquity BEH C18 column (50×2.1 mm i.d., 1.7 μm): initial conditions at 80 % B, increasing from 80 to 100 % B in 3 min,

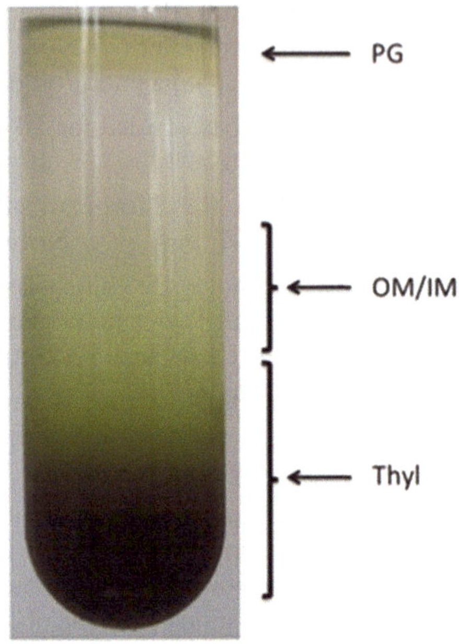

Fig. 1 Sucrose gradient floatation of plastoglobules (PG) from total membranes of isolated Arabidopsis chloroplasts. The figure shows a sucrose step gradient after overnight centrifugation at 100,000 × *g* in a SW28 rotor. PG are contained in the *yellowish* layer on the top of the gradient. Outer and inner envelope membranes (OM/IM) also appear *yellowish*; thylakoid membranes (Thyl) are intensely *green*

holding at 100 % B for 2 min, and re-equilibrating at 80 % B for 1 min (*see* **Note 30**). The flow rate is set to 800 μL/min. The column and autosampler temperatures are kept at 60 and 15 °C, respectively. The injection volume is of 2.5 μL. After injection, 700 μL of strong and weak wash solvents are employed to clean the injection needle (*see* **Note 31**). Retention times for all prenylquinones are given in Table 1.

2. The Q-TOF is operated in negative APCI mode (*see* **Note 32**). Optimized parameters for the Synapt G2 QTOF (Waters) are as follows: Corona current −18 μA, cone voltage −40 V, extraction cone −4.5 V, desolvation gas flow and temperature 800 L/h and 475 °C, respectively, source temperature 120 °C, curtain gas (cone gas) flow 20 L/h. In case another QTOF system is used, source parameters should be optimized by infusing at low flow rate the infusion solution containing prenylquinone standards (*see* **Note 33**). Data are acquired over the range *m/z* 225–1200 Da with a scan time of 0.4 s. The mobile phase is diverted to waste between 0 and 0.5 min (*see* **Note 34**). Internal calibration (Lockspray™) is performed by infusing a 500 ng/mL solution of the pentapeptide leucin-enkephalin in the mass spectrometer at a flow rate of 10 μL/min. The Lockspray™ scan time and frequency are set to 0.5

Table 1
Chromatographic and mass spectrometric characteristics of prenylquinones identified in whole leaf extracts

No.	RT (min)[a]	Ion species	*m/z* (Da)	Formula	Compound
1	1.89	$(M)^{-\bullet}$	446.3763	$C_{29}H_{50}O_3$	α-Tocopherol-quinone
2	1.96	$(M-H)^-$	401.3418	$C_{27}H_{46}O_2$	δ-Tocopherol
3	2.12	$(M-H)^-$	415.3575	$C_{28}H_{48}O_2$	γ-Tocopherol
4	2.26	$(M-H)^-$	429.3734	$C_{29}H_{50}O_2$	α-Tocopherol
5	2.65	$(M)^{-\bullet}$	450.3504	$C_{31}H_{46}O_2$	Phylloquinone
6	3.00	$(M-H)^-$	765.6183	$C_{53}H_{82}O_3$	Hydroxyplastochromanol
7	3.20	$(M)^{-\bullet}$	764.6102	$C_{53}H_{80}O_3$	Hydroxyplastoquinone
8	3.21	$(M-2H)^-$	748.6158	$C_{53}H_{82}O_2$	Plastoquinol-9
9	3.40	$(M-2H)^-$	794.6214	$C_{54}H_{84}O_4$	Ubiquinol-9
10	3.64	$(M-H)^-$	749.6227	$C_{53}H_{82}O_2$	Plastochromanol-8
11	3.68	$(M)^{-\bullet}$	794.6212	$C_{54}H_{82}O_4$	Ubiquinone-9
12	3.71	$(M-2H)^-$	862.6827	$C_{59}H_{92}O_4$	Ubiquinol-10
13	3.96	$(M)^{-\bullet}$	748.6152	$C_{53}H_{80}O_2$	Plastoquinone-9
14	4.11	$(M)^{-\bullet}$	862.6830	$C_{59}H_{90}O_4$	Ubiquinone-10
IS	0.96	$(M)^{-\bullet}$	276.2088	$C_{18}H_{28}O_2$	Decylplastoquinone

RT retention time, *IS* internal standard
[a]Column dead time = 0.16 min

and 15 s, respectively, and data are averaged over five scans for mass correction (*see* **Note 35**). Molecular or pseudo-molecular ions found for all prenylquinones are provided in Table 1.

3. To determine the presence and identity of the different prenylquinones in the leaf extracts, generate extracted ion chromatograms (EIC) using a mass window of ±0.005 Da (*see* **Note 36**) from the total ion chromatogram (TIC) (Fig. 2). Retention times for peaks corresponding to prenylquinones should be identical or close to values provided in Table 1 provided that identical chromatographic conditions are used. Extract mass spectra for the various peaks and check the mass accuracy of the molecular or pseudo-molecular ions (Table 1) to confirm the identity of the compounds of interest (*see* **Note 37**).
4. To quantify the different prenylquinones, run calibration solutions 1–5 and build calibration curves for each prenylquinone (*see* **Note 38**). For this, integrate peak areas obtained from

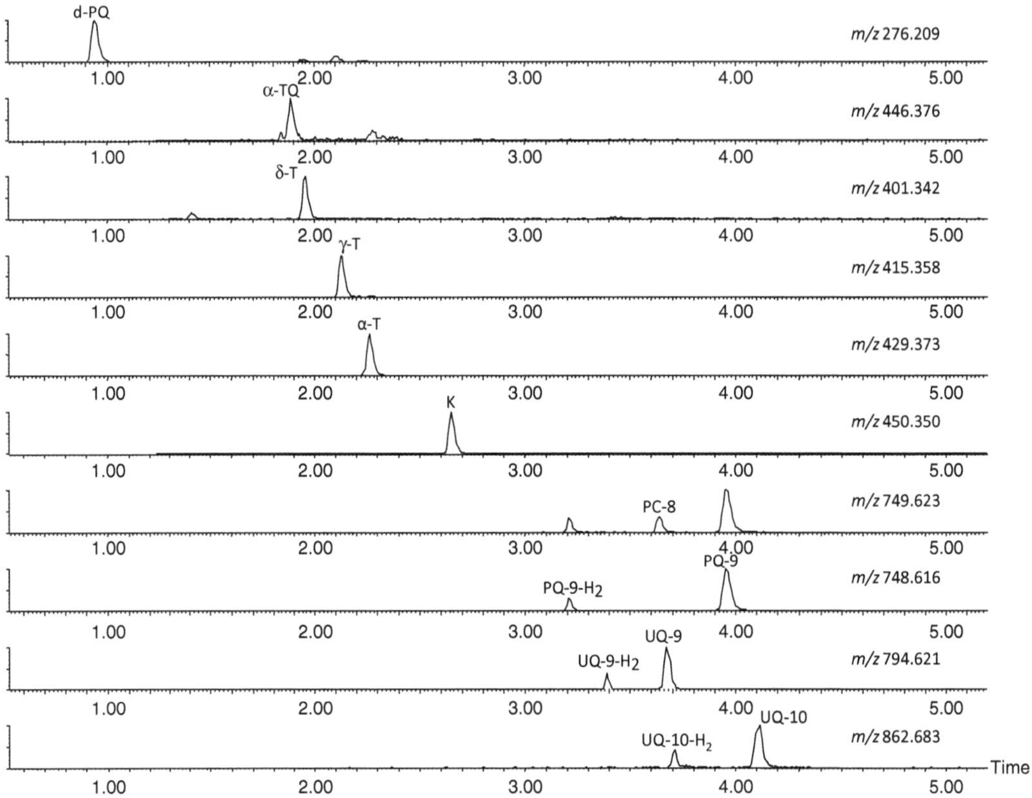

Fig. 2 Typical extracted ion chromatograms (EICs) for predominant prenylquinones present in whole leaf extracts. Windows of ±0.005 Da were used to generate EICs. d-PQ, decylplastoquinone (internal standard), α-TQ, α-tocopherol-quinone, δ-T, δ-tocopherol, γ-T, γ-tocopherol, α-T, α-tocopherol, K, phylloquinone, PC-8, plastochromanol-8, PQ-9-H_2, plastoquinol-9, PQ-9, plastoquinone-9, UQ-9-H_2, ubiquinol-9, UQ-9, ubiquinone-9, UQ-10-H_2, ubiquinol-10, UQ-10, ubiquinone-10

EICs for each prenylquinone at given concentrations and divide them by that of the corresponding internal standard (IS). Using a linear regression model, the calibration equation is calculated as follows (*a* represents the slope and *b* the *y*-intercept):

$$\frac{\text{Area}_{\text{prenylquinone}}}{\text{Area}_{\text{IS}}} = a \times \frac{\text{Concentration}_{\text{prenylquinone}}}{\text{Concentration}_{\text{IS}}} + b$$

Integrate and normalize the peaks obtained from plant samples in the same manner and calculate prenylquinone concentrations in μg/mL using the calibration equations. Normalize prenylquinone concentrations to the mass of fresh leaf tissue weighed before extraction to obtain concentrations in μg/g fresh weight.

4 Notes

1. Analytical or HPLC grade solvents should be used for whole leaf extraction.
2. Decylplastoquinone is used as internal standard (IS) to compensate for extraction and analytical variability.
3. Ascorbate (powder), BSA (powder), and PMSF (100 mM stock solution in isopropanol) should be added immediately before use.
4. This solution should be made fresh occasionally as acetone evaporates over time.
5. Decylplastoquinone, phylloquinone, α-, γ-, and δ-tocopherols were obtained from certified suppliers. Plastochromanol-8, plastoquinone-9, and α-tocopherol quinone standards were provided by Prof. Jerzy Kruk (Jagiellonian University, Kraków, Poland). Calibration standards were prepared in THF containing 15 % of water to take into account the water content of fresh leaf samples.
6. Solvents for UHPLC-APCI-QTOFMS analyses should be of LC-MS grade.
7. Alternatively, single TOF-MS or Orbitrap-type instruments able to acquire data at a compatible frequency may be employed. Care should be taken that prenylquinone concentrations fall within the linear dynamic range of the mass spectrometer used.
8. Freezing in liquid nitrogen quenches plant metabolism. Freeze-clamping might be an efficient alternative for thick tissues.
9. Leaf tissues can be stored for several weeks at this temperature.
10. During this process it is essential that tissues remain always frozen since thawing, even partial and localized, may alter prenylquinone composition. In particular, the redox ratio of plastoquinone-9 is highly sensitive to thawing. Grinding may be achieved using a mortar and a pestle previously dipped into liquid nitrogen. The use of a tissue lyser is not recommended since localized heating has been observed using such device. Microcentrifuge tubes and spatula may also be dipped in liquid nitrogen to facilitate weighing and avoid thawing. Samples should be stored in liquid nitrogen or at −80 °C immediately after weighing.
11. Avoid pipetting any particle from the pellet into HPLC vials since it might damage the UHPLC injector. If the supernatant is not perfectly clear, transfer it to another microcentrifuge tube and recentrifuge.
12. In principle, samples should be immediately analyzed after extraction to avoid possible alteration of the redox state of

prenylquinones. In practice, we observed that grinding and weighing are the most critical steps that may affect redox state and that storage for several hours after extraction under appropriate conditions does not significantly affect the redox state. However, oxidation of plastoquinol and ubiquinol is likely to occur after several days even at −80 °C.

13. During centrifugation high-density starch granules may tear through the envelope and disrupt chloroplast integrity.
14. This will allow contaminating soil particles to sediment.
15. If the leaf volume exceeds that of the Waring blender, carry out in two or three portions.
16. Do not force otherwise unwanted larger debris may enter the fine suspension.
17. Chloroplast are large and fragile organelles. Try to apply as little force as possible by swirling (rather than vortexing), invert (rather than shake), using plastic Pasteur pipets with wide openings, etc.
18. Inactivate the centrifuge break to not disturb the Percoll gradient.
19. Add 5 μL of chloroplast suspension to 1 mL of 80 % acetone. Vortex hard and centrifuge at maximum speed for 2 min in an Eppendorf-type microfuge. Transfer supernatant to 1 mL quartz cuvette (resists acetone) and determine absorbance at 652 nm (A652) using 80 % acetone as the background. Calculate chlorophyll concentration using the following formula:

$$[\text{chlorophyll}](\text{mg} / \text{mL}) = \text{A652} \times \text{dilution factor} / 36$$

20. Using the vortex here is possible because chloroplast integrity is no longer an issue.
21. Freezing and thawing helps rupture the chloroplasts.
22. Store an aliquot of the stromal fraction at −20 °C for future reference.
23. Membrane suspension can be stored in the freezer at this stage.
24. This will physically detach plastoglobules from the thylakoid membrane.
25. To avoid "overloading" the sucrose floatation gradient, do not add more than the equivalent of approximately 30 mg of chlorophyll of the total membrane fraction per tube.
26. Carefully layer the gradient using glass pipettes.
27. Inactivate the ultracentrifuge break to not disturb the sucrose gradient.
28. However, the chloroplast membrane fractions overlap and considerable variation from one experiment to another is not unusual.

Therefore, quality control experiments such as western blotting using specific marker antibodies can be carried out [11].

29. During the partition process, prenylquinones pass in the organic phase whereas sucrose stays in the aqueous phase. It is important to get rid of sucrose which may strongly interfere during UHPLC-APCI-QTOFMS analyses and prevent the detection of prenylquinones.
30. Thanks to the speed of analysis, the proposed method consumes little water and methanol and may thus be considered as a green alternative to traditional prenylquinone analysis involving the use of long chromatographic runs and toxic solvents such as chloroform or hexane. The choice of methanol as strong solvent instead of more hydrophobic and toxic solvents is rendered possible by an increase of the column temperature to 60 °C. At this temperature, prenylquinones remain stable over the time of the analysis.
31. Isopropanol or THF should be employed as strong wash solvent. It is recommended to install a stainless steel needle on the injector instead of a PEEK needle to avoid swelling of the needle. Weak wash solvent should be water or methanol 80 %.
32. Among electrospray and APCI positive and negative modes, it has been demonstrated that negative APCI yields the highest overall sensitivity for prenylquinones [10].
33. UHPLC and infusion flows should be combined to provide realistic chromatographic conditions for the optimization of MS source parameters. On the Synapt G2 QTOF, this can be performed using the in-built fluidics system. Alternatively, a low dead-volume tee may be used to combine flows.
34. This will prevent polar molecules to build in and contaminate the APCI source.
35. Using the Synapt G2 QTOF, mass accuracy should be equal or better than 2 ppm for all measured compounds.
36. Depending on the mass precision and accuracy of the mass spectrometer employed, this value may be adjusted (e.g., ±0.02 Da for old-generation single-TOF).
37. It should be noted that, using the present method based on fast and sensitive non-targeted scanning by QTOF-MS, not only prenylquinones may be detected in leaf extracts, but also a certain number of other hydrophobic compounds such as carotenoids (e.g., β-carotene, lutein) and galactolipids (e.g., MGDG-18:3-16:3, DGDG-18:3/18:3, DGDG-18:3/16:0). Thanks to the high mass accuracy and fragmentation capabilities of QTOF, these molecules may be identified and quantified in a relative manner in plant samples. In addition, multivariate statistical tools may be applied on the datasets to

obtain a global picture of lipidome changes arising from various plant treatments or mutations [3].

38. The number of prenylquinones that can be absolutely quantified is limited by the availability of pure standards. In the course of our analyses, eight prenylquinones (α-tocopherol, α-tocopherol quinone, γ-tocopherol, δ-tocopherol, phylloquinone, plastochromanol-8, plastoquinol-9, and plastoquinone-9) could be absolutely quantified based on calibration curves built from standard solutions, while six others (hydroxyplastochromanol, hydroxyplastoquinone, ubiquinol-9, ubiquinone-9, ubiquinol-10, and ubiquinone-10) were relatively quantified. It should be noted that both reduced and oxidized forms of plastoquinone-9 can be quantified using plastoquinone-9 standard curve since it has been demonstrated that both forms provide identical signals under the conditions employed [10].

References

1. Vidi PA, Kanwischer M, Baginsky S, Austin JR, Csucs G, Dormann P, Kessler F, Brehelin C (2006) Tocopherol cyclase (VTE1) localization and vitamin E accumulation in chloroplast plastoglobule lipoprotein particles. J Biol Chem 281:11225–11234
2. Zbierzak AM, Kanwischer M, Wille C, Vidi PA, Giavalisco P, Lohmann A, Briesen I, Porfirova S, Brehelin C, Kessler F, Dormann P (2010) Intersection of the tocopherol and plastoquinol metabolic pathways at the plastoglobule. Biochem J 425:389–399
3. Piller LE, Besagni C, Ksas B, Rumeau D, Brehelin C, Glauser G, Kessler F, Havaux M (2011) Chloroplast lipid droplet type II NAD(P)H quinone oxidoreductase is essential for prenylquinone metabolism and vitamin K(1) accumulation. Proc Natl Acad Sci U S A 108:14354–14359
4. Ytterberg AJ, Peltier JB, van Wijk KJ (2006) Protein profiling of plastoglobules in chloroplasts and chromoplasts. A surprising site for differential accumulation of metabolic enzymes. Plant Physiol 140:984–997
5. Lundquist PK, Poliakov A, Bhuiyan NH, Zybailov B, Sun Q, van Wijk KJ (2012) The functional network of the Arabidopsis plastoglobule proteome based on quantitative proteomics and genome-wide coexpression analysis. Plant Physiol 158:1172–1192
6. Gaude N, Brehelin C, Tischendorf G, Kessler F, Dormann P (2007) Nitrogen deficiency in Arabidopsis affects galactolipid composition and gene expression and results in accumulation of fatty acid phytyl esters. Plant J 49:729–739
7. Lippold F, vom Dorp K, Abraham M, Holzl G, Wewer V, Yilmaz JL, Lager I, Montandon C, Besagni C, Kessler F, Stymne S, Dormann P (2012) Fatty acid phytyl ester synthesis in chloroplasts of Arabidopsis. Plant Cell 24:2001–2014
8. Kruk J, Karpinski S (2006) An HPLC-based method of estimation of the total redox state of plastoquinone in chloroplasts, the size of the photochemically active plastoquinone-pool and its redox state in thylakoids of Arabidopsis. Biochim Biophys Acta Bioenerg 1757:1669–1675
9. Yoshida K, Shibata M, Terashima I, Noguchi K (2010) Simultaneous determination of in vivo plastoquinone and ubiquinone redox states by HPLC-based analysis. Plant Cell Physiol 51:836–841
10. Martinis J, Kessler F, Glauser G (2011) A novel method for prenylquinone profiling in plant tissues by ultra-high pressure liquid chromatography-mass spectrometry. Plant Methods 7:23
11. Besagni C, Piller LE, Bréhélin C (2011) Preparation of plastoglobules from Arabidopsis plastids for proteomic analysis and other studies. Methods Mol Biol 775:223–239
12. Vidi PA, Kessler F, Bréhélin C (2007) Plastoglobules: a new address for targeting recombinant proteins in the chloroplast. BMC Biotechnol 7:4

Chapter 16

Confocal Laser Scanning Microscopy Detection of Chlorophylls and Carotenoids in Chloroplasts and Chromoplasts of Tomato Fruit

Lucio D'Andrea, Montse Amenós, and Manuel Rodríguez-Concepción

Abstract

Plant cells are unique among eukaryotic cells because of the presence of plastids, including chloroplasts and chromoplasts. Chloroplasts are found in green tissues and harbor the photosynthetic machinery (including chlorophyll molecules), while chromoplasts are present in non-photosynthetic tissues and accumulate large amounts of carotenoids. During tomato fruit development, chloroplasts are converted into chromoplasts that accumulate high levels of lycopene, a linear carotenoid responsible for the characteristic red color of ripe fruit. Here, we describe a simple and fast method to detect both types of fully differentiated plastids (chloroplasts and chromoplasts), as well as intermediate stages, in fresh tomato fruits. The method is based on the differential autofluorescence of chlorophylls and carotenoids (lycopene) detected by Confocal Laser Scanning Microscopy.

Key words Chloroplast, Chlorophylls, Chromoplast, Carotenoids, Lycopene, Confocal microscopy, Tomato fruit, Fluorescence

1 Introduction

Plastids are organelles ubiquitously found in plant cells but absent from animal or fungal cells. Based on their color, structure, and metabolic profile, plastids can be categorized into different types [1]. Proplastids, the progenitors of other plastid types, are colorless plastids with limited internal membrane vesicles which are typically found in meristematic cells. Etioplasts, the plastids of dark-grown (etiolated) seedlings, are yellow plastids that contain low levels of carotenoids associated to prolamellar bodies and prothylakoid membranes. Chloroplasts are green, chlorophyll-accumulating photosynthetic organelles with distinctive internal thylakoid membranes and grana. Chromoplasts are plastids specialized in the production and accumulation of carotenoids in many flowers and fruits. Other plastids found in non-photosynthetic tissues are leucoplasts, a general term for colorless plastids that

Manuel Rodríguez-Concepción (ed.), *Plant Isoprenoids: Methods and Protocols*, Methods in Molecular Biology, vol. 1153, DOI 10.1007/978-1-4939-0606-2_16, © Springer Science+Business Media New York 2014

include elaioplasts (those accumulating oil) and amyloplasts (those accumulating starch granules) [1, 2].

Plastids fulfill different functions, serving as the main sites for photosynthesis (chloroplasts) and other important primary and secondary pathways [2, 3]. Among non-photosynthetic plastids, chromoplasts have been best studied due to their capacity to store massive levels of health-promoting carotenoid pigments and the derived effect on the coloration of plant-derived foods with red (lycopene), orange (carotenes), and yellow (xanthophylls) colors [4–6]. A well-characterized system for the study of chromoplast biogenesis is fruit ripening in tomato (*Solanum lycopersicum*), when the chloroplasts present in mature (i.e., full-size) green fruit differentiate into lycopene-accumulating chromoplasts [7–9]. The chloroplast to chromoplast transition during tomato ripening can be visualized by the change in fruit color from green to orange and red. Color changes are due to the degradation of chlorophylls and the accumulation of carotenoids (particularly lycopene) as ripening progresses. Both types of isoprenoid metabolites are autofluorescent but have different emission spectra, and this property has been exploited to monitor the presence of chloroplasts (chlorophyll-rich), chromoplasts (carotenoid-rich), and intermediate plastids in tomato fruit by Confocal Laser Scanning Microscopy (CLSM) [9, 10]. Here, we present an optimized CLSM-based protocol that virtually eliminates interference between chlorophyll and carotenoid (lycopene) fluorescence signals (Fig. 1). This protocol allows to record and quantify

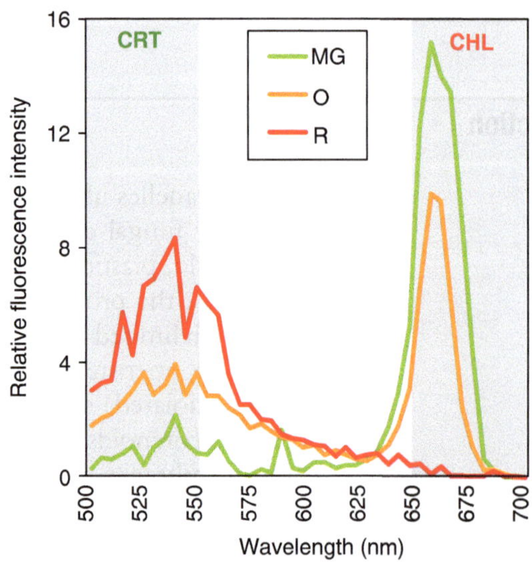

Fig. 1 Fluorescence emission spectra of tomato fruit samples at three stages of fruit development. Pericarp tissue obtained from tomatoes at the *mature green* (MG), *orange* (O), and *red ripe* (R) stages was analyzed by CLSM to generate fluorescence emission spectra after excitation at 488 nm. Representative spectra were obtained from single plastids. Fluorescence intensity is represented relative to the total fluorescence of the sample. The fluorescence emission range used to detect carotenoids (CRT, 500–550 nm) and chlorophylls (CHL, 650–700 nm) is marked

the levels of these isoprenoid pigments in the plastids present in fresh hand-cut sections of tomato fruit pericarp at different developmental stages (Fig. 2). At the mature green stage, all plastids are chloroplasts, which emit red fluorescence due to the presence of chlorophylls (Figs. 1 and 2). At the ripe stage, only fully developed chromoplasts devoid of chlorophylls and rich in lycopene are present. These chromoplasts only emit green fluorescence (Figs. 1 and 2). By contrast, a heterogeneous population of chloroplasts (red fluorescence), chromoplasts (green fluorescence), and intermediate plastids that contain high levels of both chlorophylls and lycopene (yellowish color due to the merging of red and green fluorescence) is found at the breaker and orange stages (Figs. 1 and 2). Although we describe the method for tomato fruit, it can be used (with some optimization) with any other plant material.

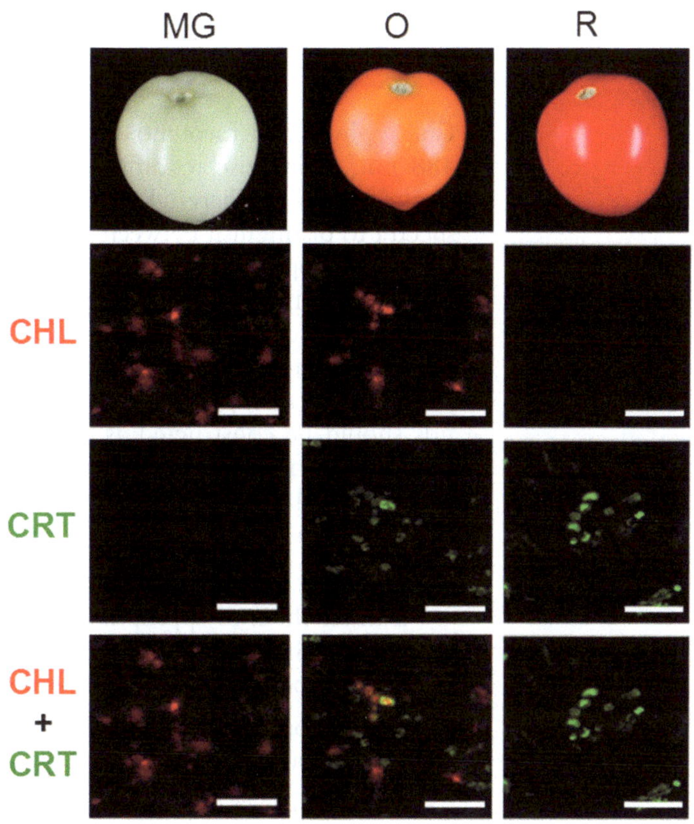

Fig. 2 Images of tomato fruit development stages and the corresponding chlorophyll and carotenoid (lycopene) autofluorescence. Fresh pericarp tissue from tomatoes at the *mature green* (MG), *orange* (O), and *red ripe* (R) stages (*upper panels*) was analyzed by CLSM. Overlay images of autofluorescence emitted at 650–700 nm (chlorophylls, CHL) or 500–550 nm (carotenoids, CRT) after excitation with the 488 nm ray line of an argon laser were obtained. Lower panels correspond to merged images (CHL + CRT). Plastids containing chlorophylls appear *red*, those containing carotenoids appear *green*, and those containing both isoprenoid pigments appear *orange*/*yellow*. Scale bars, 10 μm

2 Materials

1. Greenhouse or plant growth chambers at 22–24 °C at night, and 26–28 °C during the day.
2. Tomato seeds.
3. Soil (vermiculite).
4. Trays and pots.
5. Plastic wrap.
6. Microscope slides and coverslips.
7. Surgical blades and tweezers.
8. Olympus FV 1000 Confocal Laser Scanning Microscope or a similar equipment.

3 Methods

Chlorophylls and carotenoids can be excited with blue light at 488 nm, giving rise to different emission spectra (Fig. 1). The method described here takes advantage of the differences in such spectra to distinguish between organelles that accumulate chlorophyll (chloroplasts), lycopene (chromoplasts), or both in fresh tomato fruit tissue. To improve resolution and avoid overlapping of fluorescence signals, we restricted the detection window to 650–700 nm for chlorophyll and 500–550 nm for lycopene (Fig. 1). For other plant tissues, emission spectra of the target plastids should be constructed as described in **steps** 7 and **8** below and, based on these data, appropriate fluorescence emission windows should be selected for signal detection. **Steps** 7 and **8** can be skipped when analyzing tomato fruit samples.

1. Sow tomato seeds in pots filled with wet vermiculite and transfer them to appropriate trays in the greenhouse or plant growth chamber. Cover the pots with plastic wrap until true leaves appear. Grow the plants until fruits develop.
2. Sample tomato fruits at different developmental stages: mature green, orange, and red ripe (Fig. 2).
3. Cut a thin layer of tomato pericarp tissue using a surgical blade (*see* **Note 1**).
4. Using appropriate tweezers transfer the tissue to a glass slide (*see* **Note 2**) with a drop of water (*see* **Note 3**).
5. Cover the sample with a coverslip (*see* **Note 4**).
6. Place the sample on the microscope stage and focus progressively with the different objectives. Once the region of interest has been selected, use a water-immersion 60× objective (such as U-PlanSApo AN:1,2) to focus the plastids (*see* **Note 5**).

7. Using a zoom factor of 2.5 and the 488 nm ray line of an argon laser for excitation, scan the region of interest in lambda mode to generate emission spectra. Record the emitted fluorescence from 500 to 700 nm using a bandwidth of 10 nm and a stepsize of 5 nm (*see* **Note 6**).
8. Select representative plastids in the scanned region and plot their corresponding emission spectra data using Olympus FV10-ASW or the corresponding CLSM software (Fig. 1).
9. Based on the fluorescence emission spectra obtained, select appropriate fluorescence emission windows for signal detection. For tomato fruit, set the channel for carotenoid (lycopene) detection between 500 and 550 nm and the channel for chlorophyll detection between 650 and 700 nm (*see* **Note 7**).
10. For signal detection, fix the photomultiplier (PMT) settings as follows: PMT High Voltage (HV) ca. 720 V for carotenoids (channel 1) and 770 V for chlorophylls (channel 2); PMT Offset 12 in both channels (*see* **Note 8**).
11. Scan the region of interest taking a *z*-stack of images composed of 8–13 optical sections separated 1 μm. We recommend a resolution of 512 × 512 pixels for digital images. To reduce background noise, we suggest to use a Kalman filter to average the signal over four frames. Set the scanning speed at 4 μs/pixel.
12. Overlay the images of the *z*-stack on a maximum projection to form a single image using the CLSM software (Fig. 2).

4 Notes

1. It is important to minimize tissue damage as much as possible. Damaged cells/tissues can produce false positive signals due to autofluorescence, which typically displays a yellowish color.
2. Although the pericarp sample can be placed on the microscopy slide in any orientation, we recommend laying the sample with the internal (pulp) side facing the slide and put the coverslip on the external (cuticle) side for optimal observation in the bright field.
3. Do not allow the sample to dry. If that occurs, it is recommended to discard it and use a new sample.
4. Pay attention to not generate bubbles, as they can interfere during the focusing process.
5. Focusing can be done directly with the 60× water-immersion objective.
6. An emission wavelength range from 500 to 700 nm includes autofluorescence from chlorophylls and carotenoids in tomato fruit pericarp (Fig. 1) and it should also work for other plant

tissues. If the signal is weak, laser power or bandwidth settings can be increased.

7. It is recommended to first compare samples harboring only one type of plastid (green and red fruit pericarp, in the case of tomato) to make sure that there is no emission fluorescence overlap.
8. The pinhole aperture can be increased if photodamage is observed due to laser illumination or if electronic noise occurs when the photomultiplier HV is increased.

Acknowledgements

Our work is funded by grants from the Catalan AGAUR (2009SGR-26 and XRB), Spanish DGI (BIO2011-23680 and PIM2010IPO-00660), and European Union FP7 (TiMet, contract 245143). We are members of the IBERCAROT network funded by CYTED (112RT0445). L.D. received a predoctoral fellowship of the Spanish Ministerio de Educación FPU program.

References

1. López-Juez E, Pike K (2005) Plastids unleashed: their development and their integration in plant development. Int J Dev Biol 410:557–577
2. Neuhaus HE, Emes MJ (2000) Nonphotosynthetic metabolism in plastids. Annu Rev Plant Physiol Plant Mol Biol 51:111–140
3. López-Juez E (2007) Plastid biogenesis, between light and shadows. J Exp Bot 58: 11–26
4. Lu S, Li L (2008) Carotenoid metabolism: biosynthesis, regulation, and beyond. J Integ Plant Biol 50:778–785
5. Cazzonelli C, Pogson B (2010) Source to sink: regulation of carotenoid biosynthesis in plants. Trends Plant Sci 15:1360–1385
6. Ruiz-Sola MA, Rodriguez-Concepción M (2012) Carotenoid biosynthesis in arabidopsis: a colorful pathway. Arabidopsis Book 10:e0158
7. Waters M, Pyke K (2005) Plastid development and differentiation. Annu Plant Rev Plastids 13:30–59
8. Bian W, Barsan C, Egea I, Purgatto E, Chervin C, Zouine M, Latché A, Bouzayen M, Pech JC (2011) Metabolic and molecular events occurring during chromoplast biogenesis. J Bot 2011:289859
9. Egea I, Barsan C, Bian W, Purgatto E, Latche A, Chervin C, Bouzayen M, Pech JC (2010) Chromoplast differentiation: current status and perspectives. Plant Cell Physiol 51:1601–1611
10. Egea I, Bian W, Barsan C, Jauneau A, Pech JC, Latche A, Li Z, Chervin C (2011) Chloroplast to chromoplast transition in tomato fruit spectral confocal microscopy analyses of carotenoid and chlorophylls in isolated plastids and time lapse recording on intact live tissue. Ann Bot 108:2101–2107

Part IV

Genetic, Pharmacological, and Bioinformatic Tools

Chapter 17

Heterologous Expression of Triterpene Biosynthetic Genes in Yeast and Subsequent Metabolite Identification Through GC-MS

Ery Odette Fukushima, Hikaru Seki, and Toshiya Muranaka

Abstract

Heterologous expression of plant metabolic enzymes in microorganisms is extensively used for identifying genes involved in their pathways and producing useful compounds. Here, we describe a plasmid-based yeast expression system that easily allows the expression of different triterpene biosynthetic genes in wild-type yeast, providing a useful platform for identifying their functions and facilitating combinatorial biosynthesis of triterpenoids.

Key words Combinatorial biosynthesis, Cytochrome P450 monooxygenases (CYPs), Cytochrome P450 reductase (CPR), 2,3-Oxidosqualene, Oxidosqualene cyclases (OSCs), Triterpenoid, Triterpenol, Yeast

1 Introduction

Triterpenoids are a diverse group of specialized metabolites produced by many plant species. Interest in triterpenoids has recently increased owing to data showing their diverse biological activities and beneficial properties. Triterpenoid saponins are synthesized via the mevalonate pathway by cyclization of 2,3-oxidosqualene (a branch point in the sterol pathway) to primarily produce oleanane (β-amyrin), lupane (lupeol), ursane (α-amyrin), or dammarane triterpenoid skeletons. This first cyclization is catalyzed by oxidosqualene cyclases (OSCs). Subsequent modifications that impart functional properties and diversify the basic triterpenoid backbone include the addition of small functional groups, including hydroxyl, keto, aldehyde, and carboxyl moieties, generally followed by glycosylation reactions [1]. Cytochrome P450 monooxygenases (CYPs) perform many of these prior-to-glycosylation modifications [2].

The availability of triterpenoids depends mostly on their direct extraction from plants. Because plant tissue extractions typically yield low triterpenoid concentrations, many research groups have

Manuel Rodríguez-Concepción (ed.), *Plant Isoprenoids: Methods and Protocols*, Methods in Molecular Biology, vol. 1153, DOI 10.1007/978-1-4939-0606-2_17, © Springer Science+Business Media New York 2014

sought methods to increase their production by generating transgenic plants overexpressing biosynthetic genes or by producing them in microbial hosts. One of the most commonly employed hosts is budding yeast (*Saccharomyces cerevisiae*), the oldest industrially exploited microorganism. Molecular tools for cloning and expression of yeast genes were already developed in the 1970s and 1980s, and the expression of heterologous genes in yeast was explored early [3].

Here, we describe a plasmid-based yeast expression system for triterpene biosynthetic genes, namely, OSC (β-amyrin synthase [bAS]), CYPs, and cytochrome P450 reductases (CPRs). This is an example to illustrate the system potential allowing the production of sufficient yields of catalytic products in wild-type yeast, being useful in the functional identification of the abovementioned enzymes and in the combinatorial biosynthesis of triterpenoids [4–7].

2 Materials

2.1 Plasmid Vectors and Bacterial Strains

1. pAUR123 (TaKaRa Bio) for subcloning OSC. It carries the *ADH1* constitutive promoter sequence. For its multiplication, use *Escherichia coli* DH5α competent cells.
2. pYES3/CT (Invitrogen) for the heterologous expression of OSC in yeast under the control of *ADH1* constitutive promoter. For its multiplication, use One shot® TOP10 chemically competent *Escherichia coli* cells (Invitrogen).
3. pENTR™/D-TOPO® vector (Invitrogen) for directional cloning and subsequent transfer to corresponding destination vectors. For its multiplication, use One shot® TOP10 chemically competent *Escherichia coli* cells (Invitrogen).
4. pYES-DEST52 vector (Invitrogen) for combination assays. For its multiplication, use One Shot® ccdB Survival™ 2 T1R competent cells (Invitrogen).
5. pESC-LEU vector (Stratagene) modified for the dual expression of CPR and CYP. For its multiplication, use One Shot® ccdB Survival™ 2 T1R competent cells (Invitrogen).

2.2 Yeast Strain and Growth Media

1. *Saccharomyces cerevisiae* INVSc1 strain (*MATa his3D1 leu2 trp1-289 ura3-52*) from Invitrogen.
2. Luria–Bertani (LB) medium and plates with antibiotics (50 μg/mL of kanamycin or 100 μg/mL of ampicillin) (*see* **Note 1**).
3. Synthetic dextrose (SD) dropout medium and plates (6.7 g of minimal SD base medium and appropriate amounts of dropout supplement in 1 L of deionized H_2O). Important: stir to dissolve

and verify that the medium has a pH 5.8 (adjust if necessary). Autoclave at 121 °C for 15 min. Prepare 20 % of D-glucose or D-galactose solution separately, sterilize by filtration, and add to the SD medium before use. For making plates, add 2 % of agar before autoclaving. Then, add the D-glucose or D-galactose solution before plating.

2.3 Reagents and Chemicals

1. Standards. All sapogenins purchased were at least of analytical grade. β-Amyrin (≥98.5 % purity), α-amyrin (≥98.5 % purity), erythrodiol (≥97 % purity), uvaol (≥95 % purity), oleanolic acid (≥97 % purity), and ursolic acid (≥98.5 % purity) were purchased from Extrasynthese. Lupeol (≥94 % purity), betulin (≥98 % purity), and betulinic acid (≥98 % purity) were purchased from Sigma-Aldrich.
2. In-Fusion Dry-Down PCR cloning kit (Clontech).
3. Gateway LR Clonase Enzyme Mix (Invitrogen).
4. Frozen-EX yeast transformation II Kit™ (Zymo Research).
5. Ethyl acetate, Chloroform, and Methanol (Wako).
6. Silica-gel columns (Sep-pak, Waters).
7. *N*-Methyl-*N*-(trimethyl-d_9-silyl) trifluoroacetamide (Sigma-Aldrich).

2.4 Equipment

1. GC-MS: JMS-AMSUN200 mass spectrometer (JEOL) connected to a gas chromatograph (6890A; Agilent Technologies).
2. Capillary columns DB1-MS (J&W) (30 m × 0.25 mm, 0.25 μm film thickness) and DB1-HT (30 m × 0.25 mm, 0.1 μm film thickness).

3 Methods

3.1 Generation of Plasmid Vectors

1. To generate the pYES3-OSC expression vector for the expression of the OSC (β-amyrin synthase) driven by the constitutive *ADH1* promoter, the β-amyrin synthase cDNA from *Lotus japonicus* was subcloned into the *Kpn*I and *Xba*I sites of pAUR123 and the resultant P_{ADH1}-cOSC1-T_{ADH1} expression cassette was used to replace the P_{GAL1} to CYC1TT region of pYES3/CT using In-Fusion Dry-Down PCR cloning kit with primer combinations 1 and 2 (for the P_{ADH1}-cOSC1-T_{ADH1} cassette) and 3 and 4 (for the pYES3/CT vector fragment without the P_{GAL1} to CYC1TT region) (*see* Table 1).
2. To generate the pELC-CPR-CYP[1] expression vector for the dual expression of CPR and CYP716A12 enzymes driven by the galactose-inducible *GAL10* and *GAL1* promoters, respectively, two cloning strategies were carried out. First, a *L. japonicus*

Table 1
Oligonucleotides used for PCR assays

No.	Target seq	Sequence (5′–3′)
1	ADH1 promoter	GGATGATCCACTAGTGGATCCTCTAGCTCCCTAACATGTAGGTGG
2	ADH1 terminator	TAATGCAGGGCCGCAGGATCCGTGTGGAAGAACGATTACAACAGG
3	pYES3/CT	TGCGGCCCTGCATTAATGAATCGGCCAACG
4	pYES3/CT	TTAAGCTTTGTGTGGATAAAGGCGA
5	CPR	GGGCGGCCGCACTAGTATCGATGGAAGAATCAAGCTCCATGAAG
6	CPR	TTAATTAATCACCATACATCACGCAAATAC
7	CYP716A12	CACCATGGAGCCTAATTTCTATCTCTCCCT
8	CYP716A12	TTAAGCTTTGTGTGGATAAAGGCGA
9	CYP93E2	CACCATGCTTGAAATCCAAGGCTACGTAGTATT
10	CYP93E2	TTAGGCAGAAGAGAATGGAACAAAATGTGGAAC

The underlined sequences were added to facilitate either restriction enzyme digestion-mediated cloning or unidirectional cloning of the product into pENTR/D-TOPO (Invitrogen)

cDNA encoding a cytochrome P450 reductase (named LjCPR1, GenBank accession no. AB433810) was amplified using primers 5 and 6 (*see* Table 1) and cloned into the *Not*I and *Pac*I sites of pESC-LEU (MCS1). Subsequently, a fragment containing the single GATEWAY™ conversion cassette was excised from pAM-PAT-GW (a kind gift from Dr. Bekir Ülker, Max Planck Institute for Plant Breeding, Cologne, Germany) as the *Xho*I/*Spe*I fragment and cloned into the *Sal*I and *Nhe*I sites of pESC-LEU (MCS2) to obtain pELC-CPR. Second, the full-length CYP^1 cDNA (CYP716A12, GenBank accession no. ABC59076.1) was amplified using primers 7 and 8 (*see* Table 1) and cloned into pENTR™/D-Topo® to produce pENTR-CYP^1. The cDNA was then transferred to pELC-CPR using Gateway LR Clonase Enzyme Mix to generate pELC-CPR-CYP^1.

3. To generate the pDEST52-CYP^2 vector for the galactose-inducible expression of CYP^2, the full-length CYP^2 (CYP93E2, GenBank accession no. DQ335790) cDNA was amplified using primers 9 and 10 (*see* Table 1) and cloned into pENTRTM/D-Topo® to produce pENTR-CYP^2. The cDNA was then transferred to pYES-DEST52 using Gateway LR Clonase Enzyme Mix.

3.2 Yeast Transformation and Processing

The transformation procedure is performed in different steps, with the number of steps depending on the number of genes to be expressed. What follows is the example proposed in Fig. 1:

1. Transform the *S. cerevisiae* INVSc1 strain with pYES3-OSC or pYES3 empty vector as a control, and plate the transformed cells on SD-Trp medium agar plates (*see* **Notes 2** and **3**). Make respective glycerol stocks (*see* **Note 4**) and competent cells (*see* **Note 5**).
2. Transform *S. cerevisiae* INVSc1 harboring pYES3-OSC with pELC-CPR-CYP[1] or pELC empty vector as a control, and plate the transformed cells on SD-Trp-Leu medium agar plates. Make respective glycerol stocks and competent cells.
3. Transform *S. cerevisiae* INVSc1 harboring pYES3-OSC and pELC-CPR-CYP[1] with pDEST52-CYP[2] or pYES2/CT empty vector as control. Plate the transformed cells on SD-Trp-Leu-Ura medium agar plates. Make respective glycerol stocks.
4. Prepare yeast cultures for subsequent GC-MS analysis:
   (a) Add 200 μL of fresh culture of each sample to 5 mL of SD-Trp-Leu medium supplemented with 2 % glucose (preferably use 50 mL Falcon tubes).

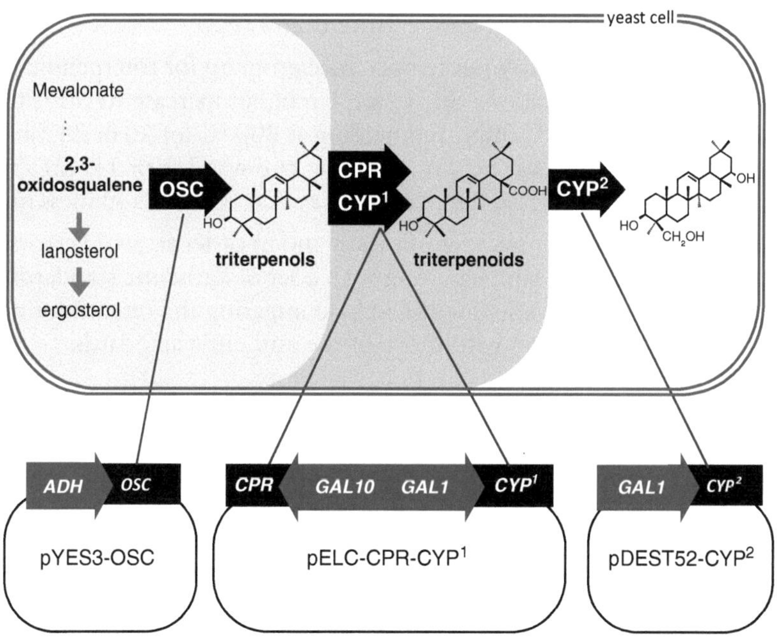

Fig. 1 Generation of OSC/CPR/CYPs-expressing yeast. OSC was introduced under the control of the *ADH1* promoter via pYES3-OSC. Using 2,3-oxidosqualene (an intermediate in the endogenous yeast sterol pathway) as the substrate, OSC (bAS) accumulated β-amyrin endogenously. Next, CPR and CYP[1] were coexpressed under the control of the galactose-inducible promoters *GAL10* and *GAL1*, respectively (via pELC-CPR-CYP[1]). Finally, CYP[2] was expressed under the control of the galactose-inducible promoter *GAL1* via pDEST52-CYP[2]

(b) Shake them (185 rpm) at 30 °C overnight or until the OD_{600} is approximately 2.

(c) After centrifugation, wash the cell pellet with 5 mL of SD-Trp-Leu medium supplemented with 2 % galactose solution (SD-Gal) and transfer to 100 mL Erlenmeyer flasks containing 10 mL of SD-Gal medium.

(d) Incubate at 30 °C with shaking at 185 rpm for 48 h.

3.3 Extraction of Triterpenoids

1. Add 3 mL of ethyl acetate to 5 mL of the culture (*see* **Note 6**), vortex at high speed for 1 min, and sonicate the samples for 30 min.
2. Centrifuge at 35 ×*g* for 5 min and transfer the supernatant to clean assay tubes. Evaporate to dryness. Repeat this process three times.
3. Filter through previously conditioned silica-gel columns (*see* **Note** 7), elute using 10 mL of ethyl acetate, collect in a separate tube, and evaporate to dryness.
4. Dissolve in 300 μL of chloroform–methanol 1:1 (v/v). Transfer 100 μL into a vial tube and evaporate to dryness.
5. For derivatization, add around 50 μL of *N*-methyl-*N*-(trimethyl-silyl) trifluoroacetamide into vials and heat at 80 °C for 30 min.

3.4 GC-MS Analysis

1. Injection temperature is 250 °C.
2. The column temperature program for triterpenoid sapogenins is as follows: 80 °C for 1 min, an increase to 300 °C at a rate of 20 °C/min, and holding at 300 °C for 20 or 28 min. The carrier gas was helium, the flow rate was 1.2 or 1.0 mL/min, and the interface temperature was 300 °C with a splitless injection.
3. For peak identification and in order to avoid errors, we analyze our samples along with a set of authentic standards (*see* Fig. 2). Peaks are identified by comparing the retention times and mass spectra with those of the authentic standards.

4 Notes

1. Prepare all solutions using ultrapure water and analytical grade reagents. Prepare and store all reagents at room temperature (unless indicated otherwise). Diligently follow all waste disposal regulations when disposing of waste materials.
2. Yeast transformations are to be performed using Frozen-EX Yeast Transformation IITM (Zymo Research) as follows:

   (a) Mix 25 μL of competent cells with 0.2–1 μg of DNA (in less than 5 μL volume).

   (b) Add 250 μL of EZ 3 solution, and mix thoroughly (vortex at low speed for 1–2 s).

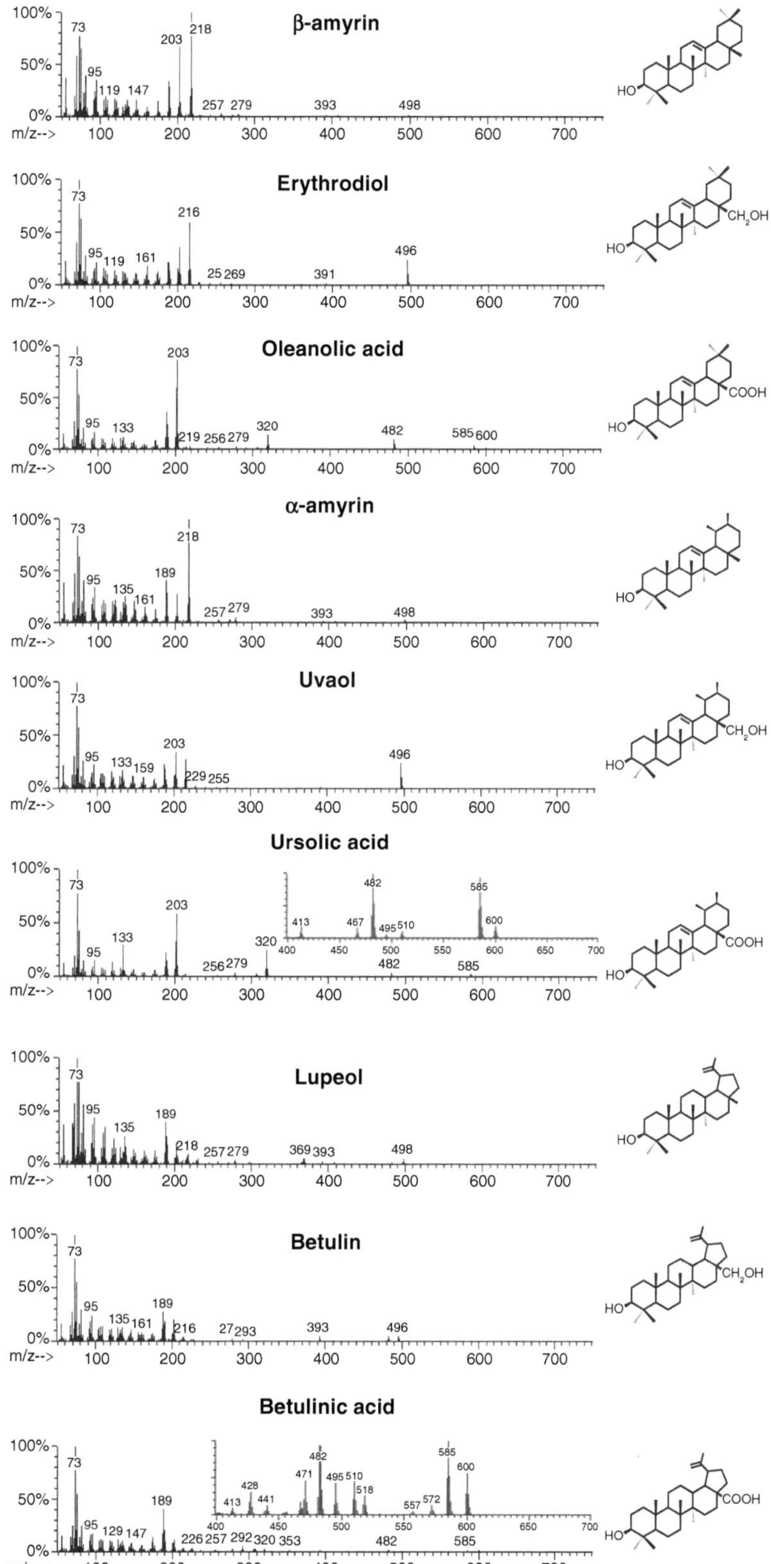

Fig. 2 Mass spectra of common authentic TMS triterpenoids

(c) Incubate at 30 °C for 1 h. Mix vigorously 2–3 times during incubation (vortex at low speed for 1–2 s).

(d) Spread 50–150 μL on appropriate SD plates. Incubate the plates at 30 °C for 2–4 days.

(e) Pick 3-4 colonies (independent clones) using a sterile toothpick and transfer to culture tubes with 2 mL of appropriate SD medium and grow at 30 °C for 2 days.

(f) Verify successful transformation by PCR (*see* **Note 3**).

3. Although yeast transformation rate is very high (nearly 100 %), it is recommended to verify successful transformation by PCR as follows:

(a) Prepare template by mixing 20 μL from each culture with 10 μL of 425–600 μm glass beads (Sigma) into 1.5 mL Eppendorf tubes and vortexing at 1,500 rpm for 45 min.

(b) Set up PCR reaction mix: 10 μL of Quick Taq™ SS Dye Mix (Toyobo, Japan), 7 μL of Milli-Q water, 1 μL of 3′ primer, 1 μL of 5′ primer, 1 μL of DNA template.

(c) Run PCR: 1 cycle at 94 °C for 2 min; 35 cycles at 94 °C for 30 s, 50 °C for 30 s, and 68 °C for 2 min; and 1 cycle at 68 °C for 10 min; followed by a pause at 4 °C.

If no band is detected, try increasing the vortexing time for the template and/or increase the template volume added to the reaction mix.

4. For long-term storage of transformed cells, make glycerol stocks for at least 3 clones per transformed gene. Add equal volumes of sterile 40 % glycerol and yeast culture to cryotubes and store below −70 °C.

5. Competent cells for all yeast strains are prepared using Frozen-EX Yeast Transformation IITM (Zymo Research) as follows:

(a) Grow yeast cells until mid-log phase (OD_{600} within 0.8 and 1.0).

(b) Pellet the cells at 500×*g* for 5 min and discard the supernatant.

(c) Add 5 mL of EZ 1 solution to wash the pellet. Re-pellet the cell and discard the supernatant.

(d) Add 0.5 mL of EZ 2 solution to resuspend the pellet.

(e) Aliquot 25–50 μL into 1.5 mL Eppendorf tubes. Competent cells can be used for transformations directly or stored frozen at or below −70 °C for future use. It is important to freeze cells slowly (wrap the aliquoted cells in 2–6 layers of paper towels or place in a Styrofoam box before placing in the freezer).

6. Of the total 10 mL culture, 5 mL is kept at −20 °C in case a new analysis is required.
7. For conditioning the silica-gel columns, add 3–5 mL of ethyl acetate to the column and let it pass through, repeating twice.

Acknowledgements

This work was financially aid in part by Frontier Research Base for Global Young Researchers, Osaka University, supported by the Ministry of Education, Culture, Sports, Science, and Technology (MEXT).

References

1. Augustin JM, Kuzina V, Andersen SB et al (2011) Molecular activities, biosynthesis and evolution of triterpenoid saponins. Phytochem 72:435–457
2. Croteau R, Kutchan TM, Lewis NG (2000) Natural products (secondary metabolites). In: Buchanan B, Gruissem W, Jones R (eds) Biochemistry and molecular biology of plants. American Society of Plant Physiologists, Rockville, pp 1250–1318
3. Bettiga M, Gorwa-Grauslund MF, Han-Hägerdal B (2010) Metabolic engineering in yeast. In: Smolke CD (ed) The metabolic pathway engineering handbook: fundamentals. CRC Press, Taylor & Francis Group, Boca Raton, FL, pp 22-1–22-48
4. Seki H, Ohyama K, Sawai S et al (2008) Licorice β-amyrin 11-oxidase, a cytochrome P450 with a key role in the biosynthesis of the triterpene sweetener glycyrrhizin. Proc Natl Acad Sci U S A 105:14204–14209
5. Fukushima EO, Seki H, Ohyama K et al (2011) CYP716A subfamily members are multifunctional oxidases in triterpenoid biosynthesis. Plant Cell Physiol 52:2050–2061
6. Seki H, Sawai S, Ohyama K et al (2011) Triterpene functional genomics in licorice for identification of CYP72A154 involved in the biosynthesis of glycyrrhizin. Plant Cell 23:4112–4123
7. Fukushima EO, Seki H, Sawai S et al (2013) Combinatorial biosynthesis of legume natural and rare triterpenoids in engineered yeast. Plant Cell Physiol 54:740–749

Chapter 18

High-Throughput Testing of Terpenoid Biosynthesis Candidate Genes Using Transient Expression in *Nicotiana benthamiana*

Søren Spanner Bach, Jean-Étienne Bassard, Johan Andersen-Ranberg, Morten Emil Møldrup, Henrik Toft Simonsen, and Björn Hamberger

Abstract

To respond to the rapidly growing number of genes putatively involved in terpenoid metabolism, a robust high-throughput platform for functional testing is needed. An *in planta* expression system offers several advantages such as the capacity to produce correctly folded and active enzymes localized to the native compartments, unlike microbial or prokaryotic expression systems. Two inherent drawbacks of plant-based expression systems, time-consuming generation of transgenic plant lines and challenging gene-stacking, can be circumvented by transient expression in *Nicotiana benthamiana*. In this chapter we describe an expression platform for rapid testing of candidate terpenoid biosynthetic genes based on *Agrobacterium* mediated gene expression in *N. benthamiana* leaves. Simultaneous expression of multiple genes is facilitated by co-infiltration of leaves with several engineered *Agrobacterium* strains, possibly making this the fastest and most convenient system for the assembly of plant terpenoid biosynthetic routes. Tools for cloning of expression plasmids, *N. benthamiana* culturing, *Agrobacterium* preparation, leaf infiltration, metabolite extraction, and automated GC-MS data mining are provided. With all steps optimized for high throughput, this *in planta* expression platform is particularly suited for testing large panels of candidate genes in all possible permutations.

Key words *Nicotiana benthamiana*, *In planta*, Transient expression, *Agrobacterium*, High throughput, Enzyme characterization, Terpenoid pathway discovery, Metabolic engineering

1 Introduction

New and increasingly affordable sequencing techniques have become an important tool for discovery of terpenoid pathways. Key biosynthetic steps are cyclization of the terpenoid backbone by terpene synthases (TPS) and oxidative decoration through cytochromes P450 (P450s). TPS are typically encoded by small to medium sized gene families, while genes encoding P450s can reach 1 % of the total genes in vascular plants [1, 2]. The resulting large

Manuel Rodríguez-Concepción (ed.), *Plant Isoprenoids: Methods and Protocols*, Methods in Molecular Biology, vol. 1153, DOI 10.1007/978-1-4939-0606-2_18, © Springer Science+Business Media New York 2014

number of candidate genes presents a major challenge in discovery and reconstitution of multi-enzyme terpenoid pathways.

Heterologous expression in bacteria, yeast, or insect cells is the most common approach for functional characterization of terpenoid enzymes from plants. Nevertheless, expression of TPS and P450s in non-plant hosts is no trivial task, which is reflected by the range of strategies employed to optimize expression. These include codon optimization of the candidate genes and the use of host strains overexpressing tRNAs for rare plant codons. Furthermore, monoterpene and diterpene synthases are plastid targeted enzymes and the N-terminal plastid transit peptide can impair the catalytic activity of these enzymes or cause formation of inclusion bodies [3, 4]. Thus, pseudomature proteins, where the plastid transit peptide has been removed, can be characterized instead [4]. Expression and characterization of plant P450s is also challenging in non-plant systems. Bacteria lack the endoplasmic reticulum (ER), where eukaryotic P450s are localized. However, N-terminal amino acid truncations or modifications can lead to expression of functional P450s in bacterial systems [5]. Although yeast has an ER compartment, N-terminal modifications such as membrane anchor swapping has been shown to improve the chance of obtaining functional P450s from vascular plants [6]. Optimization of expression conditions, assay conditions, or precursor concentrations, has also been shown to yield functional enzymes. All the mentioned engineering and optimization steps can be time demanding and are undesirable in a high-throughput context. Moreover, the enzymatic activity can be influenced by the cellular environment of the expression host. For example, the enzymatic products of a Norway spruce diterpene synthase were shown to differ between in vitro and in vivo expression systems [7, 8], whereas the products of a monoterpene synthase from sweet basil expressed in grapevine, *Arabidopsis* or tobacco differed from the products observed when the same enzyme was expressed in microbial systems [9].

The protein synthesis machinery, posttranslational modifications, cellular compartments, and the function of transit peptides for protein targeting are conserved among higher plants. Thus, plant expression systems have the capacity to express native plant terpenoid genes without any modifications of the coding sequence. Additionally, *in planta* expression systems contain the machineries for posttranslational modification and protein targeting, the relevant cellular compartments (plastids and endoplasmic reticulum), and the universal C_5 building blocks for isoprenoid biosynthesis. However, most plant based systems are not suitable for high-throughput approaches due to time consuming generation of transgenic lines and challenges of gene stacking. In this chapter we present an *in planta* expression platform suitable for high-throughput screening of terpenoid candidate genes using *Agrobacterium tumefaciens* mediated transient expression in

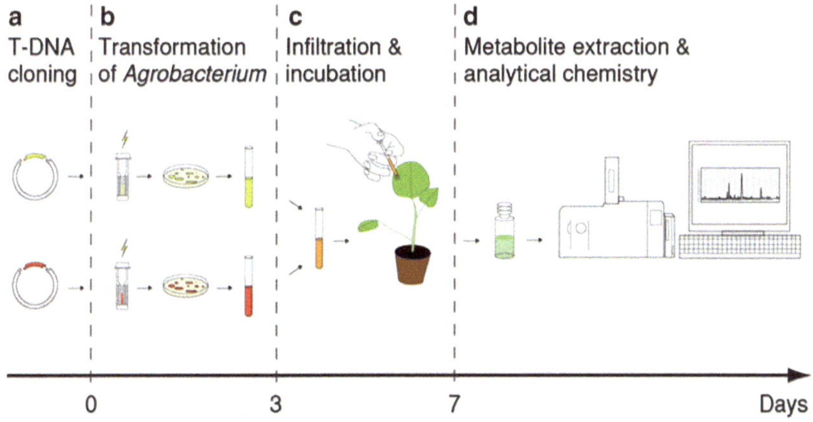

Fig. 1 Workflow illustrating co-expression of two candidate genes. (**a**) Transfer DNA (T-DNA) expression plasmids are created by cloning of native and full-length candidate gene PCR products. (**b**) Expression plasmids are transformed into *Agrobacterium* by electroporation and overnight cultures are subsequently inoculated. (**c**) Different *Agrobacterium* strains, carrying different genes of interest are mixed and co-infiltrated into leaves of *N. benthamiana*. This facilitates co-expression. (**d**) Following 4 days of incubation, metabolites are extracted and analyzed by GC-MS. After the T-DNA expression plasmids have been created, transformation, infiltration, and metabolite analysis can be performed within 7 days

Nicotiana benthamiana leaves. This system facilitates testing of hundreds of native full-length gene combinations in 7 days with minimal hands-on time (Fig. 1). In terms of workload and experimental time-line, this is comparable with microbial based heterologous systems (*see* Chapter 17) but with the advantages of a plant-based expression system. In our laboratory we have successfully expressed mono-, sesqui-, di-, and tri-terpene synthases as native full-length genes. Detection of the corresponding terpenoid products did not require supplementation of substrates, indicating availability of endogenous precursor molecules. In addition, a simple terpenoid biosynthetic pathway has been reconstituted by co-expression of a TPS and a P450. The "plug and play" nature of this transient expression platform is suitable for high-throughput screening of candidate genes, and the option of easy co-expression of multiple genes makes it a very valuable tool in discovery and reconstitution of multi-enzyme terpenoid pathways from vascular plants. The protocol presented here provides all necessary information for establishing this transient *in planta* expression platform in any molecular biological laboratory with access to plant growth and analytical facilities.

2 Materials

2.1 Growth of Biological Materials

1. *N. benthamiana* seeds can be obtained from the Tobacco Institute of Bergerac in France (http://www.imperial-tobacco-bergerac.com).

2. 5.5 cm plastic pots and a tray.
3. Greenhouse soil (e.g., Pindstrup substrate no. 2, blonde sphagnum peat with granulated clay, medium fertilizer level, pH 6.0).
4. Biological larvicide (e.g., Vectobac®).
5. Toothpicks.
6. Transparent film and Parafilm.
7. Common garden NPK 5-1-4 fertilizer.
8. Greenhouse with 16 h light (24 °C) and 8 h dark (17 °C) period, 80 % humidity (*see* **Note 1**).
9. *Agrobacterium* strain AGL-1—GV3850 [10] (*see* **Note 2**).
10. T-DNA expression plasmid encoding the anti-posttranscriptional gene silencing protein p19 (35S:p19) [11].
11. LB Medium: LB Broth low salt, adjusted to pH 7.2 with NaOH, with and without 50 μg/mL kanamycin, 34 μg/mL rifampicin, or 50 μg/mL carbenicillin. For solid media add 1.5 % agar.
12. TE buffer: 10 mM Tris–HCl pH 8.0, 1 mM EDTA.
13. 10 % (v/v) sterile glycerol solution.

2.2 General Laboratory Equipment

1. Incubators at 28 °C with and without agitation.
2. Tabletop centrifuge.
3. Spectrophotometer and UV cuvette.
4. Liquid nitrogen.
5. –80 °C freezer.
6. Electroporation cuvette (Bio-Rad; 2 mm) and Gene Pulser (Bio-Rad; Capacity 25 μF; 2.5 kV; 400 Ω).
7. Water spray atomizer.
8. Razor blade.

2.3 Metabolite Analysis

1. Silanized 2 and 20 mL glass vials with screw top including screw caps with Teflon-coated silicone septa.
2. Solid-phase micro-extraction (SPME) fiber made of polydimethylsiloxane (PDMS). Grey fiber.
3. GC-MS: Shimadzu GC-2010GC, with a 30 m Agilent HP-5MS column (250 μm i.d., 0.25 μm film) and helium as carrier gas. CTC AOC-5000 auto injector and a GCMS-QP2010 PlusMS Detector using electrical ionization mode. Similar GC-MS systems can be applied.
4. Internal standards (e.g., limonene, valencene or 1-eicosene). C7-C30 alkane standard for calculation of retention index.

5. Solvent for extraction e.g., GC-grade hexane.
6. Methanol (GC-grade) and trimethylsilyl (TMS) diazomethane for derivatization.
7. N_2 gas (grade 2 or higher).

3 Methods

Transient expression in *N. benthamiana* has been used for numerous research applications; however this protocol is optimized for rapid characterization of terpene genes.

3.1 Growth of N. benthamiana Plants

1. Fill the tray with pots containing greenhouse soil. Use water supplemented with a biological larvicide (e.g., Vectobac®) to wet the soil and a toothpick to sow a single seed in each pot.
2. Cover the tray with transparent film, to keep the seeds moist, and move the tray into the greenhouse under 16 h light (24 °C) and 8 h dark (17 °C) regime and 80 % humidity (*see* **Note 1**).
3. When seedlings start to emerge (normally after 3–4 days), remove the film and cover the bottom of the tray with water three times per week using general NPK fertilized water once per week.
4. Plants are ready for infiltration when they have developed 5–6 leaves. This usually takes 4–6 weeks.

3.2 Construction of T-DNA Expression Plasmids

T-DNA vectors for *Agrobacterium* infiltration can be created using many different cloning strategies. We find the Uracil-Specific Excision Reagent (USER™) cloning technique robust and suitable for high-throughput cloning. However, other rapid-cloning techniques can be used (e.g., In-Fusion HD® cloning or Gibson Assembly™). A subset of T-DNA expression vectors can conveniently be obtained from pCAMBIA (http://www.cambia.org/daisy/cambia/585). This protocol is intended for T-DNA expression plasmids encoding a bacterial kanamycin resistance marker (e.g., pCAMBIA130035Su) [12].

3.3 Preparation of Electrocompetent Agrobacterium

Electrocompetent *Agrobacterium* cells can also be acquired from commercial providers or prepared using standard procedures (*see* **Note 2**).

1. *Agrobacterium* is stored as a 50 % glycerol stock at −80 °C.
2. Inoculate a petri dish containing LB medium and appropriate antibiotics with the *Agrobacterium* strain AGL-1 (rifampicin resistant) harboring the pGV3850 Ti plasmid (providing carbenicillin resistance) and incubate at 28 °C for 1–2 days until colonies are visible.

3. Inoculate 5 mL of LB liquid medium, including the antibiotics and incubate overnight at 28 °C under 200 rpm agitation.
4. Use the starter culture to inoculate 500 mL of LB medium including antibiotics, and incubate approximately 15–18 h at 28 °C with agitation (170 rpm) until the OD_{600} reaches 0.5–0.8.
5. Cool the culture to 4 °C (30 min) and centrifuge for 10 min at 3,000 × *g* at 4 °C.
6. Discard the supernatant and resuspend the pellet in 5 mL of ice-cold water. Adjust the volume to 500 mL with ice-cold water.
7. This washing step is repeated twice by resuspending in 250 and 50 mL of ice-cold water, respectively.
8. Finally resuspend in 5 mL of 10 % (v/v) ice-cold glycerol solution and freeze 50 μL aliquots in liquid nitrogen.
9. The electrocompetent *Agrobacterium* cells are stored at −80 °C.

3.4 Transformation of Agrobacterium Cells

1. Mix 50 ng of expression vector with 50 μL of electrocompetent *Agrobacterium* cells thawed on ice.
2. Transfer the mixture to a pre-cooled 2 mm electroporation cuvette and electroporate using a Gene Pulser (Capacity 25 μF; 2.5 kV; 400 Ω).
3. Let the transformed bacteria recover in 950 μL of LB media for 4 h at 28 °C under agitation (170 rpm).
4. Spin the cells 3 min at 3,000 × *g* and remove 900 μL of the supernatant. Resuspend the cells in the residual liquid and spread them on LB solid medium supplemented with kanamycin, rifampicin, and carbenicillin. If necessary, leave the lid open for 10 min to let excess water evaporate.
5. Seal the petri dish with parafilm and incubate at 28 °C. Colonies should appear after 48 h.
6. The plate can be stored at 4 °C (*see* **Note 3**).

3.5 Agroinfiltration of N. Benthamiana Leaves

1. Inoculate 10 mL of LB medium containing kanamycin, rifampicin, and carbenicillin with several *Agrobacterium* colonies from the same plate and grow for 20 h at 28 °C under vigorous shaking (200 rpm). In order to ensure sufficient airflow use a 50 mL culture tube with a loose lid taped on. This is performed for each of the constructs to be included in the experiment. Cultures of an *Agrobacterium* strain harboring a plasmid encoding the anti-posttranscriptional gene silencing protein p19 should be included to ensure high levels of transient expression [11] (*see* **Note 4**).

2. After 20 h incubation at 28 °C the bacteria should have reached an OD_{600} of 1.0 or higher. Pellet the bacteria by spinning at 4,000 × *g* for 10 min at 16 °C and resuspend in 5 mL of water.
3. Repeat twice to remove residual antibiotics.
4. Measure OD_{600} of each *Agrobacterium* culture and adjust the OD_{600} to 0.4 with water. Mix each *Agrobacterium* strain harboring the plasmids encoding your gene of interest and the *Agrobacterium* strain encoding the p19 protein in a 1:1 ratio. For co-infiltration of two *Agrobacterium* strains harboring different plasmids as well as the p19 strain, strains are mixed in a 1:1:1 ratio (*see* **Notes 5** and **6**). Three leaves can be infiltrated with approximately 5 mL of infiltration solution. This will generate experimental triplicates.
5. Spray the plants with water using a water spray atomizer 1–10 min before infiltration.
6. *N. benthamiana* leaves are infiltrated by placing a 1 mL syringe onto the abaxial side of a leaf. Support with a finger the adaxial side of the leaf while pressing the piston gently (*see* **Note 7**).
7. Avoid bubbles in the infiltration solution. Infiltrate 3 leaves per plant with the same constructs to yield experimental triplicates.
8. Return the plants to the greenhouse (*see* **Note 1**).
9. Metabolites are extracted 3–4 days after infiltration.

3.6 Metabolite Extraction

3.6.1 Analysis of Volatile Terpenoid Products

This method allows an automated high-throughput analysis of headspace for monoterpenoids and sesquiterpenoids. By using SPME fibers and auto sampling it is possible to screen several samples per hour with minimal hands-on time.

1. Harvest 1–2 leaves by cutting the petiole with the razor blade. Place leaves with the petiole down in a 20 mL GC vial containing 3 mL of sterile water.
2. Cap the vials and incubate the leaf material in the water for 24–48 h (*see* **Note 8**).
3. Inject a SPME fiber through the silicone septum of the vial and incubate the fiber, without touching the leaf, for 20 min at room temperature to adsorb volatiles and analyze by GC-MS.

3.6.2 Analysis of Nonvolatile Terpenoid Products

This method allows an automated high-throughput analysis of leaf extracts for diterpenoids and triterpenoids.

1. Transfer two leaf disks (approximately 3 cm in diameter) to a 2 mL amber GC vial containing 1 mL of hexane spiked with an internal standard (e.g., 1 mg/L 1-eicosene) (*see* **Note 9**).
2. Vortex vigorously for 15 s and leave for 1 h on an orbital shaker (150 rpm) at room temperature.

3. Spin the vials at 2,500 × *g* for 30 s to sediment leaf material and transfer the solvent into new GC vials avoiding transfer of leaf material.
4. Capped samples can be stored at −20 °C until GC-MS analysis.

3.6.3 Derivatization of Oxidized Terpenoid Acid Products

1. Transfer 400 μL of solvent to a new 2 mL GC vial and add 200 μL of methanol and 100 μL of TMS diazomethane. In the presence of methanol the TMS diazomethane reacts with the carboxylic acid and yields a more volatile methyl ester.
2. Vortex briefly and leave at room temperature for 20 min.
3. Evaporate the solvent under a gentle nitrogen-stream and resuspend the residue in 400 μL of hexane by vortexing.
4. Samples can be stored at −20 °C until GC-MS analysis (*see* **Note 10**).

3.6.4 Metabolite Analysis Using GC-MS

In this section a general GC-MS program suitable for detection of most terpenoids is provided (mono- to diterpenoids). These general conditions can be further modified to shorten runtime, improve resolution or sensitivity for specific compounds. Injection volume 1 μL, splitless injector temperature 250 °C, 30 m HP-5MS column, 2 mL/min He carrier gas, GC temperature program (50 °C for 3 min, 15 °C/min to 310 °C and hold 5 min), scan range from *m/z* 50 to *m/z* 350 with 70 eV electrical ionization.

3.7 High-Throughput Data Mining

When performing large combinatorial co-expression experiments in *N. benthamiana*, detection and identification of products can be challenging and time consuming. A general strategy on how this process can be automated in order to ease the task of elucidating patterns in compounds detected from numerous samples is provided here. Furthermore, tips are given how the quantification of peak-area for each chromatogram can be exported into for example Microsoft Excel, avoiding excessive manual cut and paste. This automated data mining is especially suitable for large experimental setups containing complex mixtures of known and unknown terpenoid products.

First a custom library of compounds is built. Starting with the empty vector control and all available authentic standards, each compound is added to the custom compound library together with the corresponding mass fingerprint and retention index. The compound is assigned a unique identifier (e.g., name or number). In the analysis of the following biological sample the newly built library is used to identify known compounds and compounds constituting the phytochemical background of *N. benthamiana*. New compounds, not found in the custom library, are added to the library. As the number of previously detected compounds in the custom library grows, the chromatogram analysis becomes faster.

Most modern GC-MS software can export information of detected compounds including their unique identifier, peak-area and other relevant information. Our software (Shimadzu GCMS Postrun Analysis Program) has the option of exporting this information as text files. Using basic data processing software (Matlab, C++, or macros in Microsoft Excel) the data files can rapidly be imported into Microsoft Excel.

4 Notes

1. Since transient expression in *N. benthamiana* is very robust, growth conditions can be varied.
2. Various *Agrobacterium* strains can be used for transient *N. benthamiana* infiltration. In our laboratory we have successfully used the strains AGL-1, GV3850, and LB4404-virGN54D [14]. Competent ElectroMAX™ *Agrobacterium* LBA4404 cells can conveniently be purchased from Invitrogen or prepared using standard procedures [13].
3. In our experience freshly transformed *Agrobacterium* is best for transient expression; however, *Agrobacterium* can be kept on plates for a few weeks, without losing the ability to transform *N. benthamiana*.
4. In order to prevent posttranslational gene silencing we recommend co-infiltrating with an *Agrobacterium* strain harboring an expression plasmid encoding the p19 protein [11]. The *Agrobacterium* strain harboring p19 is included in each leaf infiltration and, thus, several 10 mL cultures of p19 *Agrobacterium* strain should be inoculated in parallel for each experiment.
5. It is possible to leave the *Agrobacterium* solution for several hours at this stage.
6. To optimize transformation efficiency, we recommend testing different densities of the bacterial suspension when using new *Agrobacterium* strains and new constructs. Furthermore, we recommend monitoring the metabolite accumulation from 3 to 10 days to determine the optimal time for metabolite extraction.
7. Choosing leaf material and conditions before infiltration is critical. Infiltration of medium sized leaves is easier and shows better reproducibility. By following guidelines the leaf stomata will be open which will ease the infiltration procedure and give best reproducibility in metabolite yield: Do not water the plants the last 24 h before infiltration. Avoid wind, temperature shifts, direct sunlight, and hygrometric shocks before and during infiltration. If possible, move the plants more than 2 h before infiltration for acclimation.

8. The incubation time in the GC vials can be extended if only minor levels of products are detected. A detached leaf can be incubated up to 1 week in a GC vial.
9. For normalization of the biomass a 50 mL falcon tube, cork borer or wad punch can be used to generate uniform leaf disks.
10. Glycosylation of transiently produced terpenoids at a hydroxyl-, or carboxyl-group has been reported, probably as a part of a detoxification metabolism [15]. However, in our lab glycosylation of terpenoid acids has not yet been observed. If terpenoid acids are glycosylated by *N. benthamiana* the glucose moieties can easily be hydrolyzed and the terpenoid can be detected after derivatization by GC-MS [15]. Alternatively one can employ LC-MS methods [16].

Acknowledgements

This work was supported by the UNIK [Universitetsforskningens Investeringskapital (Investment Capital for University Research)] Center for Synthetic Biology at the University of Copenhagen, funded by the Research Initiative of the Danish Ministry of Science, Technology, and Innovation and the Novo Nordisk Foundation, grant to BH.

References

1. Hamberger B, Bak S (2013) Plant P450s as versatile drivers for evolution of species-specific chemical diversity. Phil Trans R Soc B 368:20120426
2. Chen F, Tholl D, Bohlmann J et al (2011) The family of terpene synthases in plants: a mid-size family of genes for specialized metabolism that is highly diversified throughout the kingdom. Plant J 66:212–229
3. Williams DC, McGarvey DJ, Katahira EJ et al (1998) Truncation of limonene synthase preprotein provides a fully active "pseudomature" form of this monoterpene cyclase and reveals the function of the amino-terminal arginine pair. Biochem 37:12213–12220
4. Peters RJ, Flory JE, Jetter R et al (2000) Abietadiene synthase from grand fir (*Abies grandis*): characterization and mechanism of action of the "pseudomature" recombinant enzyme. Biochemistry 39:15592–15602
5. Barnes HJ, Arlotto MP, Waterman MR (1991) Expression and enzymatic activity of recombinant cytochrome P450 17-alpha-hydroxylase in *Escherichia coli*. Proc Natl Acad Sci U S A 88:5597–5601
6. Nedelkina S, Jupe SC, Blee KA et al (1999) Novel characteristics and regulation of a divergent cinnamate 4-hydroxylase (CYP73A15) from French bean: engineering expression in yeast. Plant Mol Biol 39:1079–1090
7. Hamberger B, Ohnishi T, Hamberger B et al (2011) Evolution of diterpene metabolism: Sitka spruce CYP720B4 catalyzes multiple oxidations in resin acid biosynthesis of conifer defense against insects. Plant Physiol 157:1677–1695
8. Martin DM, Fäldt J, Bohlmann J (2004) Functional characterization of nine Norway spruce TPS genes and evolution of gymnosperm terpene synthases of the TPS-d subfamily. Plant Physiol 135:1908–1927
9. Fischer MJ, Meyer S, Claudel P et al (2012) Specificity of *Ocimum basilicum* geraniol synthase modified by its expression in different heterologous systems. J Biotechnol 163:24–29
10. Zambryski P, Joos H, Genetello C et al (1983) Ti-plasmid vector for the introduction of

DNA into plant-cells without alteration of their normal regeneration capacity. EMBO J 2: 2143–2150

11. Voinnet O, Rivas S, Mestre P et al (2003) An enhanced transient expression system in plants based on suppression of gene silencing by the p19 protein of tomato bushy stunt virus. Plant J 33:949–956
12. Nour-Eldin HH, Hansen BG, Nørholm MHH et al (2006) Advancing uracil-excision based cloning towards an ideal technique for cloning PCR fragments. Nucleic Acids Res 34(18):e122
13. Weigel D, Glazebrook J (2006) Transformation of agrobacterium using electroporation. Cold Spring Harb Protoc. doi:10.1101/pdb.prot4665.
14. van der Fits L, Deakin EA, Hoge JHC et al (2000) The ternary transformation system: constitutive *virG* on a compatible plasmid dramatically increases Agrobacterium-mediated plant transformation. Plant Mol Biol 43:495–502
15. van Herpen TW, Cankar K, Nogueira M et al (2010) *Nicotiana benthamiana* as a production platform for artemisinin precursors. PLoS One 5:e14222
16. De Vos RCH, Moco S, Lommen A et al (2007) Untargeted large-scale plant metabolomics using liquid chromatography coupled to mass spectrometry. Nat Protoc 2:778–791

Chapter 19

Heterologous Stable Expression of Terpenoid Biosynthetic Genes Using the Moss *Physcomitrella patens*

Søren Spanner Bach, Brian Christopher King, Xin Zhan, Henrik Toft Simonsen, and Björn Hamberger

Abstract

Heterologous and stable expression of genes encoding terpenoid biosynthetic enzymes *in planta* is an important tool for functional characterization and is an attractive alternative to expression in microbial hosts for biotechnological production. Despite improvements to the procedure, such as streamlining of large scale Agrobacterium infiltration and upregulation of the upstream pathways, transient *in planta* heterologous expression quickly reaches limitations when used for production of terpenoids. Stable integration of transgenes into the nuclear genome of the moss *Physcomitrella patens* has already been widely recognized as a viable alternative for industrial-scale production of biopharmaceuticals. For expression of terpenoid biosynthetic genes, and reconstruction of heterologous pathways, *Physcomitrella* has unique attributes that makes it a very promising biotechnological host. These features include a high native tolerance to terpenoids, a simple endogenous terpenoid profile, convenient genome editing using homologous recombination, and cultivation techniques that allow up-scaling from single cells in microtiter plates to industrial photo-bioreactors. Beyond its use for functional characterization of terpenoid biosynthetic genes, engineered *Physcomitrella* can be a green biotechnological platform for production of terpenoids. Here, we describe two complementary and simple procedures for stable nuclear transformation of *Physcomitrella* with terpenoid biosynthetic genes, selection and cultivation of transgenic lines, and metabolite analysis of terpenoids produced in transgenic moss lines. We also provide tools for metabolic engineering through genome editing using homologous recombination.

Key words *Physcomitrella patens*, *In planta*, Functional characterization, Terpenoid biosynthesis, Terpene synthases, Heterologous expression, Metabolite analysis

1 Introduction

A relatively small and publicly available genome of *Physcomitrella* of just over 500 Mb, together with efficient homologous recombination in the nuclear genome, which is not feasible in any other plant system, allows convenient genome editing, including targeted removal of endogenous genes and introduction of transgenes. The higher time requirement for stable integration of genes into the genome of *Physcomitrella*, when compared to short-term

Manuel Rodríguez-Concepción (ed.), *Plant Isoprenoids: Methods and Protocols*, Methods in Molecular Biology, vol. 1153, DOI 10.1007/978-1-4939-0606-2_19, © Springer Science+Business Media New York 2014

transient systems, is clearly offset by maintenance of expression, and consequently production over time. Cost-effective and simple growth has helped establishing *Physcomitrella* as the most widely used system in plant photo-bioreactors [1]. The potential for engineering terpenoid biosynthesis was recently demonstrated by successful expression of a diterpene synthase from *Taxus brevifolia* in *Physcomitrella* which resulted in accumulation of taxadiene, the C_{20} precursor diterpene of the lighthouse anticancer drug paclitaxel [2].

Physcomitrella represents an ancient lineage of land plants, and its metabolic and chemical diversity is low compared to higher plants. This is illustrated by the number of cytochromes P450 (P450s) and UDP glycosyltransferases (UGTs) found in the genome. The genomes of *Arabidopsis thaliana* and *Oryza sativa* contain 246 and 343 P450s respectively, while the genome of *Physcomitrella* only contains 71 [3]. Similarly the genome of *Physcomitrella* contains a low number of UGTs compared to other land plants [4]. The low number of P450s and UGTs found in *Physcomitrella* and the correspondingly lower chemical diversity reduces the risk of unspecific modifications by endogenous enzymes and products through pathways used in higher plants for detoxification of xenobiotics. In particular, in the *Nicotiana benthamiana* transient expression system, (*see* Chapter 18) those pathways can interfere with production and detection of hydroxylated terpenoids in form of conjugation of terpenoids [5]. In addition, *Physcomitrella* has a simple terpenoid profile and a genome that only contains a single diterpene synthase (TPS) [6]. This gene encodes a bifunctional copalyl-diphosphate synthase/kaurene synthase (PpCPS/KS) responsible for producing *ent*-kaurene. This product is used for the biosynthesis of the phytohormones gibberellins (GAs) in higher land plants [7]. However, GAs have not been detected in *Physcomitrella* [8, 9]. Although the role of kaurene-type molecules in *Physcomitrella* is unclear, they are produced in impressively large quantities, indicating a high native capacity to produce terpenoids [10, 11].

Gene editing by efficient homologous recombination in *Physcomitrella* provides a very powerful tool for metabolic engineering [12]. Targeted knockouts of PpCPS/KS result in viable kaurene-free moss lines, which have been created independently in our laboratory and in other laboratories [13]. In addition to having a terpenoid free-background, these transgenic moss lines may have the capacity to provide geranylgeranyl diphosphate (the precursor for the biosynthesis of kaurene-type diterpenes and other plastidial isoprenoids) that could be redirected into engineered heterologous terpenoid pathways. A clean metabolic background and a strong potential for metabolic engineering by homologous recombination are unique among plants and are key strengths of this *in planta* heterologous expression system. Additionally, stable transgenic lines can be generated within 2–3 months and are easily maintained on solid media or grown in liquid photo-bioreactors. These features make *Physcomitrella* a very attractive system for producing high levels of terpenoids for basic research (purification and subsequent characterization of metabolites using NMR) as well as for the biotechnology industry, where this moss is currently used for large scale production of therapeutic antibodies in 300 L reactors [1, 4]. In this chapter we provide techniques for establishing *Physcomitrella* as a heterologous host for terpenoid biosynthesis. This includes two protocols for transformation and recovery of transgenic *Physcomitrella* lines as well as methods for metabolic profiling and characterization. Additionally, methods for medium-scale metabolite production using simple and inexpensive photo-bioreactors as well as the basic tools for performing gene editing by homologous recombination are provided.

2 Materials

2.1 Cultivation of Physcomitrella

1. Wild-type *Physcomitrella patens* (Gransden ecotype) can be obtained from the International Moss Stock Center at the University of Freiburg (http://www.moss-stock-center.org/).
2. Growth chamber with a 16 h light and 8 h dark cycle at 25 °C. Light intensities from 20 to 50 W/m^2 are used as standard growth conditions.
3. Phy B media, modified from minimal media previously described [14]. For 1 L, mix 800 mg of $Ca(NO_3)_2$, 250 mg of $MgSO_4{\cdot}7H_2O$, 12.5 mg of $FeSO_4{\cdot}7H_2O$, 0.5 g of $(NH_4)_2C_4H_4O$, 10 mL of KH_2PO_4 buffer (25 g of KH_2PO_4 per liter, adjust to pH 6.5 with 4 M KOH) and 0.25 mL of trace element solution (110 mg of $CuSO_4{\cdot}5H_2O$, 110 mg of $ZnSO_4{\cdot}7H_2O$, 1228 mg of H_3BO_3, 778 mg of $MnCl_2.4{\cdot}H_2O$, 110 mg of $CoCl_2{\cdot}6H_2O$, 53 mg of KI, and 50 mg of $Na_2MoO_4{\cdot}2H_2O$ per liter). The medium can be solidified with 0.7 % (w/v) agar A and is sterilized by autoclaving at 121 °C.
4. BCD media. For 1 L, mix 12.5 mg of $FeSO_4{\cdot}7H_2O$, 10 mL of solution B (25 g/L $MgSO_4{\cdot}7H_2O$), 10 mL of solution C (25 g/L KH_2PO_4 adjusted to pH 6.5 with 4 M KOH), 10 mL of solution D (101 g/L KNO_3), 1 mL of trace element solution (614 of mg H_3BO_3, 389 mg of $MnCl_2{\cdot}4H_2O$, 110 mg of $AlK(SO_4)_2{\cdot}12H_2O$, 55 mg of $CoCl_2{\cdot}6H_2O$, 55 mg of $CuSO_4{\cdot}5H_2O$, 55 mg of $ZnS0_4{\cdot}7H_2O$, 28 mg of KBr, 28 mg of KI, 28 mg of LiCl, and 28 mg of $SnCl_2{\cdot}2H_2O$ per liter). The medium can be solidified with 0.7 % (w/v) agar A and is sterilized by autoclaving at 121 °C. After autoclaving, add 1 mL of sterile 1 M $CaCl_2$ solution (1 mM final concentration) (*see* **Note 1**).

5. Tabletop centrifuge and spin column based plasmid purification, PCR purification and gel extraction kits.
6. PCR thermocycler and a DNA electrophoresis system for purification and PCR product verification.
7. Liquid nitrogen.
8. Sterile culturing facility (e.g., laminar air flow bench or biological safety cabinet) (*see* **Note 2**).
9. Glass bead sterilizer (250 °C) or flame for sterilization of tools.
10. Polytron (PT 1200 E, Kinematic AG) with an autoclavable Ø 12 mm dispersing aggregate (PT-DA 12/2 EC-E123, Kinematic AG) (*see* **Note 3**).
11. Common laboratory equipment: sterile petri dishes, parafilm, vortex mixer, serological pipettor and sterile pipettes (5–50 mL), 3 M surgical tape 1.25 cm (3M 1533-0).
12. Clear plastic kitchen film (e.g., Clingfilm, Vita Wrap, or Saran Wrap) cut into 2 cm wide strips.
13. Autoclaved cellophane disks (type 325P AA Packaging Ltd.) (*see* **Note 4**).
14. BugStopper (Whatman) or similar breathable flask closures for liquid culturing.

2.2 Protoplast Transformation

1. Isopropyl alcohol.
2. Driselase from *Basidiomycetes* sp. (Sigma-Aldrich).
3. Syringe filters (0.22 μm) for filter sterilization.
4. D-Mannitol (8.5 %, sterile).
5. MMM solution: 8.85 mL of 10.3 % mannitol, 150 μL of 1 M $MgCl_2$, 1 mL of 2-[*N*-morpholino]-ethanesulfonic acid (1 %, adjusted to pH 5.6 with HCl).
6. Polyethylene glycol (PEG) solution: 2 g of PEG (MW 6000), 118 mg of $Ca(NO_3)·4H_2O$, 5 mL of 8.5 % D-mannitol, and 50 μL of Tris–HCl pH 8.
7. Protoplast wash solution: 8.5 % D-mannitol, 10 mM $CaCl_2$.
8. Protoplast regeneration medium (bottom layer; PRMB): BCD media (*see* Subheading 2.1, **item 4**) supplemented with 6 % D-mannitol, 5 mM $(NH_4)_2C_4H_4O$, and 10 mM $CaCl_2$, and solidified with 0.7 % agarose (*see* **Note 1**).
9. Protoplast regeneration medium (top layer; PRMT): BCD media (*see* Subheading 2.1, **item 4**) supplemented with 8 % D-mannitol, 5 mM diammonium tartrate, and 10 mM $CaCl_2$, and solidified with 0.4 % agarose (*see* **Note 1**).
10. Hemocytometer.
11. Stainless steel filters (100 μm mesh).
12. Heated water bath.

2.3 Biolistic Transformation

1. Helios Gene Gun system (VWR) including the Helios Gene Gun kit, helium hose assembly, helium regulator, tubing prep station, syringe kit, tubing cutter, Helios Gene Gun optimization kit, and all instructions.
2. 99.8 % Ethanol.
3. Spermidine (Sigma-Aldrich). The 1 M stock solution can be kept at −20 °C.
4. 1 M $CaCl_2$ sterile solution.
5. Nitrogen and helium gas tank (grades 4.8 and 2.0, respectively).

2.4 Metabolite Analysis

1. Amber and silanized 2 mL GC glass vials including screw caps with Teflon-coated silicone septa.
2. 20 mL Solid-phase micro-extraction (SPME) vials with screw neck and magnetic caps with Teflon-coated silicone septa.
3. SPME fiber appropriate for volatile absorption, for example one based on divinylbenzene/carboxen/polydimethylsiloxane as carrier material (DVB/CAR/PDMS; Supelco).
4. GC-MS: Shimadzu GC-2010GC, with a 30 m Agilent HP-5MS column (250 μm i.d., 0.25 μm film) and helium as carrier gas. CTC AOC-5000 auto injector and a GCMS-QP2010 PlusMS Detector using electrical ionization mode. Similar GC-MS systems can be applied.
5. Solvent for extraction e.g., GC grade *n*-hexane or inhibitor-free diethyl ether and internal standards e.g., 1-eicosene (Sigma Aldrich).
6. 15 mL centrifugation tubes for 3,000 × *g* centrifugation.
7. Dodecane (Sigma-Aldrich) for volatile trapping.
8. 65 °C oven and an extra fine analytical balance (±0.01 mg) for dry-weight determination.

2.5 DNA and RNA Isolation

1. Microcentrifuge tube pestle.
2. Edwards buffer: 200 mM Tris–HCl pH 7.5, 250 mM NaCl, 25 mM EDTA, 0.5 % SDS [15].
3. TE buffer: 10 mM Tris–HCl pH 8.0, 1 mM EDTA.
4. RNAqueous kit (Ambion).

3 Methods

3.1 Cultivation of Physcomitrella

Cultivation of *Physcomitrellla* is performed according to standard protocols [16]. All work with *Physcomitrella* is done under sterile conditions using sterile materials and standard sterile techniques (*see* **Note 2**). All moss handling is performed with sterile forceps.

3.1.1 Cultivation on Solid Media Using Petri Dishes

1. Place a lump (approximately 2–5 mm in diameter) of *Physcomitrella* gametophyte tissue on a petri dish with Phy B solid media. Seal the plates with 3 M surgical tape (*see* **Note 5**).
2. Incubate the cultures at standard conditions in a 25 °C growth chamber.

3.1.2 Cultivation on Solid Media Overlaid with Cellophane Disks

By growing *Physcomitrella* on top of cellophane the tissue does not adhere to the solid agar and is therefore easier to handle.

1. Overlay a 9 mm petri dish filled with Phy B solid media with a sterilized disk of cellophane. Close the lid and let the cellophane disk absorb moisture for 10 min. Straighten the disk to avoid air bubbles and wrinkles.
2. Add a lump of approximately 1-week-old tissue (around 10 mm in diameter) to 10 mL of sterile water in a 50 mL tube and homogenize using a Polytron (*see* **Note 3**).
3. Use a serological pipette to transfer 1–3 mL of homogenized suspension to petri dishes with cellophane overlaid Phy B media and let the plates dry uncovered (*see* **Note 6**).
4. Seal the plates carefully with surgical tape and incubate the cultures 1 week in a growth chamber at standard conditions.
5. Harvest the tissue by scraping it off the cellophane.
6. For further biomass generation repeat **steps 1–5** using one plate of protonemal tissue blended in 10 mL of water.

3.1.3 Cultivation in Liquid Medium Using Medium-Scale Photo-bioreactors

1. Add tissue from a plate of 1-week-old protonemal tissue to 5–10 mL of sterile water and homogenize by a Polytron (*see* **Note 3**).
2. Use 5–10 mL to inoculate 50–100 mL BCD liquid media in a 500 mL Erlenmeyer flask (*see* **Note** 7).
3. Seal the flask using a sterile BugStopper and incubate for up to 3 months on a rotary shaker in a growth chamber.
4. In order to keep the tissue in the haploid stage and to enhance growth rates, the tissue should be homogenized every 1–2 weeks.

3.2 Protoplast Transformation

Two approaches can be applied for integration of foreign genes into the genome of *Physcomitrella*. One is targeted integration to a specific locus via homologous recombination and the other is non-targeted or random integration. For random integration of DNA fragments into the genome of *Physcomitrella* any plant expression vector can be utilized (*see* **Note 8**). The major requirement is that the vector contains an appropriate selection marker (*see* **Note 9**). *Physcomitrella* has a uniquely high rate of homologous recombination among plants, and it is possible to perform metabolic engineering by gene editing or targeted integration. For this, 1 kb flanking regions homologous to the targeted locus are added on each side of the DNA to be integrated (e.g., the expression cassette of a selection marker and the gene of interest) (*see* **Note 10**). Protoplast transformation is the most commonly used method for *Physcomitrella* transformation, and is well described in the literature [12]. The method requires careful handling and regeneration of fragile protoplasts and must be done under sterile conditions. This method can be very efficient and yield a large number of stable transformants. However, several attempts may be needed to successfully recover stable *Physcomitrella* lines.

3.2.1 Preparation of DNA for Transformation

1. Prepare plasmid DNA from *E. coli* cultures using mini- or maxiprep kits.
2. Linearize 100 μg of DNA by overnight digestion with an appropriate restriction enzyme. Inactivate restriction enzyme by heating at 60–80 °C for 20 min. Verify digestion by electrophoresis.
3. Precipitate DNA by addition of sodium acetate (pH 5.2, final concentration 0.3 M) and 0.7 volumes of isopropyl alcohol. Mix well and centrifuge at 15,000 × *g* for 15 min. Carefully decant the supernatant, and wash the pellet with 70 % ethanol. Centrifuge again at 15,000 × *g* for 5–15 min and decant the supernatant. Air-dry the pellet and redissolve the DNA in 50 μL of water. Final DNA concentration should be between 1 and 2 μg/μL.

3.2.2 Preparation and Transformation of Protoplasts

Protoplasts are highly sensitive to external force, so handle with great care.

1. On the day of transformation, first prepare the PEG solution, and allow it to sit for 2–3 h before use. After this, filter-sterilize the solution by passing it through a 0.22 μm syringe filter. Prepare petri dishes with PRMB, overlaid with sterile cellophane.
2. Collect 5–7-day-old moss protonema tissue by scraping it off cellophane-overlaid agar plates. Place in a sterile 50 mL plastic tube, and weigh the tissue. Prepare 1 mL of a 0.5 % driselase solution for every 40 mg tissue. Dissolve the Driselase powder in 8.5 % D-mannitol, and sterilize using a syringe filter, adding it directly to *Physcomitrella* tissue. Incubate the mixture for 30–60 min with occasional inversion of tube.
3. Pour the Driselase treated tissue through a sterile 100 μm stainless steel mesh screen, and recover the protoplasts in a sterile 100 mL beaker. Most undigested tissue and cellular debris will not pass through the mesh.
4. Centrifuge the protoplasts at 200 × *g* for 5 min, with gentle acceleration and breaking.
5. Carefully remove the supernatant using a serological pipette, being careful not to disturb the protoplast pellet.

6. Resuspend the pellet in protoplast wash solution using the same volume as driselase solution in **step 2**.
7. Repeat **steps 4** and **5**.
8. Resuspend the protoplasts in half the original volume of 8.5 % D-mannitol and estimate the protoplast density using a hemocytometer.
9. Centrifuge at 200 × *g* for 5 min, remove supernatant, and resuspend the pellet in sterile MMM solution to give a protoplast concentration of 1.5–2 × 10^6 protoplasts/mL (*see* **Note 11**).
10. Add 20 μg of linearized DNA to the bottom of a 15 mL conical tube. Total volume of DNA should be less than 30 μL. Add 300 μL of protoplast suspension to the DNA and then add 300 μL of sterile PEG prepared in **step 1**. Mix by flicking the tube gently.
11. Incubate the mixture in a 45 °C water bath for 5 min followed by 5 min at room temperature.
12. Dilute protoplast suspension 5 times with 300 μL of 8.5 % D-mannitol, waiting for 1 min between dilutions.
13. Dilute an additional 5 times with 1 mL of 8.5 % D-mannitol.
14. Centrifuge the transformed protoplasts at 200 × *g* with gentle acceleration and braking for 5 min, and remove the supernatant with a pipette.
15. Resuspend the protoplasts in 500 μL of 8.5 % D-mannitol, and add 2.5 mL of molten PRMT, and mix (*see* **Note 12**).
16. Evenly disperse 1 mL of the protoplast suspension using cut pipet tips to fresh PRMB petri dishes, overlaid with cellophane. Each transformation will result in 3 plates.
17. Seal plates with 3 M tape and incubate in the growth chamber under standard conditions.
18. Allow protoplasts to regenerate their cell walls for 5–7 days, and transfer the cellophane with regenerated protoplasts to Phy B media containing an appropriate selection agent for 2 weeks (*see* **Note 9**). Proceed with selection of positive transformants (*see* Subheading 3.4).

3.3 Biolistic Transformation

Biolistic transformation is a robust alternative to PEG-mediated transformation of protoplasts. This protocol requires more specialized equipment, but it can be performed with a minimal number of tissue handling steps.

3.3.1 Preparation of DNA-Coated Gold Particles

1. Mix 1.8 mg of PVP with 90 μL of 99.8 % ethanol, vortex and leave for 5 min until the PVP is fully dissolved. Transfer 4.5 μL to a 15 mL tube and add 3 mL of 99.8 % ethanol (final PVP concentration: 0.03 mg/mL).

2. Mix 25 mg of gold particles (1 μm in diameter) and 100 μL of 0.05 M spermidine. Vortex vigorously for 1 min and add 100 μL of transformation plasmid (diluted to 100 ng/μL) to the gold/spermidine solution. Vortex vigorously for 1 min.
3. To precipitate DNA onto the gold particles, add 100 μL of sterile 1 M $CaCl_2$ dropwise while vortexing. Leave at room temperature for 10 min.
4. Vortex and pellet the gold particles by a brief centrifugation. Remove the supernatant and wash with 1 mL of 99.8 % ethanol. Repeat this washing step two more times.
5. Resuspend the gold particles in 3 mL of 0.03 mg/mL PVP solution prepared in **step 1**.

3.3.2 Tubing Coating

1. Connect the Bio-Rad Tubing Prep Station to a nitrogen gas cylinder according to the manufacturer's manual (*see* **Note 13**). Gently insert 50 cm of Gold-Coat Tubing into the tubing support cylinder. Turn the nitrogen flow on (0.4 L/min) for 10 min. Dismount the tubing.
2. Connect the Gold-Coat Tubing to a 10 mL syringe using 5 cm of syringe adaptor tubing. Vortex the DNA–gold suspension from the previous section. Immediately after this, use the 5 mL syringe to gently suck the DNA–gold suspension into the rubber tube. Avoid air bubbles in the Gold-Coat Tubing. Mark where the DNA–gold suspension starts and stops on the gold-coat tubing.
3. Immediately following this insert the Gold-Coat Tube containing the DNA–gold suspension into the tubing support cylinder. Wait for 2 min to let the gold particles sediment in the tube.
4. Slowly suck out the PVP solution using the 10 mL syringe, leaving the sedimented gold particles in the Gold-Coat Tubing. This should take 1 min.
5. After removing the PVP solution, remove the syringe and start the rotation of the Tubing Prep Station. Rotate for 2 min to ensure even distribution of the gold particles.
6. Turn the nitrogen flow on (0.4 L/min) for 5–10 min to dry the DNA–gold particles.
7. Cut the Gold-Coat Tubing into 130 mm pieces using a BioRad Tubing Cutter. These pieces of Gold-Coat Tubing with DNA-coated gold particles on the inner wall are referred to as "cartridges" for the Helios Gene Gun. You will get around 30 cartridges from this, which should be enough for 7–8 rounds of transformation.
8. The cartridges are stored in tightly capped storage vials (e.g., a 20 mL polyethylene scintillation vial) wrapped with Parafilm and kept at 4 °C. The cartridges can be used for 12 months after date of preparation. Silica gel can be added to absorb any condensation and extend storage time.

3.3.3 Particle Shooting

The following steps need to be carried under sterile conditions.

1. Attach the Helios Gene Gun to the helium regulator according to the manufacturer's guidelines and set the pressure to 120 psi.
2. Scrape 7-day-old moss protonema from cellophane-overlaid agar into a clump in the center of a petri dish. The diameter of the moss clump should be approximately 2 cm.
3. Load 4 cartridges into a sterile cartridge holder using sterile forceps, and assemble the loaded cartridge holder, Helios Gene Gun, battery, and a sterile barrel liner.
4. Hold the Helios Gene Gun vertically and aim the gun at the lump of moss. The distance between the barrel liner and the moss should be 1–2 cm. Hold in the safety interlock switch and press the trigger down. Shoot helium through the cartridge twice.
5. Repeat **step 5** three more times for the remaining 3 cartridges.
6. Close the petri dish with surgical tape and let the moss recover in a growth chamber for 2 days.
7. Transfer the moss lump to 10 mL sterile water in a 50 mL tube and blend with a Polytron.
8. Transfer the solution to three cellophane-overlaid Phy B solid media plates including appropriate antibiotics for selection of transformed cells, and let the solution dry until the tissue remains moist, but without excess water.
9. Seal the plates with surgical tape and incubate the cultures for 2 weeks in the growth chamber. To obtain stable transformants follow the selection procedure decribed below (*see* Subheading 3.4).

3.4 Selection of Transformants

1. After 2 weeks on Phy B selective media, transfer the cellophane disks with recovered transformants to solid Phy B media and incubate for 2 weeks under standard conditions. Unstable or transient transformants will lose the transformed DNA and the ability to survive on selective media during this period.
2. Cellophane disks with transformed *Physcomitrella* are transferred onto solid Phy B media supplemented with an appropriate selective agent and incubated for another 2 weeks (*see* **Note 9**).
3. Repeat **step 1** and **2** two more times and the moss lines obtained after three rounds of selective pressure are considered to be stably transformed moss lines. Verify transformants using metabolic profiling and PCR amplification of transgenes.

3.5 Metabolite Profiling

It is possible to obtain moss lines that do not express the gene of interest despite surviving the selection. We recommend performing metabolic and genetic screening in order to confirm the existence of heterologously produced metabolites before further characterization of the moss lines. Always use glass equipment for handling solvents to avoid plasticizer contaminants.

3.5.1 Automated Analysis of Volatile Terpenoid Products

This method is preferred for initial screening of volatiles in stable moss lines. Many lines can be screened in parallel with minimal hands-on time using auto sampling.

1. Sterilize 20 mL GC vials, caps, and septa separately. Pipet approximately 4 mL of solid Phy B media into each vial and inoculate by placing a lump of moss on top of the agar.
2. Cap the GC vials and incubate cultures 1–4 weeks in the growth chamber.
3. Sample the headspace volatiles by incubating a Solid Phase Microextraction (SPME) fiber in the vials at room temperature for 30 min and analyze by GC-MS.

3.5.2 Quantification of Volatile Terpenoid Products

1. Inoculate a 500 mL Erlenmeyer flask containing 100 mL of BCD media with at least 2.5 mL of 1-week-old plate blended for 30 s in 5 mL sterile water. Seal the flasks using sterile BugStopper or other breathable flask cap.
2. Incubate the moss cultures with 70 rpm agitation under standard conditions.
3. After 4 days add 5–10 mL of dodecane, under sterile conditions, and continue cultivation for up to 3–5 weeks.
4. Transfer approximately 10 mL of culture containing dodecane to a 15 mL glass centrifuge tube and centrifuge at 4,000 rpm for 5 min.
5. Take 100 μL of dodecane from the upper phase, dilute 10 times in hexane spiked with an internal standard, and analyze by GC-MS (*see* **Note 14**).
6. The concentration of the metabolite of interest can be calculated based on the internal standard and an authentic standard run in parallel.

3.5.3 Analysis of Semi- or Non-volatile Terpenoid Products

1. Transfer a lump of moss into a 2 mL GC glass vial containing 500–1,000 μL of organic solvent such as n-hexane including an appropriate internal standard with a concentration of 0.2–1 mg/L.
2. Cap the vial and vortex the samples briefly.
3. Extract at room temperature for 1 h or overnight at 4 °C while mixing.
4. Transfer the extract to a new vial and analyze by GC-MS.

3.5.4 Dry Weight Determination for Quantification of Terpenoid Products

Determining the dry weight is the most accurate way to quantify the biomass. When performing small-scale extractions the dry weight of the moss samples is around 5–15 mg and therefore an extra fine analytical balance is needed (±0.01 mg).

1. After extraction of terpenoid products and removal of organic solvent, the tissue is dried at 65 °C for 24 h.
2. Tare the balance using the vial containing the dried moss.
3. Remove the dry moss from the glass vial and reweigh the empty vial.

3.5.5 GC-MS Analysis Conditions

The following conditions provide a typical starting point for separation of terpenoids using GC-MS. These settings and conditions allow separation of mixtures of mono-, sesqui-, and di-terpenes, as well as further oxidized products: Injection volume 1 μL, splitless injector temperature 250 °C, 30 m HP-5MS column, 2 mL/min He carrier gas, GC temperature program (50 °C for 3 min, 15 °C/min to 310 °C and hold 5 min), scan range from *m/z* 50 to *m/z* 350 with 70 eV electrical ionization. These general conditions can be modified to shorten runtime, improve resolution or sensitivity for specific compounds.

3.6 DNA and RNA Isolation for Further Characterization

3.6.1 Rapid Small Scale DNA Isolation

1. Snap-freeze 100 mg of *Physcomitrella* tissue in a 2 mL microcentrifuge tube in liquid nitrogen.
2. Grind the frozen tissue using a microcentrifuge tube pestle, add 500 μL of Edwards buffer and vortex vigorously (*see* **Note 15**).
3. Centrifuge the sample at 15–20,000 × *g* for 10 min at room temperature.
4. Transfer 400 μL of the supernatant to a new tube and add 400 μL of isopropanol. Mix gently.
5. Centrifuge the sample 15–20,000 × *g* for 10 min at room temperature and discard the supernatant.
6. Invert the tubes and let them dry on a paper towel.
7. Once the tube is completely dry resuspend the pellet in 300 μL of TE buffer at 50 °C for 30 min and store at −20 °C until use (*see* **Note 16**).
8. For PCR use 1 μL of DNA as template.

3.6.2 Rapid Small Scale RNA Isolation

1. Snap-freeze 100 mg of moss tissue in a 1.5 mL microcentrifuge tube in liquid nitrogen.
2. Add 300 μL of Lysis Solution from the RNAqueous kit and grind with a microcentrifuge pestle while the tissue is defrosting.

3. Spin the tissue at 20,000 × *g* for 3 min and grind the tissue further. Repeat this two times.
4. Add 150 μL of 100 % ethanol and vortex briefly and proceed with the extraction according to the RNAqueous kit protocol.
5. Elute in two times 15 μL of Elution Buffer, preheated to 75 °C.
6. DNase I treat and inactivate the enzyme according to the RNAqueous kit protocol.

4 Notes

1. Adding $CaCl_2$ after autoclaving minimizes precipitation of the media.
2. Since *Physcomitrella* is typically grown on Phy B media without supplemental antibiotics there is a significant risk of contamination by fungi or bacteria, which may not be obvious until after several weeks of growth. We recommend using a sterile facility e.g., a LAF bench dedicated for work with plant tissue culturing to limit the risk of microbial contamination.
3. Many different Polytrons or tissue homogenizers can be used. Most importantly, the blender tip needs to be autoclavable. Excessive blending will lead to poor regeneration.
4. Prevent the cellophane disks from sticking together autoclave with filter paper in between each cellophane disk.
5. Plates sealed with 3 M surgical tape have a high growth rate, but will dry up within 2–4 weeks, depending on the starting volume of media. Sealing the plates with kitchen film or Parafilm facilitates storage for 6 months or more, however, this will also slow the growth of the cultures significantly.
6. Do not overdry the plates. The cellophane will curl and the moss will not survive.
7. In addition to BCD media, liquid Phy B and KNOP media can be used for liquid growth. In order to obtain very high growth rates approximately 1.5 % of CO_2 can be supplemented to the moss culture and the light intensity can be increased.
8. In our lab, pCAMBIA vectors (http://www.cambia.org/daisy/cambia/585) are used as starting point for generating non-targeted *Physcomitrella* expression vectors. Since most of the vector backbone of the pCAMBIA binary vectors is nonessential for *Physcomitrella* transformation and expression, we recommend cloning the essential parts of pCAMBIA vectors (expression cassettes of the gene of interest and the selection marker) into a high copy number subcloning vector with a

smaller backbone. The use of such simpler expression vectors facilitates downstream cloning and DNA propagation. Furthermore, a viral 2A linker for expression of several genes under the control of a single promoter can be used [17].

9. We have successfully used four different selection agents for recovery of stable tranformants, namely hygromycin (30 µg/mL), G418 (kanamycin) (50 µg/mL), sulfadizine sodium (150 µg/mL), or gentamicin sulfate (100 µg/mL).
10. We recommend using at least 750 base pairs of homologous genomic sequence flanking the desired insertion site. However, transformation of DNA that does not contain targeting sequences can also lead to long-term maintenance and expression of foreign DNA. Many integration sites have successfully been established for heterologous expression in *Physcomitrella*. A very useful subset of moss vectors for targeted integration can be found at the homepage of Michael Prigge (Mark Estelle's Lab http://labs.biology.ucsd.edu/estelle/moss2.html).
11. If transformations yield few surviving lines, the protoplast concentration can be increased up to tenfold. However, this increases the risk of recovering a lawn of transformants, resulting in an inability to distinguish individual transformants due to high colony density.
12. Keep the PRMT in a 45 °C water bath. This will keep the agar melted, but cool enough to ensure protoplast survival.
13. The manual for connecting the Bio-Rad Tubing Prep Station can be found online (http://www.bio-rad.com/webroot/web/pdf/lsr/literature/Bulletin_9541.pdf).
14. Depending on your metabolite abundance, the dilution can be varied from 3 to 1,000 times.
15. DNA isolation can also be performed in a 96-well high-throughput format using metal beads and a mixer mill for 2 × 30 s. After being frozen in liquid nitrogen the sample lids are more fragile and can break after excessive mixing in the mixer mill.
16. Small amounts of pigments might be left in the DNA extract. However, this does not typically inhibit PCR. For a cleaner DNA template include one or two steps of ethanol wash: Add 100 µL of 70 % ethanol, vortex briefly and spin the sample for 15–20,000 × *g* for 10 min at room temperature. Discard supernatant and dry the tubes by inverting them on a paper towel. Resuspend in 300 µL of TE buffer as described.

References

1. Viana AAB, Pelegrini PB, Grossi-de-Sá MF (2012) Plant biofarming: novel insights for peptide expression in heterologous systems. Pept Sci 98:416–427
2. Anterola A, Shanle E, Perroud PF et al (2009) Production of taxa-4(5),11(12)-diene by transgenic *Physcomitrella patens*. Transgenic Res 18:655–660
3. Hamberger B, Bak S (2013) Plant P450s as versatile drivers for evolution of species-specific chemical diversity. Phil Trans R Soc B 368: 20120426
4. Yonekura-Sakakibara K, Hanada K (2011) An evolutionary view of functional diversity in family 1 glycosyltransferases. Plant J 66:182–193
5. Liu Q, Majdi M, Cankar K et al (2011) Reconstitution of the costunolide biosynthetic pathway in yeast and *Nicotiana benthamiana*. PLoS One 6:e23255
6. Chen F, Tholl D, Bohlmann J et al (2011) The family of terpene synthases in plants: a mid-size family of genes for specialized metabolism that is highly diversified throughout the kingdom. Plant J 66:212–229
7. Aya K, Hiwatashi Y, Kojima M et al (2011) The Gibberellin perception system evolved to regulate a pre-existing GAMYB-mediated system during land plant evolution. Nat Commun 22:544
8. Anterola A, Shanle E (2008) Genomic insights in moss gibberellin biosynthesis. Bryologist 111:218–230
9. Johri MM (2008) Hormonal regulation in green plant lineage families. Physiol Mol Biol Plants 14:23–38
10. Simonsen HT, Drew DP, Lunde C (2009) Perspectives on using *Physcomitrella patens* as an alternative production platform for thapsigargin and other terpenoid drug candidates. Perspect Medicin Chem 3:1–6
11. von Schwartzenberg K, Schultze W, Kassner H (2004) The moss *Physcomitrella patens* releases a tetracyclic diterpene. Plant Cell Rep 22(10): 780–786
12. Schaefer DG, Zrÿd JP (1997) Efficient gene targeting in the moss *Physcomitrella patens*. Plant J 11:1195–1206
13. Hayashi K, Horie K, Hiwatashi Y et al (2010) Endogenous diterpenes derived from *ent-daurene*, a common gibberellin precursor, regulate protonema differentiation of the moss *Physcomitrella patens*. Plant Physiol 153:1085–1097
14. Ashton NW, Grimsley NH, Cove DJ (1979) Analysis of gametophytic development in the moss, *Physcomitrella patens*, using auxin and cytokinin resistant mutants. Planta 144:427–435
15. Edwards K, Johnstone C, Thompson C (1991) A simple and rapid method for the preparation of plant genomic DNA for PCR analysis. Nucleic Acids Res 19:1349
16. Cove DJ, Perroud PF, Charron AJ et al (2009) The moss Physcomitrella patens: a novel model system for plant development and genomic studies. Cold Spring Harb Protoc 2009(2):69–104
17. Halpin C, Cooke SE, Barakate A et al (1999) Self-processing 2A-polyproteins—a system for co-ordinate expression of multiple proteins in transgenic plants. Plant J 17:453–459

Chapter 20

Quantification of Plant Resistance to Isoprenoid Biosynthesis Inhibitors

Catalina Perelló, Manuel Rodríguez-Concepción, and Pablo Pulido

Abstract

Plants use two pathways for the production of the same universal isoprenoid precursors: the mevalonic acid (MVA) pathway and the methylerythritol 4-phosphate (MEP) pathway. Inhibitors of the MVA pathway prevent the activity of the shoot apical meristem and the development of true leaves in seedlings, whereas those inhibiting the MEP pathway show an additional bleaching phenotype. Here, we describe two methods to quantify plant resistance to inhibitors of the MVA pathway or the MEP pathway based on seedling establishment and photosynthetic pigment measurements. Although the methods are presented for Arabidopsis, they are valid for other plant species. These methods can be used as inexpensive and high-throughput alternatives to in vitro assays to estimate the activity of the corresponding target enzymes and to screen for mutants with altered levels or activities of these enzymes.

Key words MEP pathway, Inhibitors, Mevinolin, Clomazone, Fosmidomycin, Resistance, Quantification

1 Introduction

All free-living organisms synthesize their isoprenoids from the same building units, isopentenyl diphosphate (IPP) and dimethylallyl diphosphate (DMAPP). But unlike most organisms, plants have two unrelated pathways for the biosynthesis of these universal isoprenoid precursors (Fig. 1): the methylerythritol 4-phosphate (MEP) pathway produces IPP and DMAPP for monoterpenes, photosynthesis-related isoprenoids (chlorophylls, carotenoids, and prenylquinones such as plastoquinone, phylloquinone, and tocopherol), and hormones (gibberellins, abscisic acid, strigolactones), and the mevalonic acid (MVA) pathway provides precursors for the production of ubiquinone, sterols and derived hormones (brassinosteroids), triterpenes, sesquiterpenes, and polyterpenes [1, 2]. All the MEP pathway enzymes are encoded by the nuclear genome and imported into plastids, whereas the MVA pathway enzymes have been found in different subcellular compartments, including

Manuel Rodríguez-Concepción (ed.), *Plant Isoprenoids: Methods and Protocols*, Methods in Molecular Biology, vol. 1153, DOI 10.1007/978-1-4939-0606-2_20,

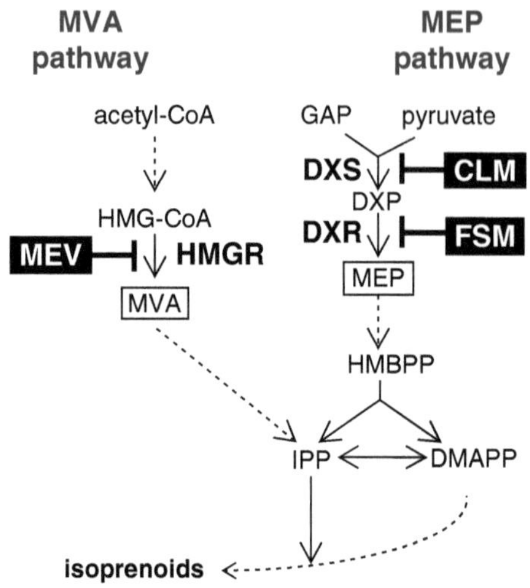

Fig. 1 Schematic representation of the MVA and MEP pathways. The steps inhibited by mevinolin (MEV), clomazone (CLM), and fosmidomycin (FSM) and the corresponding target enzymes are indicated. *DMAPP* dimethylallyl diphosphate, *DXP* deoxyxylulose 5-phosphate, *DXR* DXP reductoisomerase, *DXS* DXP synthase, *GAP* glyceraldehyde 3-phosphate, *HMBPP* hydroxymethylbutenyl 4-diphosphate, *HMG-CoA* hydroxymethylglutaryl CoA, *HMGR* HMG-CoA reductase, *IPP* isopentenyl diphosphate, *MEP* methylerythritol 4-phosphate, *MVA* mevalonic acid

peroxisomes, the endoplasmic reticulum, and the cytosol [1]. In particular, the proposed main rate-determining enzyme of the MVA pathway, hydroxymethylglutaryl CoA reductase (HMGR; EC 1.1.1.34), is anchored to the endoplasmic reticulum exposing the catalytic domain of protein towards the cytosol [3]. The production of MEP by the sequential activity of the stromal enzymes deoxyxylulose 5-phosphate (DXP) synthase (DXS; EC 4.1.3.37) and DXP reductoisomerase (DXR; EC 1.1.1.267) has also been proposed as rate determining in the MEP pathway [4]. Although methods for the determination of HMGR, DXS, and DXR activities are available (*see* Chapters 2 and 3), they are time consuming, require specialized equipment, and cannot be used for genetic screens. An inexpensive and high-throughput alternative to enzymatic assays that has the added advantage of estimating HMGR, DXS, or DXR activities in vivo is to quantify the resistance to competitive inhibitors specifically targeting these enzymes (Fig. 1) [5–7].

A limited exchange of isoprenoid precursors takes place among different subcellular compartments. However, the complete genetic or pharmacological block of either the MVA pathway or the MEP pathway in null mutants or inhibitor-treated wild-type plants is lethal, indicating that the loss of one of the two pathways cannot be compensated by the remaining pathway. For example, growth of *Arabidopsis thaliana* plants in medium supplemented with mevinolin (MEV) or other statins that inhibit HMGR activity

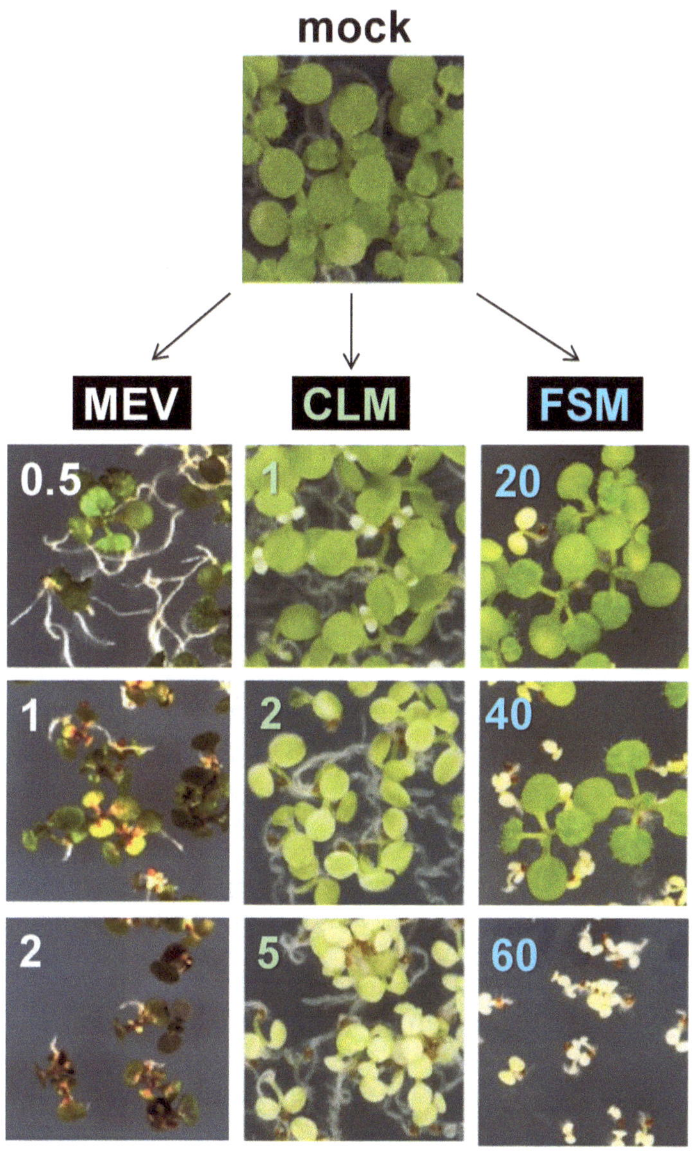

Fig. 2 Phenotype of Arabidopsis seedlings grown in the presence of isoprenoid inhibitors. Wild-type plants (Columbia) were grown on MS plates either supplemented or not (mock) with mevinolin (MEV), clomazone (CLM), or fosmidomycin (FSM) for 2 weeks under long-day conditions. *Numbers* indicate the concentration (μM) of inhibitor in the corresponding plate

[8, 9] causes inhibition of root growth, formation of adventitious roots, and, at higher concentrations of the inhibitor, an arrest of seedling establishment (SE, defined as the production of true leaves that can support further plant development) due to a blockage of shoot apical meristem development (Fig. 2) [6, 10, 11]. Inhibition of the MEP pathway in plants germinated and grown in the presence of the DXS inhibitor clomazone (CLM) [12, 13] or

the DXR inhibitor fosmidomycin (FSM) [14, 15] also causes a concentration-dependent inhibition of SE [5–7, 16]. Therefore, a quantification of SE rates can be used as a common estimator of resistance to inhibition of the MVA pathway or the MEP pathway. In the case of the MEP pathway, a bleached phenotype caused by a blockage in the production of photosynthetic pigments is also obvious in inhibitor-treated plants (Fig. 2) and can be easily quantified by measuring chlorophyll and carotenoid levels [5–7, 16, 17]. A difference between the phenotype of plants grown in media supplemented with CLM or FSM is that CLM only causes bleaching of true leaves (but not cotyledons) at low concentrations of the inhibitor, whereas the reduction in the levels of photosynthetic pigments is similar in cotyledons and true leaves from FSM-treated plants (Fig. 2). The reason behind this differential phenotype might be that CLM is not the actual molecule that inhibits DXS but it needs to be converted in the plant tissues to keto-CLM, the biologically active inhibitor [12, 13].

Based on the competitive nature of MEV, CLM, and FSM mode of action, resistance to these inhibitors (i.e., increased SE rates or, in the case of the MEP pathway inhibitors, reduced bleaching) can be accomplished by increasing the levels or the activity of HMGR, DXS, or DXR, respectively [6, 7]. Thus, transgenic lines with higher HMGR levels show resistance to MEV (but not to MEP pathway inhibitors), whereas lines overexpressing DXS show resistance to both CLM and FSM (due to a higher production of DXP, which competes with FSM for the active site of DXR) and those overexpressing DXR are resistant to FSM but not to CLM or MEV [6, 7]. Screening for Arabidopsis mutants with increased resistance to these isoprenoid inhibitors has led to identification of posttranscriptional mechanisms that control the levels and/or activity of the target enzymes but also other regulatory pathways modulating isoprenoid biosynthesis [6, 16, 18–21]. Other studies have used inhibitor resistance assays to estimate the in vivo activities of the target enzymes [5, 22]. Here we present two simple methods to quantify plant resistance to MEV, CLM, and FSM based on SE and photosynthetic pigment measurements. Although the methods are presented for Arabidopsis, they are valid for other plant species. However, concentrations of inhibitors should be adapted to the particular species under study since it is known that the SE block and bleaching phenotypes require different concentrations in different plant species [19, 23].

2 Materials

2.1 Plant Germination and Growth

1. Equipment: Scale, autoclave, laminar airflow hood, water bath, plant growth chamber.
2. Containers for in vitro plant growth (Petri dishes for Arabidopsis).

3. Ethanol, absolute.
4. MEV (Sigma): Prepare a 1 mM stock solution in ethanol (*see* **Note 1**).
5. CLM (Sigma): Prepare a 5 mM stock solution in methanol (*see* **Note 1**).
6. FSM (Sigma): Prepare a 100 mM stock solution in Tris–HCl 10 mM pH = 8.5 and sterilize by filtration with a 20 μm microfilter (*see* **Note 1**).
7. Murashige and Skoog (MS) medium (*see* **Note 2**) and plates: For 1 L, mix 4.4 g of MS salts without vitamins (Duchefa) and 0.5 g of 2-(*N-morpholino*) ethanesulfonic acid (MES), and stir until the solution becomes clear. Adjust to pH = 5.7 with 1 M KOH, take to 1 L with deionized water, and add 8 g of plant agar (Duchefa). Sterilize in autoclave at 121 °C for 20 min, and then cool in a water bath at 50–60 °C. Add the inhibitors (*see* **Note 3**) before pouring the medium into the appropriate plates (Petri dishes for Arabidopsis) under a laminar airflow hood (*see* **Note 4**).

2.2 Pigment Extraction

1. Equipment: Scale, vortex, multi-tube shaker, microcentrifuge, and spectrophotometer.
2. Liquid nitrogen.
3. Mortar and pestle.
4. Organic solvents: Methanol, chloroform, ethyl acetate, acetone (*see* **Note 5**).
5. TB solution: 50 mM Tris–HCl pH = 7.5, 1 M NaCl.
6. Nitrogen stream.

3 Methods

3.1 Seedling Establishment (SE)

SE rate is defined as the percentage of seedlings producing green true leaves that are photosynthetically active and therefore able to support full plant development. In this method, seedlings are germinated and grown on MS plates supplemented with a range of concentrations of MEV, CLM, or FSM and visually inspected to monitor the presence of true leaves.

1. Prepare MS plates with and without inhibitors at the appropriate concentrations.
2. Surface-sterilize seeds (*see* **Note 6**), and plate them on the solid medium (*see* **Note 7**). Wrap the plates in aluminum foil and transfer to 4 °C.
3. After the appropriate stratification period at 4 °C (from 2 to 4 days for Arabidopsis), place plates in a growth chamber (*see* **Note 8**).

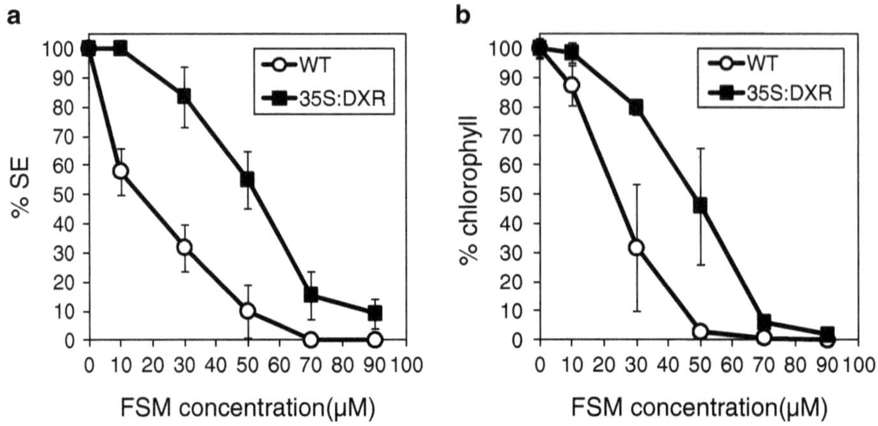

Fig. 3 Plots comparing the results from the two methods for quantification of FSM resistance. Wild-type (WT) and transgenic Arabidopsis lines overexpressing DXR (35S:DXR) were grown on MS plates either supplemented or not with the indicated concentrations of FSM for 2 weeks under long-day conditions. Resistance to the inhibitor was quantified as the percentage of seedling establishment (% SE, panel **a**) or chlorophyll levels (**b**) relative to those measured for each genotype on mock plates (0 μM FSM). Data represent mean and standard deviation of triplicates

4. Incubate the plates until the second set of true leaves are clearly visible in plants grown on plates without inhibitors (about 2 weeks for Arabidopsis).
5. Count the total number of germinated seedlings (t) and the number of seedlings with a green first set of true leaves (n) in each plate (*see* **Note 9**).
6. Calculate the proportion of seedlings with true leaves (a) in mock plates ($a_0 = n_0/t_0$) and in each of the plates supplemented with inhibitor ($a_i = n_i/t_i$). Then, calculate the mean of the a_0 values obtained from all replicates ($\bar{a}_0$), and recalculate each of the a_i values relative to $\bar{a}_0$. The percentage of SE for each concentration of inhibitor (i) is calculated with the formula $SE = 100 \times a_i/\bar{a}_0$ (*see* **Note 10**).
7. Plot the generated values (means and errors) to construct dose–response curves for each genotype (Fig. 3).

3.2 Pigment Levels

This method is only valid to quantify resistance to inhibitors of the MEP pathway (CLM and FSM), since inhibition of the MVA pathway with MEV does not significantly interfere with the biosynthesis of chlorophylls and carotenoids. Although this method is more time consuming and requires more seedlings than the SE method, it is a valid alternative to quantify more subtle differences in resistance. In particular, determination of pigment levels appears to be more robust than SE calculation to quantify CLM resistance.

1. Follow **steps 1–3** in Subheading 3.1.
2. Incubate the plates until growth of albino seedlings is completely blocked on plates with the highest concentration of inhibitors (about 10 days for Arabidopsis) (*see* **Note 11**).

3. Collect samples of whole seedlings (between 25 and 50 mg of fresh weight) in ice-cold 2 mL microcentrifuge tubes, carefully note the exact weight of the tissue (*w*), and immediately freeze in liquid nitrogen.
4. Grind individual samples in liquid nitrogen to a fine powder in a precooled mortar and pestle.
5. Transfer the powder to a 2 mL microcentrifuge tube precooled in liquid nitrogen. It is important to work quickly so that the plant material does not thaw.
6. Immediately add 400 μL of methanol to each powdered sample, vortex, and transfer the samples to an ice bucket covered with aluminum foil (*see* **Note 12**).
7. When enough samples have been processed, briefly vortex the tubes with the methanol extract and transfer them to a multi-tube shaker in a 4 °C chamber. Shake for 30 min in darkness (e.g., covered with aluminum foil).
8. Add 400 μL of TN solution to each sample, and shake for 10 min at 4 °C in the dark.
9. Add 400 μL of chloroform to the mixture, and shake for 3 min at 4 °C in the dark.
10. Remove the tubes from the multi-tube shaker, and centrifuge them at 15,000 × *g* for 5 min at 4 °C.
11. Recover the organic (upper) phase containing the pigments, and transfer it to a clean tube.
12. Dry under a stream of nitrogen or by centrifugal evaporation (*see* **Note 13**).
13. Add 50 μL of ethyl acetate to each dried sample, resuspend, and centrifuge at 15,000 × *g* for 1 min.
14. Transfer 25 μL of the supernatant to a microcentrifuge tube containing 775 μL of acetone. Mix by inverting the tube several times until all the solution is uniformly colored. Keep protected from light on ice. Prepare a blank sample with 25 μL of ethyl acetate in 775 μL of acetone.
15. Transfer to a quartz cuvette to measure absorbance at 470.0, 644.8, and 661.6 nm in a spectrophotometer.
16. Calculate chlorophyll and carotenoid contents (μg/mL) in the solution using the following equations [24]:

   Chlorophyll *a*: $Ca = 11.24 \times [A661.6] - 2.04 \times [A644.8]$

   Chlorophyll *b*: $Cb = 20.13 \times [A644.8] - 4.19 \times [A661.6]$

   Total chlorophylls: $Ca + b = 7.05 \times [A661.6] - 18.09 \times [A644.8]$

   Total carotenoids: $Cc = (1{,}000 \times [A470.0] - 1.90 \times Ca - 63.14 \times Cb)/214$

17. Calculate the concentration of chlorophylls and carotenoids in the tissue samples as follows (w = weight of seedlings collected in **step 3**):

    μg chlorophylls/mg of fresh weight = $1.6 \times Ca + b/w$

    μg carotenoids/mg of fresh weight = $1.6 \times Cc/w$

18. Make the values for inhibitor-treated samples relative to those measured in mock plates (*see* **Note 14**).
19. Plot the generated values (means and errors) to construct dose–response curves (Fig. 3).

4 Notes

1. Store aliquots of stock solutions of inhibitors in sealed vials at −20 °C (they should be stable for at least 1 month). Before use, let the solutions to thaw on ice.
2. Vitamins and sucrose are not required. Adding sucrose to the growth media actually results in a strong increase in plant resistance to isoprenoid inhibitors [21]. The amount of MS salts can be reduced to half, but this will also affect resistance to inhibitors.
3. The recommended final concentrations of inhibitor for Arabidopsis (Columbia ecotype) are as follows:

    MEV: 0.1, 0.5, 1, 2, and 5 μM

    CLM: 0.5, 1, 2, 5, and 10 μM

    FSM: 10, 20, 40, 60, and 80 μM

    Concentrations should be experimentally determined for other plant species or Arabidopsis accessions (e.g., Landsberg *erecta* and Wassilewskija are more resistant to MEV and FSM than Columbia). A set of concentrations ranging in phenotypic effect from low (only a few individuals show symptoms) to high (all individuals are affected) should be established to generate dose–response curves. Mock plates are prepared by adding the same amount of inhibitor solvent used to prepare the plates with the highest concentration of the corresponding inhibitor.
4. Once the medium temperature decreases to 50–60 °C, it is safe to add the inhibitors without the risk of inactivating them. At this temperature, however, the medium tends to become solid relatively fast when removed from the water bath to room temperature. It is recommended to pour the medium into the plates as fast as possible under sterile conditions (hood). The plates should be left open in the hood for 30 min (or until the medium becomes solid) and then sealed and stored at 4 °C until used.

5. Follow appropriate regulations (fume hood, protective clothing) when handling and disposing organic solvents.
6. For Arabidopsis, place the seeds in a sterile microfuge tube and add three volumes of ethanol. Shake for 5 min, remove the ethanol, and let the seeds dry leaving the tube open in a laminar airflow sterile hood.
7. For meaningful statistical analysis, triplicate plates with at least 50 individuals each should be prepared for every concentration to be tested. It is very useful to distribute the seeds on the medium like a grid because it facilitates counting and prevents shade avoidance responses and other neighbor-related effects. When comparing two genotypes (e.g., wild type vs. mutant or transgenic lines), they should be grown in the same plate.
8. Use the same growth conditions (light, temperature, humidity) for all the experiments and replicates. In particular, differences in light intensity or photoperiod result in dramatic changes in resistance to CLM and FSM.
9. Since the counting is by visual inspection, different experimenters might have different criteria to count positive hits (i.e., to define when a seedling has established). To reduce this possible bias, it is important to establish clear objective criteria beforehand.
10. Different genetic backgrounds or seed stocks might show defects in germination or seedling development that need to be taken into account to estimate the real influence of isoprenoid inhibitors on SE. Therefore, the proportion of seedlings producing green true leaves in media with inhibitor should be represented relative to that in mock plates.
11. Depending on the genotype or the plant species used and the incubation time, the number of individuals needed for each replicate will need to be calculated to produce between 25 and 50 mg of fresh tissue for pigment extraction. Take into account that higher concentrations of inhibitor typically result in lower amounts of tissue per seedling.
12. Chlorophylls and carotenoids are highly sensitive to light and heat. To avoid degradation, the tubes containing methanol extracts should be maintained in darkness at low temperature while processing other samples.
13. Dried residues should have a green-brownish color, corresponding to the chlorophyll and carotenoid pigments. They can be stored protected from light at −80 °C for a few days until used for pigment analysis.
14. Different genetic backgrounds or seed stocks might produce different levels of photosynthetic pigments in the absence of

inhibitors. To compare resistance to CLM or FSM of lines with different pigment contents, it is recommended that chlorophyll and carotenoid contents are presented relative to those measured in samples grown without the inhibitors.

Acknowledgements

Our work is funded by grants from the Catalan AGAUR (2009SGR-26 and XRB), Spanish DGI (BIO2011-23680 and PIM2010IPO-00660), and European Union FP7 (TiMet, contract 245143). We are members of the IBERCAROT network funded by Programa Iberoamericano de Ciencia y Tecnologia para el Desarrollo (CYTED, code 112RT0445). C.P. received a predoctoral fellowship of the JAE-PRE program (CSIC).

References

1. Pulido P, Perello C, Rodriguez-Concepcion M (2012) New insights into plant isoprenoid metabolism. Mol Plant 5:964–967
2. Vranova E, Coman D, Gruissem W (2013) Network analysis of the MVA and MEP pathways for isoprenoid synthesis. Annu Rev Plant Biol 64:665–700
3. Campos N, Boronat A (1995) Targeting and topology in the membrane of plant 3-hydroxy-3-methylglutaryl coenzyme A reductase. Plant Cell 7:2163–2174
4. Rodríguez-Concepción M (2006) Early steps in isoprenoid biosynthesis: multilevel regulation of the supply of common precursors in plant cells. Phytochem Rev 5:1–15
5. Pulido P, Toledo-Ortiz G, Phillips MA, Wright LP, Rodriguez-Concepcion M (2013) Arabidopsis J-protein J20 delivers the first enzyme of the plastidial isoprenoid pathway to protein quality control. Plant Cell 25:4183–4194
6. Rodríguez-Concepción M, Forés O, Martínez-García JF, González V, Phillips MA, Ferrer A, Boronat A (2004) Distinct light-mediated pathways regulate the biosynthesis and exchange of isoprenoid precursors during Arabidopsis seedling development. Plant Cell 16:144–156
7. Carretero-Paulet L, Cairo A, Botella-Pavia P, Besumbes O, Campos N, Boronat A, Rodriguez-Concepcion M (2006) Enhanced flux through the methylerythritol 4-phosphate pathway in Arabidopsis plants overexpressing deoxyxylulose 5-phosphate reductoisomerase. Plant Mol Biol 62:683–695
8. Alberts AW, Chen J, Kuron G, Hunt V, Huff J, Hoffman C, Rothrock J, Lopez M, Joshua H, Harris E, Patchett A, Monaghan R, Currie S, Stapley E, Albers-Schonberg G, Hensens O, Hirshfield J, Hoogsteen K, Liesch J, Springer J (1980) Mevinolin: a highly potent competitive inhibitor of hydroxymethylglutaryl-coenzyme A reductase and a cholesterol-lowering agent. Proc Natl Acad Sci U S A 77:3957–3961
9. Istvan ES, Deisenhofer J (2001) Structural mechanism for statin inhibition of HMG-CoA reductase. Science 292:1160–1164
10. Re EB, Jones D, Learned RM (1995) Co-expression of native and introduced genes reveals cryptic regulation of HMG CoA reductase expression in *Arabidopsis*. Plant J 7: 771–784
11. Kasahara H, Hanada A, Kuzuyama T, Takagi M, Kamiya Y, Yamaguchi S (2002) Contribution of the mevalonate and methylerythritol phosphate pathways to the biosynthesis of gibberellins in *Arabidopsis*. J Biol Chem 277:45188–45194
12. Zeidler J, Schwender J, Mueller C, Lichtenthaler HK (2000) The non-mevalonate isoprenoid biosynthesis of plants as a test system for drugs against malaria and pathogenic bacteria. Biochem Soc Trans 28:796–798
13. Matsue Y, Mizuno H, Tomita T, Asami T, Nishiyama M, Kuzuyama T (2010) The herbicide ketoclomazone inhibits 1-deoxy-D-xylulose 5-phosphate synthase in the 2-C-methyl-D-erythritol 4-phosphate pathway and shows antibacterial activity against Haemophilus influenzae. J Antibiot (Tokyo) 63:583–588

14. Kuzuyama T, Shimizu T, Takahashi S, Seto H (1998) Fosmidomycin, a specific inhibitor of 1-deoxy-D-xylulose 5-phosphate reductoisomerase in the nonmevalonate pathway for terpenoid biosynthesis. Tetrahedron Lett 39: 7913–7916
15. Steinbacher S, Kaiser J, Eisenreich W, Huber R, Bacher A, Rohdich F (2003) Structural basis of fosmidomycin action revealed by the complex with 2-C-methyl-D-erythritol 4-phosphate synthase (IspC). Implications for the catalytic mechanism and anti-malaria drug development. J Biol Chem 278:18401–18407
16. Flores-Pérez U, Sauret-Güeto S, Gas E, Jarvis P, Rodríguez-Concepción M (2008) A mutant impaired in the production of plastome-encoded proteins uncovers a mechanism for the homeostasis of isoprenoid biosynthetic enzymes in Arabidopsis plastids. Plant Cell 20:1303–1315
17. Laule O, Furholz A, Chang HS, Zhu T, Wang X, Heifetz PB, Gruissem W, Lange M (2003) Crosstalk between cytosolic and plastidial pathways of isoprenoid biosynthesis in Arabidopsis thaliana. Proc Natl Acad Sci U S A 100:6866–6871
18. Crowell DN, Packard CE, Pierson CA, Giner JL, Downes BP, Chary SN (2003) Identification of an allele of CLA1 associated with variegation in Arabidopsis thaliana. Physiol Plant 118:29–37
19. Sauret-Güeto S, Botella-Pavia P, Flores-Perez U, Martinez-Garcia JF, San Roman C, Leon P, Boronat A, Rodriguez-Concepcion M (2006) Plastid cues posttranscriptionally regulate the accumulation of key enzymes of the methylerythritol phosphate pathway in Arabidopsis. Plant Physiol 141:75–84
20. Kobayashi K, Suzuki M, Tang J, Nagata N, Ohyama K, Seki H, Kiuchi R, Kaneko Y, Nakazawa M, Matsui M, Matsumoto S, Yoshida S, Muranaka T (2007) Lovastatin insensitive 1, a Novel pentatricopeptide repeat protein, is a potential regulatory factor of isoprenoid biosynthesis in Arabidopsis. Plant Cell Physiol 48:322–331
21. Flores-Perez U, Perez-Gil J, Closa M, Wright LP, Botella-Pavia P, Phillips MA, Ferrer A, Gershenzon J, Rodriguez-Concepcion M (2010) Pleiotropic regulatory locus 1 (PRL1) integrates the regulation of sugar responses with isoprenoid metabolism in Arabidopsis. Mol Plant 3:101–112
22. Leivar P, Antolin-Llovera M, Ferrero S, Closa M, Arro M, Ferrer A, Boronat A, Campos N (2011) Multilevel control of Arabidopsis 3-hydroxy-3-methylglutaryl coenzyme A reductase by protein phosphatase 2A. Plant Cell 23:1494–1511
23. Rodríguez-Concepción M, Boronat A (2002) Elucidation of the methylerythritol phosphate pathway for isoprenoid biosynthesis in bacteria and plastids. A metabolic milestone achieved through genomics. Plant Physiol 130: 1079–1089
24. Lichtenthaler HK (1987) Chlorophylls and carotenoids: pigments of photosynthetic biomembranes. Methods Enzymol 148:351–382

Chapter 21

A Flexible Protocol for Targeted Gene Co-expression Network Analysis

Diana Coman, Philipp Rütimann, and Wilhelm Gruissem

Abstract

The inference of gene co-expression networks is a valuable resource for novel hypotheses in experimental research. Routine high-throughput microarray transcript profiling experiments and the rapid development of next-generation sequencing (NGS) technologies generate a large amount of publicly available data, enabling in silico reconstruction of regulatory networks. Analysis of the transcriptome under various experimental conditions proved that genes with an overall similar expression pattern often have similar functions. Consistently, genes involved in the same metabolic pathway are found in co-expressed modules. In this chapter, we describe a detailed workflow for analyzing gene co-expression networks using large-scale gene expression data and explain critical steps from design and data analysis to prediction of functionally related modules. This protocol is platform independent and can be used for data generated by ATH1 arrays, tiling arrays, or RNA sequencing for any organism. The most important feature of this workflow is that it can infer statistically significant gene co-expression networks for any number of genes and transcriptome data sets and it does not involve any particular hardware requirements.

Key words Co-expression, Isoprenoids, Circadian clock, Gene module, Network, Transcriptome

1 Introduction

1.1 Principle and Purposes of Co-expression Network Analysis

The vast volume of transcriptome data generated by microarray studies and more recently by RNA sequencing (RNA-seq) technology has facilitated the development and application of advanced statistical computational analysis tools for data mining, interpretation, and visualization, aiming at disentangling the complex regulation of biological systems. One popular data analysis method describing the interaction of biomolecules at system level is the gene co-expression network (GCN) analysis, which is based on the graph theory [1–3]. The graph theory concepts and properties are useful for describing, inferring, and visualizing relationships between biomolecules as part of larger biological systems [1, 2, 4]. In essence, a graph is a generalized mathematical abstraction of relationships between entities represented as nodes, which are connected by edges. A co-expression network is

Manuel Rodríguez-Concepción (ed.), *Plant Isoprenoids: Methods and Protocols*, Methods in Molecular Biology, vol. 1153, DOI 10.1007/978-1-4939-0606-2_21, © Springer Science+Business Media New York 2014

a special case of a graph with nodes representing genes and the edges showing inferred relationships between genes. It applies the concepts and properties of a graph to facilitate the understanding of complex biological systems [2].

GCNs are constructed based upon the "guilt by association" concept; that is, genes that share a similar expression profile across multiple spatial, temporal, and/or environmental conditions are likely involved in the same biological process and regulated by common transcriptional programs [5, 6]. Consistently, genes associated with the same metabolic pathway are highly co-expressed across various experimental conditions compared to genes from different pathways in various organisms including plants [7–11]. Nevertheless, genes for cellular metabolic pathways may not always form discrete co-expressed modules because the enzymes encoded by the genes may have distinct spatial or temporal physiological functions or may be regulated at translational or posttranslational levels [12].

Basic network topology properties such as network hubs, gene modules, or network motifs can reveal information about the biological significance of network components [1, 2, 13]. For example, one of the most general observations on biological networks is their propensity for a scale-free topology [1]. In scale-free networks, few nodes tend to be densely connected (commonly referred to as hubs), whereas the majority of the nodes have few connections, thereby determining an overall sparse connectivity of the entire network. From a biological perspective, scale-free networks may constitute an evolutionary advantage by preserving the robustness to accidental node (gene) failure of the system [14, 15]. Removal of hub node/s disturbs the wiring of the network to a degree that leads to network collapse [14, 16, 17]. In complex biological systems, this property translates into essentiality of highly connected genes [11, 14, 18, 19].

One of the primary goals of GCN analysis is to identify gene modules (groups of genes with similar expression patterns across various conditions), which might reflect common regulation of the co-expressed genes. For example, the co-expressed genes found in a module could encode enzymes that are part of the same pathway (e.g., catalyze consecutive biosynthetic or biodegradation steps), or they may encode subunits of protein complexes [20, 21]. The modular organization of cellular processes, as opposite to the single-gene view, emerged as an immediate property of biological networks inferred from GCN analysis and is supported by increasing evidence [1, 4, 11, 12, 20, 22, 23].

Network motifs are yet another example of how graph properties may assist in revealing complex regulatory mechanisms of biological systems [24, 25]. Such motifs are small sets of recurring regulation patterns, which occur in GCNs much more often than would be expected in random networks [26]. Interestingly, the network motifs are conserved across species indicating their importance in maintaining basic functions of biological systems [24, 27].

1.2 Types of GCN: Global vs. Targeted

After identifying modules of co-expressed genes, the next step in a GCN analysis is the inference of biological significance of its components (genes). Based on the existing data sets, GCN can generate novel hypotheses for further testing using genetic and molecular biology tools. Global GCN approaches (also known as condition independent) can be used to identify modules in a network in a knowledge-independent manner [10, 12, 23, 28]. Alternatively, practical information on specific biological processes may be obtained by studying networks in a more focused manner by performing targeted GCN analyses (also known as condition dependent) [21, 29]. In brief, a targeted GCN analysis starts with the assembly of a set of genes relevant for a biological process based on experimental knowledge and literature information. These genes are commonly referred to as *guide genes*. To further refine a targeted GCN, custom combination of data sets can be selected by the user based on specific biological questions [30–33]. Next, pairwise similarity scores between guide genes and genes of interest (commonly referred to as *query genes*) are calculated. Significant correlations of guide and query genes are finally represented as GCN (Fig. 1).

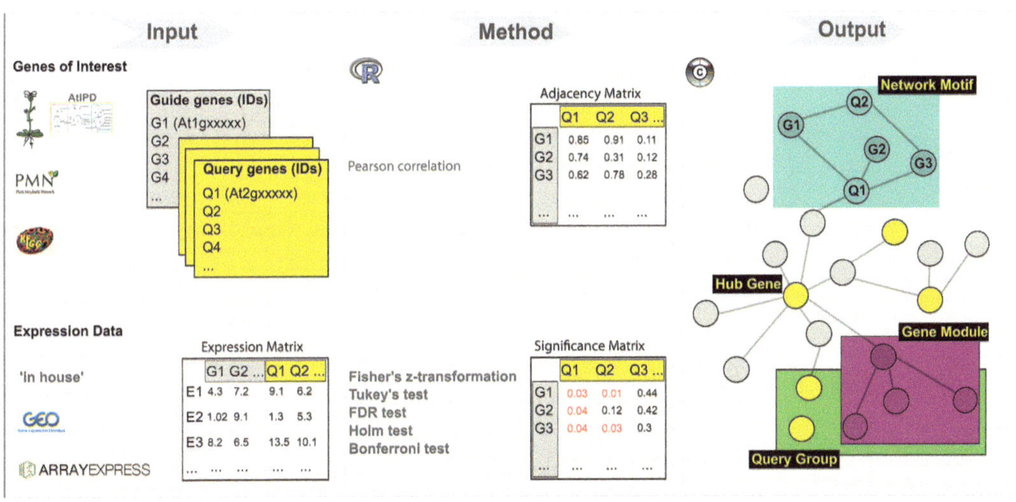

Fig. 1 Targeted GCN analysis workflow. Examples of public databases for gene identifiers (AtIPD, http://www.atipd.ethz.ch; PMN, http://www.plantcyc.org/; KEGG, http://www.genome.jp/kegg/) and for expression data (GEO, http://www.ncbi.nlm.nih.gov/geo/; ArrayExpress, http://www.ebi.ac.uk/arrayexpress/) are shown. Guide genes (e.g., G1, G2, G3) are shown in *gray*. Query genes (e.g., Q1, Q2, Q3) are shown in *yellow*. E1, E2, and E3 stand for expression data sets. Significant correlation relationships (e.g., equal or higher than 0.05 in the significance matrix) are shown in *red*. Open-source software used for data analysis and visualization (R, http://www.r-project.org/, and Cytoscape, www.cytoscape.org/) are shown as *pictograms*. The GCN is shown as undirected graph with genes shown as nodes and significant correlation relationships shown as edges. Typical network topological properties are indicated (e.g., gene module). A query group comprises genes, for which prior knowledge is available (e.g., targets of the same transcription factor, encoding proteins localized in the same subcellular compartment, encoding enzymes in the same pathway)

A targeted GCN has the advantage that it facilitates integration of the existing knowledge (e.g., subcellular localization, *cis-elements*, DNA–protein or protein–protein interactions) and therefore allows delineation of co-expression relationships specifically between genes of interest across relevant experiments. As such, it allows a straightforward interpretation of observed co-expressed modules in a biological context. However, it should be noted that a module identified using a targeted GCN approach does not reflect the complete interaction landscape and may be still part of larger modules, consistent with the notion that biological networks are not only modular but as well hierarchical [28, 29, 34].

1.3 Applicability of GCN Analysis

GCN analyses have been successfully performed in various organisms, from bacteria [35, 36], yeast [8, 37], and plants [11, 38] to humans [39, 40]. GCNs aided in identification of functionally coherent modules corresponding to major cellular processes or in assigning annotation to uncharacterized genes. As such, GCN analyses improved the understanding of biological processes or regulatory mechanisms by complementing molecular biology analyses.

The model plant *Arabidopsis thaliana* (Arabidopsis) benefits from the richest collection of transcriptome data, setting an excellent framework for GCN analysis followed by experimental validation of novel hypotheses. Several global GCN analyses have been performed in Arabidopsis and provided valuable knowledge such as the regulon organization of Arabidopsis transcriptome [12] or functional annotation of unknown genes [11, 41].

Focused targeted GCN studies in Arabidopsis have been successful in assigning putative functions to unknown genes or in associating genes with specific biological pathways and processes. For example, potential functions supported by experimental evidence were assigned to genes from the cytochrome P450 superfamily [42], brassinosteroid-related genes [30], genes involved in regulation of glucosinolate biosynthesis [31], or genes essential for secondary cell wall integrity in Arabidopsis [43, 44]. Furthermore, several studies have used targeted GCN aiming at understanding the regulation and integration in the cellular network of the isoprenoid pathway in Arabidopsis [32, 45, 46]. Particularly, the workflow described here was used by Vranová et al. [33] to identify organ-specific co-expressed modules formed by circadian clock genes and genes encoding enzymes in the isoprenoid pathway from Arabidopsis.

1.4 Future Perspectives of GCN Analysis

The understanding of transcriptional regulatory networks is one of the most significant challenges in systems biology. In this context, GCN analysis is a valuable tool for discovering modules of correlated genes as a first step towards elucidating transcriptional regulation processes. Such analyses set a solid framework for more detailed downstream analyses. For example, *cis*-element analysis is one of the most popular GCN follow-up analyses because gene co-expression is caused, at least to some extent, by co-occurrence of specific *cis*-elements in the promoters of co-expressed genes [6, 47, 48].

Several GCN studies in Arabidopsis attempted to combine transcriptome network analysis with protein interaction data, sequence information, or genetic data [21]. Noteworthy, by combining phylogenetic profiles with proteomics and comparative genomics data sets, conserved co-expressed gene modules across plant species have been identified [34, 38, 49]. Such conserved co-expressed gene modules with specific functions, once identified in model species, provide a solid framework for inferring gene functions in species with agricultural or economic value [49].

This chapter presents a user-friendly protocol to perform targeted GCN analyses for generating experimentally testable hypotheses. Furthermore, several examples of follow-up bioinformatics analyses are presented as a guideline for broadening the understanding of GCN results in a biological context. The flexibility of the workflow overcomes the typical caveats of existing online web tools such as strict limitations in the number of input genes or lack of statistical evaluation of identified correlation relationships [21]. The workflow is platform independent (ATH1 arrays, tiling arrays, or RNA-seq-generated transcriptome data sets can be used), does not pose any particular hardware requirements, and can be used with any major computer operating system.

2 Materials

1. Gene identifiers (*see* **Note 1**).
2. Normalized expression data (*see* **Notes 2–4**).
3. The R software (available at http://www.r-project.org/) and the TinnR text editor for R (http://www.sciviews.org/Tinn-R/) (*see* **Note 5**).
4. The t_GCN.r script containing the actual code to compute the GCN (*see* **Note 6**).
5. The Cytoscape software (available at www.cytoscape.org/) (*see* **Notes 5** and 7).

3 Methods

To perform a targeted GCN analysis, follow the workflow below, which uses as input gene identifiers (IDs) (*see* **Note 1**) and expression data (*see* **Notes 2** and **3**). Based on the input, the pairwise similarity of gene expression profiles is computed and the significance of the identified correlated genes is assessed within a statistical framework.

The final output of this workflow is the gene co-expression network (*see* **Note 8**). Furthermore, several useful downstream analyses for delineation of biological significance of co-expressed gene modules are proposed.

3.1 General Data Retrieval

1. Assemble a list comprising the guide gene IDs and one or more lists with the query gene IDs (Fig. 1) (*see* **Notes 1** and **9**). If prior knowledge for the query genes is available (e.g., targets of the same transcription factor, encoding proteins localized in the same subcellular compartment, encoding enzymes in the same pathway) it can be used to group the query genes according to the information available. In this case, an additional file consisting of the names of each query group should be created (*see* **Note 9**).
2. Select and retrieve relevant microarray expression data for the genes of interest (*see* **Notes 2** and **3**).

At the end of these steps, the following files, which will be used for computing the GCN, should have been created: one file with the expression profiles of the guide genes (e.g., guide_expression.csv; *see* **Note 10**) and one or more files with the expression profiles of the individual or grouped query genes (e.g., query_expression.csv, query_group1_expression.csv, query_group2_expression.csv) (*see* **Notes 9** and **10**). Each input file contains an expression matrix with row names representing experiment names and column names representing gene IDs (Fig. 1). Each entry in the table is the normalized expression of a gene in a certain experiment (*see* **Note 3**).

3.2 Calculation of the Gene Co-expression Network

In the next steps, based on the input files (*see* Subheading 3.1) the custom-designed *t_GCN.r* script implemented in R (*see* **Note 6**) calculates the pairwise similarity scores between the expression profiles of guide and query genes, evaluates the statistical significance, and computes the GCN according to significant pairwise similarity scores.

3.2.1 Calculation of the Adjacency Matrix

Open the R command line interface or any R graphical interface (e.g., RStudio, http://www.rstudio.com/), and set the working directory to the folder where the input files are stored (e.g., T:// Your_WorkingDirectory/). Open the *t_GCN.r* script in TinnR, and run the commands in the order they are written (*see* **Notes 11** and **12**). First, the input files containing the expression data for the genes of interests are read into R. Next, the pairwise similarity between gene expression profiles is estimated by calculating the correlation coefficient between each guide gene and each query gene. The code provided here calculates the pairwise similarity between gene expression profiles using the popular Pearson correlation measure (*see* **Note 13**). The Pearson correlation coefficient quantifies the strength and direction of a linear relationship between two sets of measurements (e.g., gene expression profiles) and is chosen here as similarity metric because it is relatively insensitive to the amplitude of expression levels and at the same time evaluates the similarity of the overall shape of gene expression profiles. Other similarity metrics can be used by adjusting the *t_GCN.r* script (*see* **Note 14**).

The pairwise Pearson correlation coefficients ($r(G^{(k)}Q^{(r)})$) between guide genes ($G^{(k)}$) and query genes ($Q^{(r)}$) are computed as follows:

$$r\left(G^{(k)}Q^{(r)}\right)=\frac{\sum_{i=1}^{n}\left(G_i^{(k)}-\overline{G^{(k)}}\right)\left(Q_i^{(r)}-\overline{Q^{(r)}}\right)}{\sqrt{\frac{1}{n-1}\sum_{i=1}^{n}\left(G_i^{(k)}-\overline{G^{(k)}}\right)^2}\sqrt{\frac{1}{n-1}\sum_{i=1}^{n}\left(Q_i^{(r)}-\overline{Q^{(r)}}\right)^2}}$$

where n denotes the number of experiments (i.e., number of rows in the expression matrix). Note that the correlation coefficient $r(G^{(k)}Q^{(r)})$ varies between −1 (perfect anticorrelation or negative correlation) and +1 (perfect positive correlation).

At the end of this step, an adjacency matrix is computed (Fig. 1). The elements of the matrix are the pairwise Pearson similarity scores between gene expression profiles.

3.2.2 Assessment and Selection of the Statistically Significant Co-expressed Genes

Once a similarity metric has been chosen (e.g., Pearson correlation) and all pairwise similarity scores between expression profiles have been calculated, the next step is to select a score threshold and create a GCN by linking all gene pairs with scores exceeding the respective threshold. One simple method is to choose a cutoff based on the Pearson correlation coefficients. For example, a cutoff of 0.50 could be corresponding to a moderate correlation and a cutoff of 0.80 would indicate strong positive correlation. However, because Pearson correlation coefficients are sensitive to outliers, a more robust cutoff, based on the significance level of the observed correlation, should be used [3, 21]. For example, Fisher's z-transformation is particularly useful when the number of conditions tested in a microarray study varies considerably and it has been successfully applied to construct GCNs [50]. This statistical test assigns a significance estimate to Pearson correlation coefficients based on the number of conditions across which it is calculated. Genes with strong correlations across a greater number of samples receive higher significance levels (i.e., *P*-values) indicating that the likelihood of obtaining an equal or a higher correlation coefficient value by chance alone is small.

To test if the pairwise correlations are significantly bigger than zero (*see* **Note 15**) the *z*-test is applied to the adjacency matrix containing the pairwise similarity scores. The Fisher z-transformation converts the correlation coefficients $r(G^{(k)}Q^{(r)})$ into a normal distributed z-variable using the following formula:

$$Z_{G^{(k)}Q^{(r)}} = \frac{1}{2}\log\left(\frac{1+r\left(G^{(k)}Q^{(r)}\right)}{1-r\left(G^{(k)}Q^{(r)}\right)}\right)$$

The null and the alternative hypothesis are defined as H_0: $r(G^{(k)}Q^{(r)}) \leq 0$ and H_A: $r(G^{(k)}Q^{(r)}) > 0$, respectively. P-values are obtained for each pair of guide and query genes and indicate whether they are significantly positively correlated (*see* **Note 15**).

If the query genes are grouped based on prior information (*see* **Note 9**), an additional statistical significance test—Tukey's test—has to be applied to identify guide genes which are significantly co-expressed with a group of query genes. Values of 2 or greater of the Tukey's test indicate that a guide gene is significantly correlated with a group of query genes.

3.2.3 Applying Multiple Testing Correction

An important factor to be considered when using P-values to assess the statistical significance of pairwise correlation coefficients is that the number of statistical tests applied is equal to the number of genes being tested. Thus, the P-values need to be further adjusted for multiple testing to control the number of false positives. The *t_GCN.r* script provides several types of multiple testing corrections: the false discovery rate (FDR), which estimates the proportion of false positives in the data, and the more stringent corrections, such as Holm or Bonferroni (family-wise error rate tests), which calculate the probability of having even one (or more) false positive.

After running the *t_GCN.r* script the result files are saved automatically in the working directory (*see* **Note 12**). The output contained in the result files is the significance matrix whose elements are the adjusted P-values and indicate significant correlation relationships (Fig. 1). These result files will be used for further downstream analysis.

3.2.4 Preparing the GCN for Visualization with Cytoscape

After completing the calculation of the similarity scores between gene expression profiles and assessing their statistical significance, the next step is to visualize the GCN and analyze its properties (Fig. 1). The GCN is represented as undirected graph consisting of nodes (genes) and edges. Edges are only shown between two genes (e.g., $G^{(k)}$ and $Q^{(r)}$; *see* **Notes 16** and **17**) if their expression profiles across a set of experiments are significantly correlated—that is, the corresponding P-value is smaller or equal to a chosen threshold.

To visualize the GCN in Cytoscape, prepare a specific network file, which specifies the nodes and the edges of the network (*see* **Note 18**). Use the result files containing the adjusted P-values as estimators of significant co-expression between gene pairs. For example, P-values smaller or equal to 0.05 can be selected as indicators of significantly co-expressed genes (*see* **Note 19**).

To generate a network file, transform the adjusted *P*-values as follows: assign the value 1 to significantly co-expressed genes ($P \leq 0.05$, *see* **Note 19**) or else assign the value 0. Next, import the network file into Cytoscape and display the GCN as a graph with nodes representing genes and edges representing significant correlation between pairs of genes (Fig. 1).

3.3 Basic Analysis of Network Properties

Standard analyses of gene co-expression networks can be performed using the Cytoscape tool. Nevertheless, once a GCN is computed, any software for network analysis may be used (*see* **Note 20**). To extract biological information from GCNs, three topology network properties can be analyzed: network hubs, gene modules, and network motifs (*see* Subheading 1).

3.3.1 Identification of Network Hub Genes

Hubs are defined as genes with many interaction partners in a network (Fig. 1). Hub genes can be identified by calculating the node degree and the node degree distribution. The degree of a node is one of its most prominent characteristics in a network and is defined as the number of edges incident at a node. To calculate the node degree, use the *CalculateNodeDegree* Cytoscape plug-in (*see* **Note 21**).

Another important characteristic of networks is their scale-free property (*see* Subheading 1), which can be evaluated based on of their node degree distribution. Networks with scale-free topology have power law node degree distributions (i.e., few nodes with high node degree and many nodes with low node degree). To evaluate the node degree distribution, use the *NetworkAnalyzer* Cytoscape plug-in (*see* **Note 21**).

3.3.2 Identification of Gene Modules

Gene modules are defined as highly interconnected regions of a network and are therefore likely to be co-regulated (Fig. 1). Gene modules can be detected by assessing the local network connectivity (the number of edges connecting a subgroup of nodes divided by the total possible number of edges) and the average clustering coefficient (the connectivity degree corresponding to the neighbors of a node) compared to random networks.

To identify gene modules, use the *ClusterViz* Cytoscape plug-in (*see* **Note 21**).

3.3.3 Identification of Network Motifs

Network motifs, defined as subset of recurring regulation patterns, occur significantly more often than by chance alone (*see* **Note 22**). To identify overrepresented network motifs assess the clustering coefficient with the *CytoKavosh* Cytoscape plug-in (*see* **Note 21**).

3.3.4 Further Network Parameters

To calculate further informative topological parameters of GCNs (e.g., network diameter, average number of neighbors, average clustering coefficient, shortest path length) use the *NetworkAnalyzer* Cytoscape plug-in.

To assess whether interesting topological properties occur more often than expected by chance the GCN of interest can be compared with a random network (*see* **Note 22**). To generate random networks (*see* **Note 21**) use the *RandomNetworks* Cytoscape plug-in.

3.3.5 Further Data Analysis Examples

The interpretation of GCNs requires biological knowledge. The nodes in a GCN can be labeled with available information such as the name of the gene, subcellular localization, and biological function. In this context, superimposing functional annotations (e.g., gene ontologies, GO) onto GCNs is particularly useful to assess whether topological network characteristics are linked to statistically overrepresented categories.

To identify overrepresented GO categories in co-expression networks, use the *BiNGO* Cytoscape plug-in (*see* **Note 21**). Furthermore, other omics data (expression data, proteome data, etc.) can be integrated into the GCN analysis by using the *OmicsAnalyzer* Cytoscape plug-in (*see* **Note 21**).

4 Notes

1. Online dedicated databases for metabolic pathways such as Plant Metabolic Network (PMN, http://www.plantcyc.org/) or KEGG (http://www.genome.jp/kegg/; [51]) are accessible by various software platforms and tools, allowing manual or automated gene ID retrieval. Particularly, the AtIPD (http://www.atipd.ethz.ch; [52]) is a comprehensive, manually curated database for the isoprenoid pathway from Arabidopsis.
2. Microarray expression data for the genes of interest can be retrieved from dedicated databases. For example, the eFP Browser accessible at http://bar.utoronto.ca/affydb/cgi-bin/affy_db_exprss_browser_in.cgi [53] hosts microarray expression data for various tissues of Arabidopsis during plant development and different experimental conditions (abiotic and biotic stresses, hormone treatments, etc.). Experiments specific to light and circadian regulation in plants can be found at DIURNAL (http://diurnal.mocklerlab.org/; [54]). Alternatively, any type of expression data (measured by microarrays or RNA-seq) can be retrieved from specific repositories (e.g., GEO, http://www.ncbi.nlm.nih.gov/geo/ [55], or ArrayExpress, http://www.ebi.ac.uk/arrayexpress/ [56]).
3. The normalization of microarray expression data is required to remove noise intrinsic to technical error and to obtain comparable data. One of the most popular normalization methods for microarray is the robust multi-array average (RMA) method [57], which includes background correction, normalization,

and calculation of probe set summaries and provides log 2 scale expression values as output.

4. In high-throughput gene expression measurements the number of variables (genes) is much higher than the number of experiments. To insure sufficient "degrees of freedom," multiple microarray experiments should be combined when performing GCN analyses [3, 21].
5. All software and scripts used in this book chapter are public and freely available.
6. The *t_GCN.r* script and demo input files can be downloaded from http://www.pb.ethz.ch/downloads/.
7. Cytoscape is a powerful open-source software platform designed for network visualization and analysis [58]. Furthermore, Cytoscape allows users to integrate custom networks with annotations, gene expression profiles, and other types of data and is available for all the major operating systems.
8. Several online resources such as ATTEDII (http://atted.jp; [59]), CressExpress (http://cressexpress.org; [60]), or Genevestigator Co-Expression Tool (https://www.genevestigator.com; [61]) could be used for co-expression analyses; however, they are limited regarding the number of genes, types of expression experiments available, or statistical evaluation [21].
9. Run the "guide-query GCN" code in the *t_GCN.r* script to calculate co-expression of guide and query genes. To additionally assess the co-expression of guide genes to groups of query genes, run the "guide-query groups GCN" code in the *t_GCN.r* script. The latter is particularly useful for example in assessing whether a gene with unknown function is part of a pathway based on similar expression pattern across a set of relevant experimental conditions.
10. Alternatively, plain text files can be used. In this case use the appropriate R commands (e.g., "read.table()").
11. The names of the input files must match exactly (case sensitive) the names used in the script. Duplicate entries (e.g., IDs) are not allowed.
12. Basic R commands and functions are provided in the R Reference Card available to download from http://cran.r-project.org/doc/contrib/Short-refcard.pdf.
13. Pearson correlation assumes normal distribution of data; thus, the microarray expression values must be normalized.
14. Spearman's rank correlation can be used as an alternative non-parametric similarity measure, particularly when the data does not follow a normal distribution [62]. Spearman's rank correlation

is obtained by replacing the numerical expression values with their rank across all microarray experiments and is thus more robust against outliers [63].

15. The *t_GCN.r* script evaluates by default the statistical significance of positive Pearson correlations, but it can easily be adapted to estimate the statistical significance of negative Pearson correlations.
16. Correlation between guide genes or between query genes may exist but are not evaluated in this type of targeted GCN analysis.
17. A restriction of GCNs is that they estimate correlations, which may reflect indirect biological relationships. Two genes can be co-expressed when considering them separately, but not correlated conditionally on other genes. To address this issue, partial correlation and graphical Gaussian models can be applied to distinguish direct from indirect correlations [45, 64].
18. Cytoscape supports many different standard network file formats, including Simple Interaction Format (SIF), BioPAX, PSI-MI, SBML, or tab-delimited text files. A typical SIF network file specifies nodes and interactions as follows: *node1 relationship_type node2* (http://wiki.cytoscape.org/Cytoscape_User_Manual/Network_Formats/).
19. To change the *P*-value threshold simply modify the *t_GCN.r* script with the desired value (e.g., 0.01).
20. Alternatively, GCNs can be visualized with BioLayout (http://www.biolayout.org/) or VisANT (http://visant.bu.edu/).
21. Numerous additional features are available in Cytoscape (http://apps.cytoscape.org/) as plug-ins (e.g., advanced network and molecular profiling analyses, layouts, connection with databases).
22. The Erdös–Rényi model [65], Watts and Strogatz model [66], or the Barabási–Albert model [67] can be used to generate random networks.

Acknowledgments

We thank Dr. Eva Vranová and Prof. Peter Bühlmann for helpful discussions and Philipp Ihmor for critically reading the manuscript. This work was supported by the Seventh Framework Program of the European Commission through the TiMet collaborative project (grant 245143) to W.G.

References

1. Barabási AL, Oltvai ZN (2004) Network biology: Understanding the cell's functional organization. Nat Rev Genet 5(2):101–115
2. Huber W, Carey VJ, Long L et al (2007) Graphs in molecular biology. BMC Bioinformatics 8(Suppl 6):S8
3. Lèbre S, Lelandais G (2009) Modeling a regulatory network using temporal gene expression data: why and how? In: G. Alterovitz, R. Benson and M. Ramoni (eds) Automation in proteomics and genomics: an engineering case-based approach. Wiley, Chichester UK. pp. 69–96
4. Jeong H, Tombor B, Albert R et al (2000) The large-scale organization of metabolic networks. Nature 407(6804):651–654
5. Oliver S (2000) Guilt-by-association goes global. Nature 403(6770):601–603
6. Gao F, Foat BC, Bussemaker HJ (2004) Defining transcriptional networks through integrative modeling of mRNA expression and transcription factor binding data. BMC Bioinformatics 5:31
7. Hughes TR, Marton MJ, Jones AR et al (2000) Functional discovery via a compendium of expression profiles. Cell 102(1):109–126
8. Stuart JM, Segal E, Koller D et al (2003) A gene-coexpression network for global discovery of conserved genetic modules. Science 302(5643):249–255
9. Gachon CMM, Langlois-Meurinne M, Henry Y et al (2005) Transcriptional co-regulation of secondary metabolism enzymes in Arabidopsis: functional and evolutionary implications. Plant Mol Biol 58(2):229–245
10. Wei HR, Persson S, Mehta T et al (2006) Transcriptional coordination of the metabolic network in Arabidopsis. Plant Physiol 142(2): 762–774
11. Heyndrickx KS, Vandepoele K (2012) Systematic identification of functional plant modules through the integration of complementary data sources. Plant Physiol 159(3):884–901
12. Mentzen WI, Wurtele ES (2008) Regulon organization of Arabidopsis. BMC Plant Biol 8:99
13. Tieri P, de la Fuente A, Termanini A et al (2011) Integrating Omics data for signaling pathways, interactome reconstruction, and functional analysis. Methods Mol Biol 719:415–433
14. Jeong H, Mason SP, Barabasi AL et al (2001) Lethality and centrality in protein networks. Nature 411(6833):41–42
15. Leclerc RD (2008) Survival of the sparsest: robust gene networks are parsimonious. Mol Syst Biol 4:213
16. Albert R, Jeong H, Barabasi AL (2000) Error and attack tolerance of complex networks. Nature 406(6794):378–382
17. Jalili M (2011) Error and attack tolerance of small-worldness in complex networks. J Informetrics 5(3):422–430
18. Krylov DM, Wolf YI, Rogozin IB et al (2003) Gene loss, protein sequence divergence, gene dispensability, expression level, and interactivity are correlated in eukaryotic evolution. Genome Res 13(10):2229–2235
19. Zotenko E, Mestre J, O'Leary DP et al (2008) Why do hubs in the yeast protein interaction network tend to be essential: reexamining the connection between the network topology and essentiality. PLoS Comput Biol 4(8): e1000140
20. Hartwell LH, Hopfield JJ, Leibler S et al (1999) From molecular to modular cell biology. Nature 402(6761):47–52
21. Usadel B, Obayashi T, Mutwil M et al (2009) Co-expression tools for plant biology: opportunities for hypothesis generation and caveats. Plant Cell Environ 32(12):1633–1651
22. Gavin AC, Aloy P, Grandi P et al (2006) Proteome survey reveals modularity of the yeast cell machinery. Nature 440(7084):631–636
23. Freeman TC, Goldovsky L, Brosch M et al (2007) Construction, visualisation, and clustering of transcription networks from Microarray expression data. PLoS Comput Biol 3(10):2032–2042
24. Alon U (2007) Network motifs: theory and experimental approaches. Nat Rev Genet 8(6): 450–461
25. Huang CY, Cheng CY, Sun CT (2007) Bridge and brick network motifs: identifying significant building blocks from complex biological systems. Artif Intell Med 41(2):117–127
26. Milo R, Shen-Orr S, Itzkovitz S et al (2002) Network motifs: simple building blocks of complex networks. Science 298(5594):824–827
27. Kim TH, Kim J, Heslop-Harrison P et al (2011) Evolutionary design principles and functional characteristics based on kingdom-specific network motifs. Bioinformatics 27(2):245–251
28. Mao LY, Van Hemert JL, Dash S et al (2009) Arabidopsis gene co-expression network and its functional modules. BMC Bioinformatics 10:346
29. Aoki K, Ogata Y, Shibata D (2007) Approaches for extracting practical information from gene co-expression networks in plant biology. Plant Cell Physiol 48(3):381–390

30. Lisso J, Steinhauser D, Altmann T et al (2005) Identification of brassinosteroid-related genes by means of transcript co-response analyses. Nucleic Acids Res 33(8):2685–2696
31. Hirai MY, Sugiyama K, Sawada Y et al (2007) Omics-based identification of Arabidopsis Myb transcription factors regulating aliphatic glucosinolate biosynthesis. Proc Natl Acad Sci U S A 104(15):6478–6483
32. Meier S, Tzfadia O, Vallabhaneni R et al (2011) A transcriptional analysis of carotenoid, chlorophyll and plastidial isoprenoid biosynthesis genes during development and osmotic stress responses in Arabidopsis thaliana. BMC Syst Biol 5:77
33. Vranová E, Coman D, Gruissem W (2012) Structure and dynamics of the isoprenoid pathway network. Mol Plant 5(2):318–333
34. Mutwil M, Usadel B, Schutte M et al (2010) Assembly of an interactive correlation network for the Arabidopsis genome using a novel heuristic clustering algorithm. Plant Physiol 152(1):29–43
35. Zampieri M, Soranzo N, Bianchini D et al (2008) Origin of co-expression patterns in *E. coli* and *S. cerevisiae* emerging from reverse engineering algorithms. PLoS One 3(8):e2981
36. Zare H, Sangurdekar D, Srivastava P et al (2009) Reconstruction of Escherichia coli transcriptional regulatory networks via regulon-based associations. BMC Syst Biol 3:39
37. Segal E, Shapira M, Regev A et al (2003) Module networks: identifying regulatory modules and their condition-specific regulators from gene expression data. Nat Genet 34(2): 166–176
38. Ficklin SP, Feltus FA (2011) Gene coexpression network alignment and conservation of gene modules between two grass species: maize and rice. Plant Physiol 156(3):1244–1256
39. Ma S, Shi M, Li Y et al (2010) Incorporating gene co-expression network in identification of cancer prognosis markers. BMC Bioinformatics 11:271
40. Oldham MC, Langfelder P, Horvath S (2012) Network methods for describing sample relationships in genomic datasets: application to Huntington's disease. BMC Syst Biol 6:63
41. Horan K, Jang C, Bailey-Serres J et al (2008) Annotating genes of known and unknown function by large-scale coexpression analysis. Plant Physiol 147(1):41–57
42. Ehlting J, Provart NJ, Werck-Reichhart D (2006) Functional annotation of the Arabidopsis P450 superfamily based on large-scale co-expression analysis. Biochem Soc Trans 34:1192–1198
43. Brown DM, Zeef LAH, Ellis J et al (2005) Identification of novel genes in Arabidopsis involved in secondary cell wall formation using expression profiling and reverse genetics. Plant Cell 17(8):2281–2295
44. Persson S, Wei H, Milne J et al (2005) Identification of genes required for cellulose synthesis by regression analysis of public microarray data sets. Proc Natl Acad Sci U S A 102(24):8633–8638
45. Wille A, Zimmermann P, Vranova E et al (2004) Sparse graphical Gaussian modeling of the isoprenoid gene network in Arabidopsis thaliana. Genome Biol 5(11):R92
46. Ruiz-Sola MA, Rodriguez-Concepcion M (2012) Carotenoid biosynthesis in Arabidopsis: a colorful pathway. Arabidopsis Book 10:e0158
47. Xu XJ, Wang LS, Ding DF (2004) Learning module networks from genome-wide location and expression data. FEBS Lett 578(3): 297–304
48. Vandepoele K, Quimbaya M, Casneuf T et al (2009) Unraveling transcriptional control in Arabidopsis using cis-regulatory elements and coexpression networks. Plant Physiol 150(2): 535–546
49. Movahedi S, Van Bel M, Heyndrickx KS et al (2012) Comparative co-expression analysis in plant biology. Plant Cell Environ 35(10):1787–1798
50. Weirauch MT (2011) Gene coexpression networks for the analysis of DNA microarray data. Applied statistics for network biology. Wiley-VCH Verlag GmbH & Co. KGaA, Weinheim, pp 215–250
51. Kanehisa M, Goto S, Sato Y et al (2012) KEGG for integration and interpretation of large-scale molecular data sets. Nucleic Acids Res 40(1): 109–114
52. Vranová E, Hirsch-Hoffmann M, Gruissem W (2011) AtIPD: a curated database of Arabidopsis isoprenoid pathway models and genes for isoprenoid network analysis. Plant Physiol 156(4):1655–1660
53. Toufighi K, Brady SM, Austin R et al (2005) The botany array resource: e-Northerns, expression angling, and promoter analyses. Plant J 43(1):153–163
54. Mockler TC, Michael TP, Priest HD et al (2007) The Diurnal project: Diurnal and circadian expression profiling, model-based pattern matching, and promoter analysis. Cold Spring Harb Symp 72:353–363
55. Barrett T, Troup DB, Wilhite SE et al (2011) NCBI GEO: archive for functional genomics data sets-10 years on. Nucleic Acids Res 39:1005–1010

56. Parkinson H, Sarkans U, Kolesnikov N et al (2011) ArrayExpress update-an archive of microarray and high-throughput sequencing-based functional genomics experiments. Nucleic Acids Res 39:1002–1004
57. Irizarry RA, Hobbs B, Collin F et al (2003) Exploration, normalization, and summaries of high density oligonucleotide array probe level data. Biostatistics 4(2):249–264
58. Smoot ME, Ono K, Ruscheinski J et al (2011) Cytoscape 2.8: new features for data integration and network visualization. Bioinformatics 27(3):431–432
59. Obayashi T, Nishida K, Kasahara K et al (2011) ATTED-II updates: condition-specific gene coexpression to extend coexpression analyses and applications to a broad range of flowering Plants. Plant Cell Physiol 52(2): 213–219
60. Srinivasasainagendra V, Page GP, Mehta T et al (2008) CressExpress: a tool for large-scale mining of expression data from Arabidopsis. Plant Physiol 147(3):1004–1016
61. Hruz T, Laule O, Szabo G et al (2008) Genevestigator v3: a reference expression database for the meta-analysis of transcriptomes. Adv Bioinformatics 2008:420747
62. D'haeseleer P (2005) How does gene expression clustering work? Nat Biotechnol 23:1499–1501
63. Bickel DR (2003) Robust cluster analysis of microarray gene expression data with the number of clusters determined biologically. Bioinformatics 19(7):818–824
64. Ma S, Gong Q, Bohnert HJ (2007) An Arabidopsis gene network based on the graphical Gaussian model. Genome Res 17(11):1614–1625
65. Erdős P, Rényi A (1961) On the strength of connectedness of a random graph. Acta Math Hung 12:261–267
66. Watts DJ, Strogatz SH (1998) Collective dynamics of 'small-world' networks. Nature 393(6684):440–442
67. Barabási AL, Albert R (1999) Emergence of scaling in random networks. Science 286(5439):509–512

INDEX

A

Activity ... 9–19, 21–35, 41–45, 48–52, 77, 81, 83, 85, 149, 151, 162, 198, 235, 246, 274, 276
Acylated sterol glucosides (ASGs) ... 115–119, 122, 124–126, 128, 129, 131
Agrobacterium ... 246–251, 255
Alkaloids ... 171–185
Antioxidant ... 77–95, 100, 110, 111, 153, 213
Arabidopsis thaliana ... 1, 12, 17, 24, 42, 115, 138, 144, 151, 217, 258, 274, 288
Assay ... 11–15, 17, 18, 22–27, 29–32, 35, 41–52, 102, 103, 107, 189, 236, 238, 240, 246, 276, 2774

C

Calibration ... 15, 16, 64, 68, 71, 105, 110, 138, 162–167, 177, 216, 220–223, 225
Carotenoid ... 3, 4, 77–95, 146, 198, 199, 213, 214, 225, 227–232, 273, 276, 278–282
Chlorophyll ... 3, 4, 78, 83–85, 131, 197, 218, 219, 224, 227–232, 273, 276, 278–281
Chloroplast ... 42, 100, 102, 108, 213–232
Cholesterol ... 115, 119, 121–123, 130, 171, 172, 175
Chromatography ... 10–12, 14, 26, 31, 43, 49, 57–75, 99–111, 115–132, 136, 137, 139, 140, 144, 150, 151, 154, 158, 161–167, 173, 174, 177, 194, 196, 198, 214, 217
Chromoplast ... 227–232
Clomazone (CLM) ... 2, 274–278, 280–281
Co-expression ... 247, 252, 285–296
Confocal laser scanning microscopy (CLSM) ... 227–232

D

Degradation ... 17, 22, 34, 82–85, 87, 103, 104, 109, 125, 131, 156, 162, 210, 214, 217, 228, 281, 286
1-Deoxy-D-xylulose 5-phosphate (DXP) ... 2, 9–19, 57–59, 61, 63, 64, 196, 274, 276
Deoxyxylulose ... 2, 57–75, 196, 274
Dimethylallyl diphosphate (DMAPP) ... 1–4, 9, 41, 42, 51, 57, 59, 171, 196, 273, 274
Diterpene ... 2–4, 149–158, 246, 258
DMAPP. *See* Dimethylallyl diphosphate (DMAPP)
Dolichol ... 3, 42, 135–139, 142, 143
DXP. *See* 1-Deoxy-D-xylulose 5-phosphate (DXP)
DXP reductoisomerase (DXR) ... 2, 58, 274, 276, 278
DXP synthase (DXS) ... 2, 9–14, 16–18, 274–276

E

Enzyme ... 1–4, 9–19, 21, 27, 41, 45, 49–52, 57, 136, 151, 214, 236–238, 246, 247, 258, 263, 269, 273, 274, 276, 286–288, 290
Expression ... 42–47, 49, 50, 99, 100, 204, 235–243, 245–254, 257–270, 285–296

F

Farnesyl diphosphate (FPP) ... 2, 3, 41–52, 171, 196
Fluorescence (FD) ... 10, 94, 101, 104–108, 111, 204, 214, 228–232
Fosmidomycin (FSM) ... 2, 274–278, 280, 281
FPP. *See* Farnesyl diphosphate (FPP)
Free sterols ... 115, 116, 119, 122–125, 129, 131

G

Gas chromatography – mass spectrometry (GC–MS) ... 116, 124, 126, 130–132, 150–157, 161–168
Gene ... 43, 46, 50, 99, 100, 157, 204, 242, 245–248, 250, 251, 253, 258, 259, 261–263, 265, 266, 269, 285–296
Geranyl diphosphate (GPP) ... 2, 3, 41, 42, 44, 48, 49, 51, 188, 196, 258
Geranylgeranyl diphosphate (GGPP) ... 2, 3, 188, 196, 258

H

Hemiterpene ... 2–4
High-performance liquid chromatography (HPLC) ... 10–12, 14, 26, 59, 63–65, 68, 69, 71, 74, 75, 81, 82, 85–88, 93, 94, 99–112, 128, 139, 140, 142–144, 146, 152, 155, 183, 191, 214, 215, 217, 219, 223
HMG-CoA reductase (HMGR) ... 2, 21–35, 274
Hormone ... 3, 4, 78, 81, 273, 294

Manuel Rodríguez-Concepción (ed.), *Plant Isoprenoids: Methods and Protocols*, Methods in Molecular Biology, vol. 1153, DOI 10.1007/978-1-4939-0606-2, © Springer Science+Business Media New York 2014

Hydroxymethylbutenyl diphosphate (HMBPP)....59, 274
3-Hydroxy-3-methylglutaryl coenzyme A (HMG-CoA)....2, 21–27, 31–34, 274

I

Inhibitor....2, 4, 11–13, 16, 17, 23, 26, 33, 44, 261, 273–282
Isopentenyl diphosphate (IPP)....1–4, 9, 41, 42, 44, 48, 49, 51, 52, 57, 59, 136, 171, 188, 196, 273, 274
Isoprene....1–4, 9, 100, 101, 135–137, 162, 163

L

Leaf....12, 24, 25, 51, 83, 84, 99, 103–106, 110, 131, 152, 154, 155, 161, 187, 190, 192, 193, 197, 215, 217, 221, 223–225, 251–254
Liquid chromatography – mass spectrometry (LC–MS)....11, 12, 17, 63, 74, 116, 121, 128, 174, 189, 191, 223, 254

M

Mass spectrometry (MS)....10, 26, 57–75, 121, 128, 132, 139, 150, 155, 157, 159, 162, 174, 176–178, 182, 184, 194, 225, 275
Membrane....3, 22, 63, 78, 100, 115, 116, 136, 173, 191, 205, 207, 209–211, 213, 214, 219, 220, 224, 227, 246
Methylerythritol....1, 2, 57–59, 136, 196, 273, 274
Methylerythritol cyclodiphosphate (MEcPP)....59, 196
Methylerythritol 4-phosphate (MEP) pathway....1, 2, 9–19, 57, 273–276, 278
Mevalonate....2, 9, 21, 23, 29, 31, 34, 136, 188, 235
Mevalonic acid (MVA)....1–4, 21, 22, 25, 32, 171
  pathway....1, 3, 4, 136, 273, 274, 276, 278
Mevinolin....2, 26, 274, 275
Microarray....204, 285, 290, 291, 294–296
Microdissection....204
Monoterpene....2–4, 10, 149, 161–163, 203, 246, 251, 273

N

Network....285–296
Nicotiana benthamiana....245–254, 258

O

Octaprenyl....2
Oil gland....203–211

P

Phylloquinone....3, 4, 100, 103–106, 112, 213, 214, 216, 221, 223, 226, 273
Physcomitrella patens....257–270
Pigment....3, 78, 86, 101, 107, 228, 229, 270, 276–282
Plastid....1, 3, 4, 9, 10, 57–75, 213, 227, 229–232, 246, 273
Plastoglobule (PG)....214–220, 224
Plastoquinone....3, 4, 99, 100, 102–108, 213–216, 221–226, 273
Polyisoprenoid....4, 135–146
Polyprenol....3, 135–139, 142, 143
Prenylquinone....99, 100, 108, 213–226, 273
Prenyltransferase....2, 3, 41, 50
Profiling....57–75, 151, 187–199, 213–226, 259, 266–268, 296

Q

Quantification....18, 48, 94, 106–108, 115, 116, 119–122, 124, 125, 127–132, 164, 165, 188, 214, 252, 267, 268, 273–282
Quinone....99–101, 103–106, 108, 110, 111, 199, 216, 221–223

R

Reductase....2, 21–35, 59, 236, 238, 274
Rice....150, 152, 154–156
Root....24, 25, 135, 136, 149–158, 275

S

Saccharomyces cerevisiae....236
Saponins....151, 171–185
Seedling....23–26, 33, 35, 51, 77, 138, 139, 143, 144, 227, 249, 275, 277–281
Sesquiterpene....2–4, 42, 149, 161–163, 203, 251, 273
Solanesyl....2
Solanum lycopersicum....25, 172, 178, 179, 228
Spectrometry....10, 11, 26, 57–75, 115–132, 150, 153, 155, 157, 161–167, 171–185, 194–196, 214, 221
Steroidal....171–185
Sterol....21, 22, 115–132, 235, 239
  esters....115, 119, 121, 122, 125, 129, 130
  glucosides....115, 119, 123–125
Synthase....2, 3, 9–19, 41–52, 58, 59, 157, 236, 237, 245–247, 258, 274

T

Tetraterpene....3, 4
Thin layer chromatography (TLC)....10, 22, 25, 28, 30–32, 34, 35, 116, 117, 120, 126, 131, 136–138, 140, 142, 144, 146
Tobacco....22, 23, 192, 193, 197, 246, 247
Tocopherol....3, 4, 77–95, 213, 214, 216, 221–223, 226, 273

Tomato 22, 84, 172, 174, 182, 184, 192, 193, 227–232
Transcriptomic 203–211
Trichome 187–199, 203
Triterpene 2–4, 149–158, 171, 235–243, 273

U

Ubiquinone 3, 4, 42, 106, 107, 221, 222, 226, 273
Ultra-high performance liquid chromatography (UPLC) 51, 171–185, 191, 193, 194, 199, 219

V

Vitamin 77, 81, 82, 100, 213, 277, 280
Vitamin A 77, 81, 85
Vitamin E 77, 82
Vitamin K 213
Volatile 34, 52, 59, 74, 86, 87, 116, 144, 149–151, 153, 154, 161–167, 251, 252, 261, 267
Volatile organic compounds (VOCs) 161–167

Y

Yeast 27, 34, 135, 151, 157, 158, 203, 235–243, 246, 288

MIX
Papier aus verantwortungsvollen Quellen
Paper from responsible sources
FSC® C105338

If you have any concerns about our products,
you can contact us on
ProductSafety@springernature.com

In case Publisher is established outside the EU,
the EU authorized representative is:
Springer Nature Customer Service Center GmbH
Europaplatz 3, 69115 Heidelberg, Germany

Printed by Libri Plureos GmbH
in Hamburg, Germany